GOLD-MINING BOOMTOWN

GOLD-MINING BOOMTOWN
*People of White Oaks, Lincoln County,
New Mexico Territory*

By ROBERTA KEY HALDANE

UNIVERSITY OF OKLAHOMA PRESS
Norman, Oklahoma

LIBRARY OF CONGRESS CATALOGING-IN-PUBLICATION DATA

Haldane, Roberta Key, 1935–
Gold-mining boomtown : people of White Oaks, Lincoln County, New Mexico Territory /
By Roberta Key Haldane.
     p. cm.
Includes bibliographical references and index.
ISBN 978-0-87062-410-0 (cloth)
ISBN 978-0-8061-4417-7 (paper)
1. White Oaks (N.M.)—Biography. 2. White Oaks (N.M.)—History. 3. Frontier and pioneer life—New Mexico—White Oaks I. Title.

F804.W55H35 2012
978.9'64—dc23

2011044350

The paper in this book meets the guidelines for permanence and durability of the Committee on Production Guidelines for Book Longevity of the Council on Library Resources, Inc. ∞

Copyright © 2012 by the University of Oklahoma Press, Norman, Publishing Division of the University. Paperback published 2013. Manufactured in the U.S.A.

All rights reserved. No part of this publication may be reproduced, stored in a retrieval system, or transmitted, in any form or by any means, electronic, mechanical, photocopying, recording, or otherwise—except as permitted under Section 107 or 108 of the United States Copyright Act—without the prior written permission of the University of Oklahoma Press. To request permission to reproduce selections from this book, write to Permissions, University of Oklahoma Press, 2800 Venture Drive, Norman OK 73069, or email rights.oupress@ou.edu.

To all the hardworking burros, mules, horses, and oxen, on whose backs

White Oaks and its mines and the ongoing

life of the town were built. It is time

these hoofed servants of humankind receive

due esteem, even if only in memory.

CONTENTS

ACKNOWLEDGMENTS	xi
THE LIVELIEST LITTLE GOLD CAMP IN THE SOUTHWEST *An Introduction*	3
THE AGUAYOS IN LINCOLN COUNTY *Descendants of the Spanish and Mexican Houses De Aguayo*	11
JOHN AH NUE, A.K.A "JOHN CHINAMAN" *Celestial Extraordinaire*	22
JAMES A. ALCOCK *Feisty Little Irishman and Manager of the Carrizozo Cattle Ranche Company*	27
SUSAN McSWEEN BARBER *Cattle Queen of New Mexico*	31
JAMES W. BELL *Deputy Sheriff Shot Dead by Billy the Kid during His Escape from Lincoln*	44
EDWIN R. BONNELL *Builder of the Bonnell Opera House*	49
JAMES CARLYLE, ALIAS BERMUDA CARLISLE *Gunned Down in the Greathouse-Kuch Ranch Shootout*	56
THE DE GUEVARA, LALONE, AND LACEY FAMILIES *Mexico and Texas Meet Canada in White Oaks*	60
SCOTSWOMAN JANE MALCOLM GALLACHER AND HER THREE SONS *Founders of the J + H Ranch*	66
BENJAMIN GUMM AND SONS *Builders of the Gumm House, the Hoyle House, and the White Oaks Schoolhouse*	74
JOHN A. HALEY *Printer's Ink Flowed in His Veins*	85
JOHN Y. HEWITT *First Citizen of White Oaks*	88
"JOLLY JERRY" HOCKRADLE *Age Cannot Weary the Prospector*	96
EMERSON HOUGH *Lawyer, Conservationist, Author of* Heart's Desire	100

MATTHEW WATSON "WATT" HOYLE   107
Partner in the Old Abe Mine and Builder of Hoyle House

WILLIAM AND JOHN HUDGENS   116
Double Trouble from Louisiana

ANDREW "ANDY" HUDSPETH   124
Foremost Legal Mind in the Southwest

DAVID L. "HAPPY JACK" JACKSON   129
A Lesson in Do-It-Yourself Integration

MAX H. KOCH   134
Photographer of Note

DR. ALEXANDER GALLATIN LANE   138
From Confederate Surgeon to White Oaks Physician and Pharmacist

JUDGE FRANKLIN HOUSTON LEA   144
Former Quantrill Guerilla

NEW ENGLAND SEA CAPTAIN JOHN LEE   149
Around the World aboard the Yankee Clipper

THE LESLIES   162
Settlers at Texas Park

THE ENTERPRISING MAYER BROTHERS   168
Frederick, Paul, and Charles

WILLIAM CALHOUN McDONALD   178
First Elected Governor of New Mexico

CHARLES MADISON MERRILL   193
A Prospector's Prospector

URBAIN OZANNE   196
A Frenchman with a Nose for Business

DR. MELVIN G. PADEN   208
Legendary Country Doctor

E. W. AND EMMELINE PARKER   215
Civilizers of a Raucous Mining Camp and Parents of a Mining Dynasty

COLONEL GEORGE WORTH PRICHARD   230
New Mexico's Clarence Darrow

THE QUEEN FAMILIES *Mining in Their Blood*	235
JAMES B. REDMAN/REDMOND, G. S. AND J. S. REDMAN/ READMAN, S. J./W. J. WOODLAND *Man of Many Names*	245
JOHN BURCHEM SLACK *From International Diamond Hoaxer to Coffin Builder and Undertaker*	249
LEVIN WASHINGTON STEWART *Mercantilist from Saint Louis*	256
JONES AND STANLEY TALIAFERRO *Newspapermen, Politicians, Mercantilists, Miners*	261
THE LEGEND OF MADAM VARNISH	267
WILLIAM H. WEED *Santa Fe Trail Trader and White Oaks Merchant, Dreamer, and Doer*	269
SAMUEL AND MARTHA FRANCES WELLS *Not of the "White Oaks 400" and Proud of It*	274
MARCUS WHITEMAN *Russian Jew and Founder of the Pioneer Store*	279
JOHN E. WILSON AND SON-IN-LAW JOHN WAUCHOPE *Bit Hard by the Gold Bug*	283
JOHN V. "OLD JACK" WINTERS *Owner of the North Homestake Mine*	288
GEORGE RICHARD "DICK" YOUNG *Mercantilist from Mississippi and Klondike Prospector*	291
ALBERT AND JACOB ZIEGLER *German-Jewish Merchants Chasing the American Dream*	298
AFTERWORD	305
BIBLIOGRAPHY	307
INDEX	317

ACKNOWLEDGMENTS

Because this book has been a decade in research and writing, the passage of time has meant the deaths of many good people who helped in the project, among them Mary McCourt Anderson, Janice Gnatkowski, John Hall, Jack Harkey, Nora Henn, Paul Hile, Sara Jackson, Mary Jo Jamison, Louise Kelt, Dee Ann Lee, Melvin James Lee, Pat Lee, Robert and Dorothy Leslie, Evelyn Lynch, Dell Morris, Donald M. Queen, Carol Queen Watt, and Mary Lou Townsend Welch.

Special thanks to those still living who gave most liberally of their time and treasured mementoes: Ina Hunter Dow, Salinas, California; Chris Harkey, Carrizozo, New Mexico; John Mottinger, Santa Teresa, New Mexico; and Larue Lane Wetzel, White Oaks.

I am indebted to a host of aficionados of old White Oaks who gladly furnished photos, diaries, memoirs, family scrapbooks, Bible records, genealogies, newspaper clippings, videos, artifacts, and magazine articles (from New Mexico unless otherwise noted): Rose Barton, Capitán; Martin Bittner, Redwood City, California; Tim Bonnell, Wichita, Kansas; Lorraine De Aguayo Brimberry, Ruidoso; Howard Bryan, Albuquerque; Ann Buffington, Eugene, Oregon; Rhonda Barber Burrows, Nogal; Norma Cinert, Tularosa; Betty Corn, Las Cruces; Evelyn Fitzpatrick Dobbs, Roswell; Roy N. Dow, Carrizozo; Rich Eastwood, Enseñada, Mexico; Donna Edmund, Chloride; Winnie Eoff, Roswell; Bethel Treat Grant, Yuma, Arizona; Pat and Art Guevara, Payson, Arizona; Charles Hayes, Morro Bay, California; Nisha Hoffman, Ruidoso; Jean Parker Hughes, Rolling Hills Estates, California; Marilyn Lemon Ingley, Bend, Oregon; Troy Kelley, Johnson City, New York; Parker Kemp, Hermosa Beach, California; Rebecca LoPorto, Tucson, Arizona; Bill Matthews, Redondo Beach, California; J. N. "Snooks" McDaniels, Carrizozo; Willeta Morris, Fort Thomas, Arizona; Anne New, White Oaks; Loris Norris, Abilene, Texas; Jeannette Lemon Olson, Gales Ferry, Connecticut; Polly Owen, Roswell; Morris B. Parker III, Newport Beach, California; James Perrin, Hammond, Louisiana; Verna Reed, Carlsbad; Jane Gallacher Shafer, J + H Indian Tank Ranch, Carrizozo; Marlow Sharpe, Nogal; Eleanor Shockey, Ruidoso; James Sorrells, Woodway, Texas; Steven Spencer, Río Rancho; Johnson Stearns, Carrizozo; Dave Taliaferro, Albuquerque; Bonnie Turrentine, Minneapolis, Minnesota; Don Ward, Carrizozo; and Dr. G. Richard Wheelock, Del Mar, California.

Thanks also to Frederick Nolan of Chalfont Saint Giles, England, for his generous encouragement and sound advice; deceased historian Philip J. Rasch for the invaluable heritage of his research; and Karen Mills, Lincoln County Historical Records Clerk, for her unfailing enthusiasm for the project and for furnishing countless leads.

I have many people to thank at archive collections, universities, libraries, and museums (from New Mexico unless otherwise noted): Jim Bradshaw, J. Evetts Haley History Center, Midland, Texas; Irisha Corral and her staff, Thomas C. Donnelly Library, New Mexico Highlands University, Las Vegas; Betty Crawford-Gore, Santa Fe Trail Center, Larned, Kansas; Elvis Fleming, Historical Center for Southeast New Mexico, Roswell; Michele Gillett, East Mountain Library, Tijeras; Sherrill Goken, Iroquois County Genealogical Society, Watseka, Illinois; Kathryn Hodson, University of Iowa Libraries, Iowa City, Iowa; Austin Hoover, Dennis Daily, Marah DeMeule, Charles Stanford, and Bill Boehm, Río Grande Historical Collections, Las Cruces; Tomas Jaehn, Palace of the Governors, Santa Fe; Daniel Kosharek, Photo Archives, Palace of the Governors, Santa Fe; Norma Meeks, New Mexico Bureau of Geology and Mineral Resources, Socorro; Linda Newman, Thomas C. Donnelly Library, Las Vegas; Bobbie Sago and Claudia Rivers, Special Collections, University Library of the University of Texas at El Paso, Texas; Mary Schmitt, Alamogordo Public Library, Alamogordo; Cliff Vanderpool and his staff, Panhandle-Plains Historical Museum, Canyon, Texas; and the staffs of the Las Vegas Public Library; New Mexico State Records Center and Archives, Santa Fe; Special Collections Library, Albuquerque; and Tucumcari Public Library.

I extend my profound gratitude to Robert Clark of the Arthur H. Clark Company for taking on a daunting project, and to editorial assistant Katy Oard, both of whose knowledge, diligence, and skill made the writing much better. The talented folks at the University of Oklahoma Press, manuscript editor Emily Jerman and copyeditor Kelly Parker, expertly ushered the manuscript into its finished form; they found all the things I missed, and they did it diplomatically.

The Doña Ana County Historical Society graciously gave me permission to reprint several articles in this book. The following articles are reprinted, with changes, from the *Southern New Mexico Historical Review*: "What Became of Susan McSween Barber's Diamonds?" (January 1999); "David L. 'Happy Jack' Jackson of Old White Oaks" (January 2005); "William Calhoun McDonald: First Elected Governor of the State of New Mexico" (January 2006); "Andrew 'Andy' Hudspeth: From Bookkeeper and Lawyer for the Cree VV Ranch in the Sacramento Mountains to Chief Justice of the New Mexico Supreme Court" (January 2007); and "William and John Hudgens: Double Trouble from Louisiana" (January 2008).

And last and foremost I would like to thank Robert Workhoven, dedicated partner, uncompromising editor, most generous friend.

GOLD-MINING BOOMTOWN

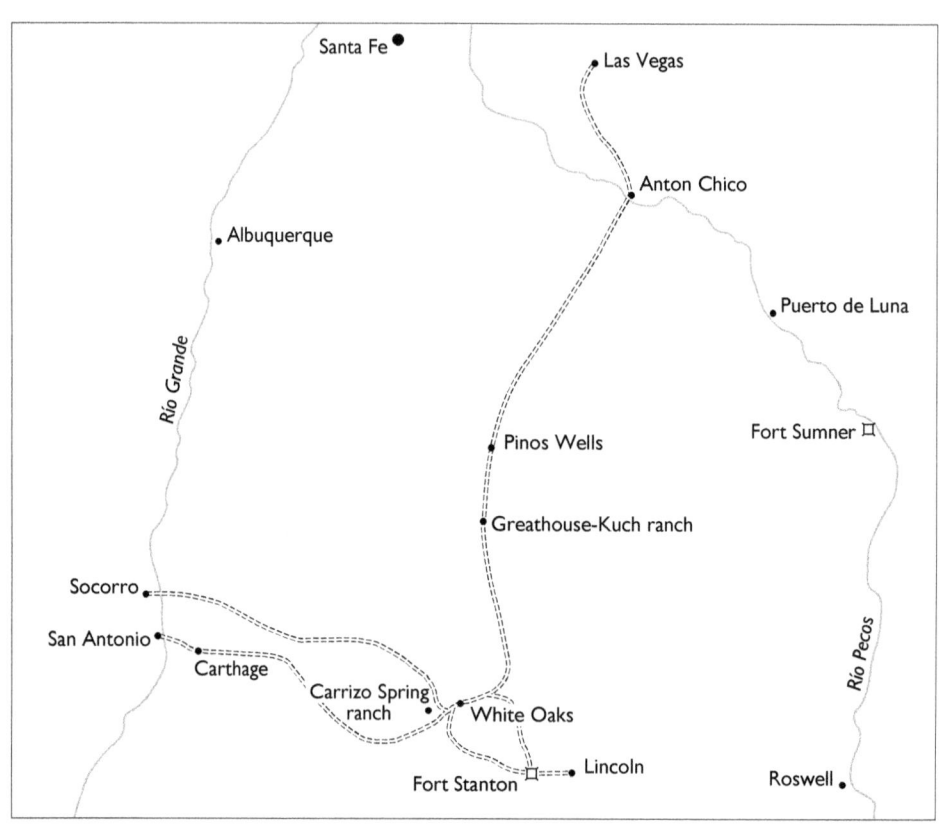

Central New Mexico, showing major stagecoach, mail, and freight routes to White Oaks, 1880–1900.

THE LIVELIEST LITTLE GOLD CAMP IN THE SOUTHWEST
An Introduction

Readers familiar with the violent historical legacy of old Lincoln County in territorial New Mexico may be surprised to learn that there is much more history in that place and time than books on the county's namesake war would suggest. The last quarter of the nineteenth century also witnessed the boom-to-bust life of the gold-mining camp of White Oaks.

Lincoln County, created in 1869 by the New Mexico territorial legislature in honor of assassinated president Abraham Lincoln, covered 27,000 square miles of land. Sprawling across the entire southeastern one-fourth of the territory of New Mexico, it measured 150 miles from north to south and 200 miles from east to west. This was old Lincoln County, an immensity that attracted men who longed for wild and wide-open spaces (or were running from the law). For twenty years, from 1869 to 1889, it remained a vast breeding ground for cattle, with millions of acres of public lands available for free grazing. One of this country's last frontiers, the county gained notoriety as one of the most lethal landscapes in the American West.

The territorial legislature of January 1889 carved two new counties, Cháves and Eddy, out of old Lincoln County. In time other counties also encroached on the land, and a new Lincoln County emerged.

Tucked away in a beautiful valley in the northernmost reaches of Lincoln County, White Oaks burst into life on the edge of the cataclysm that was the Lincoln County War. The town lies a scant twenty-eight miles as the crow flies from the Lincoln of Billy the Kid fame, a geographic fact that contributed much to its beginnings.

Mountains almost piling on each other encircle White Oaks: Baxter on the west, Lone on the northwest, pine-shrouded Patos on the southeast, and mighty Carrizo

Wide view of White Oaks, looking south, in the late 1890s. Photographer unknown. Courtesy Loris Norris.

John A. Brown, Lone Mountain, March 23, 1889. Photographer unknown. Courtesy Sara Jackson, My House of Old Things, Ancho, New Mexico.

on the south at 9,700 feet. Until the mid-1800s, isolated prospectors and the occasional sheepherder obtained gold here—if they could elude raiding Comanches and Apaches—by placer mining or by mining with burro-drawn arrastras.

The area's solitude was interrupted in the spring of 1879 when John Baxter ventured east out of the Middle Río Grande Valley. A failed forty-niner and former Union soldier, Baxter wandered southern New Mexico searching for gold. He heard rumors of the precious metal on the flanks of a certain mountain in the southeast. Pooling resources with old friends Jack Winters and John Wilson, Baxter and his partners packed east across the Jornada del Muerto to the place now known as Baxter Mountain.

In August 1879, George Wilson (no relation to John) drifted into camp and struck a deal to work and share in

Dave Jackson at the White Oaks *arrastra*, 1952. Leonard Sharpe, his son Marlow, and Jackson all helped unearth this mill on the east side of White Oaks. Today the *arrastra* stones can be seen in the White Oaks Schoolhouse Museum. Photographer unknown. Courtesy Marlow Sharpe.

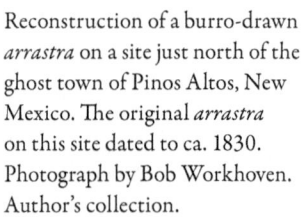

Reconstruction of a burro-drawn *arrastra* on a site just north of the ghost town of Pinos Altos, New Mexico. The original *arrastra* on this site dated to ca. 1830. Photograph by Bob Workhoven. Author's collection.

Bill of sale for the Homestake Mine, August 1879. Herman B. Weisner Papers, Archives and Special Collections, New Mexico State University Library.

any finds. One day as Wilson rested and idly chipped away at a nearby rock, he exposed a vein of gold ore. He rushed back to camp to break the news. For unknown reasons (the law at his heels?), George Wilson lost no time in selling his share of the discovery and disappeared. Wilson's find would become the Homestake Mine, one of the richest in the region.

Before long, Baxter also departed for a new strike he had heard of in southern New Mexico. Word of Wilson's find leaked out by early 1880 and the stampede was on. In May 1880, there were only fifty people in the White Oaks camp. That spring citizens laid out their town site with the help of Jimmy Dolan from Lincoln. One month later there were 350 people. Gold was the lodestone that drew adventurers to White Oaks—that magical, glittering metal that for eons has infected mankind with an incurable sickness. Many who came to "the Oaks," as the town was widely known, had already moved from continent to continent, place to place, culture to culture. It was a rough life. Wide-open in the beginning, White Oaks was, as Mark Twain once remarked about a similar town, "no place for a Presbyterian."

Claims with names like the Comstock, Black Prince, White Swan, and Old Abe—the richest of them all—pockmarked Baxter Mountain. Then came the shantytowns; Hidetown, Tent Town, and Hogtown sprang up overnight. Bars, dance halls, gambling houses, and brothels flourished. Eight saloons could not keep up with the thirst of the miners, outlaws, and cowboys, and life was a noisy din from dusk till dawn.

White Oaks became a magnet for gunslingers, gamblers, whiskey peddlers, indigents, and entrepreneurs. To add to this volatile mix, cowboys off the Block and Carrizozo cattle ranches swaggered into the saloons, where they joined vaqueros from Raventón Draw in drinking and gambling. So many colorful characters moved into and out of town that the landscape for a time resembled the set of a Hollywood Western.

The Pioneer was the saloon of saloons, with its

Twelve mule teams pulling ore wagons near the top of the ridge behind the gold mines near White Oaks. Date and photographer unknown. Archives and Special Collections, RG83-151, New Mexico State University Library.

Group of White Oaks miners showing off their tools. Date and photographer unknown. Courtesy John and Wesley Mottinger.

myriad of lights and mirrors and a long mahogany bar. The Hudgens brothers sold three grades of whiskey in the Pioneer—all from the same barrel. Down the street, Madam Varnish, so-called for the slick way she manipulated the cards while parting gamblers from their money, held sway in the Little Casino Saloon.

From 1879 to 1884, White Oaks was a dangerous place to be, even as it grew into a mining, ranching, and mercantile center for fifty miles around. Early on, the likes of Billy the Kid and his gang were familiar figures on horseback on the town's dusty streets and they hurrahed the town with gunshots on more than one occasion. Dedrick's livery and stable, run by a trio of brothers, bought stolen cattle. The men sold beef to local miners and fenced cattle as far south as Pat Coughlan's ranch in Tularosa. At the same time, a thriving counterfeiting ring, said to involve Billy the Kid, Tom Cooper, and Billy Wilson, operated out of the town.

It was no country for old men; most adventurers were young men and women in a young country. The average life expectancy was much shorter than it is today. Few of those in these pages managed to live to the age of seventy. The census of 1880, taken right after the first gold strike on Baxter Mountain, lists 268 names. It shows only four men over the age of sixty.

In 1880, twenty-six people were foreign-born. They came from Mexico, Ireland, Canada, Prussia/Germany, France, and England. Ireland claimed the greatest number with twelve. But twenty-two other people had foreign-born parents. A whopping total of forty-two people hailed from Texas. Though not enumerated in the census, Chinese miners and workmen probably lived somewhere in the area. It has been

Early street scene in White Oaks. Note the blacksmith repair shop on the left and general merchandise store on the right. Date and photographer unknown. Courtesy John and Wesley Mottinger.

White Oaks picnic, ca. 1896. Mrs. Alice Koch is in a dark skirt and white blouse, with her back to the camera, and leaning on her hand. To her right are her two young daughters, Merle and Edna. The Jones Taliaferro family is to Mrs. Koch's right, with Jones holding a plate in hand. Dr. Paden (bald), is seated in front of the two ladies sitting on boxes, and John Gumm is seated just to Paden's right. Albert Ziegler (man in white collar and seated facing the camera) is in the center of the photo in front of man standing with small boy. Photographer unknown (probably Max H. Koch). Courtesy Marilyn Lemon Ingley.

claimed that few children lived in rough and raucous mining boomtowns. This was not true of White Oaks; the census taker counted fifty-nine children fifteen years and under, or 22 percent of the population—and there were twenty-nine women.

A curiously democratic form of society arose in this frontier mining camp in which each person knew his or her own place, but believed that place to be as good as anyone else's. But certain differences did exist that made it seem unlikely that such a town could continue to exist, much less thrive. Half the people, for instance, came from the recently defeated South of the Civil War, the other half from the victorious North.

When the popular young deputy James Carlyle was gunned down in November 1880 in a battle between a White Oaks posse and Billy the Kid's gang at the Greathouse-Kuch ranch forty miles north, the citizens of White Oaks stiffened their backbones and resolved never again to allow an outlaw in or near their town. Eventually, the baby steps of law and order in Lincoln County began with the election in late 1880 of new sheriff Pat Garrett. Garrett often rode over the mountains to White Oaks from the county seat of Lincoln to deliver campaign speeches and on one such occasion was drowned out by raucous brays from the miners' burros on the edge of town.

As outlaws became scarce, the climate began to favor a more law-abiding class of citizenry, especially from the north and east. A town hall built in 1882 became the focal point for community life. People used it for church services, meetings, plays, and dances. Garrett himself swung the ladies to old-time music there. The very first church service was held in the hall early in 1884. "The Old Rugged Cross" was to have begun the service, but so few of the fledgling congregation knew it or any other hymn that songs like "Old Black Joe" and "Swanee River" had to be sung instead.

Example of a five-stamp milling machine. This example stands in front of the James Douglas mansion at Jerome, Arizona, an old copper-mining town. The Old Abe had four such five-stamp mills, each weighing 850 pounds. Photograph by Bob Workhoven. Author's collection.

By 1885 White Oaks looked like a real town, with 213 houses, a hotel, a general store, and more than fifty other established businesses. The year 1890 found White Oaks at its peak population of twelve hundred, the largest town in Lincoln County, second largest town in the territory, and the center of an important supply trade throughout southeast New Mexico.

That same year also saw a second major strike with the discovery of a large vein of pure gold in the Old Abe. But major production had to be halted in March 1895 after a disastrous fire suffocated eight miners underground. Operations soon resumed, and the Old Abe eventually yielded $875,000 in gold and at its zenith employed forty workers mining between forty-five to fifty tons of ore per day. Its working shaft at 1,450 feet made it the deepest dry hole in the United States. A total of 150,000 ounces of gold were mined at White Oaks, with gold production up to 1904 valued at $3 million, or in today's terms, more than half a billion dollars.

Grand new buildings were built in the 1890s, including the imposing Exchange Bank on White Oaks Avenue. People traveled long distances to attend elegant social affairs and celebrations hosted by the town's elite. In July 1892, Judge John Hewitt and his wife hosted one such affair for two hundred guests at which attendees from as far away as El Paso listened to light opera and arias from the old masters.

The individuals or families profiled in the pages of this book form a diverse and remarkable group. Their occupations ranged from outlaw to lawman, rancher to mercantilist, miner to doctor, saloonkeeper to stage-coach owner. And all were—or became, as they pushed

Ladies' social group seated in the Taliaferro home. Note the typical Victorian wallpaper, carpet, hangings, and pictures. Ella Taliaferro is to the right on her "fainting couch" behind son Jones, Jr. Date and photographer unknown. Courtesy Chris and Jack Harkey.

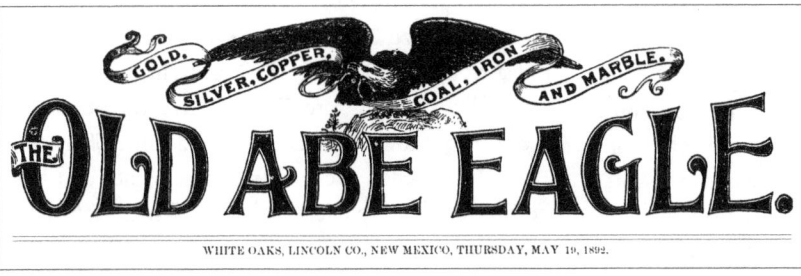

Two White Oaks newspaper mastheads: *top*, the *White Oaks Golden Era*, 1885; *bottom*, the *Old Abe Eagle*, 1892. Author's collection.

across the western frontier—uniquely American. Some of the people the reader will meet include

- A deputy sheriff from White Oaks, shot and killed by Billy the Kid during the Kid's escape from the Lincoln County Courthouse.
- A Russian Jew who found his niche as a mercantilist after emigrating to White Oaks via New York.
- The New England sea captain who sailed around the world three times and married a Samoan princess before relocating his family to White Oaks.
- Another mercantilist from New York City, who began as a Santa Fe Trail freighter before joining the California gold rush and eventually opening a highly successful "emporium" in the Oaks.
- Three men who forsook the mines of White Oaks for the fresh promise of the Yukon goldfields in the waning years of the nineteenth century.
- A black man who came to the Oaks as a barely literate youth and rose to become full partner in the Wild Cat Leasing Company and a much-loved and highly respected member of the community.
- The town's sole surviving Chinese man, who eked out an existence for decades in his little shack until dying there at ninety years of age.
- A remarkable woman who survived the Lincoln County War to reinvent herself as one of New Mexico's most successful ranchers.
- The town's undertaker, one of the all-time great swindlers of the Old West.

The list goes on. Territorial Governor Lewis Wallace, author of *Ben-Hur*, stayed in White Oaks. Pat Garrett visited often and was in town the very day Billy the Kid made his historic escape from a makeshift jail in the Lincoln courthouse. William C. McDonald arrived in 1880 from Fort Scott, Kansas, as a twenty-two-year-old lawyer. He rose to become manager of the Carrizozo Cattle Ranche Company and the El Capitán Land and Cattle Company, two of the largest cattle companies in New Mexico in the early twentieth century. He capped a long and successful political career by being elected first governor of the newly minted state of New Mexico.

Another politician, Col. George Prichard, began as a mining litigator in White Oaks and graduated to high-profile criminal cases. His sparkling eyes, silky white goatee, and animated hand gestures mesmerized juries. He too rose to prominence in New Mexico politics as district attorney for the Fifth Federal District of New Mexico and, later, state attorney general.

Writers in abundance drew inspiration from White Oaks, and some of the Southwest's finest literary talent emerged there. Emerson Hough arrived in 1880 as one of a host of eager young lawyers. He turned to writing as a career after a stint on White Oaks' first newspaper, the *Golden Era*. Later he wrote a highly successful series of Western novels. His 1905 novel *Heart's Desire*, modeled after White Oaks, immortalized the town. Eugene Manlove Rhodes, called by some the Bard of the Tularosa, walked the streets of El Dorado. And Jack Thorp wrote some of his famous cowboy songs there.

Miles from anywhere, White Oaks was always confronted by a major transportation problem. A series of stage companies carried mail, passengers, and express to and from Las Vegas, 182 miles away and the business hub of New Mexico. Freighters too were critical to the flow of goods and supplies. Later, as the railroad pushed from Las Vegas to San Antonio on the Río Grande south of Socorro, mail and passengers were brought to White Oaks by stagecoach ninety miles from that terminus. French entrepreneur Urbain Ozanne took over the stage and ran it superbly from 1886 to 1894. For those years, it was the primary means of transportation for hire between San Antonio, Fort Stanton, and Lincoln.

What White Oaks lacked from the beginning was a railroad. Charles Eddy, a promoter from southeast New Mexico, saw an opportunity. He would build a railroad from El Paso to the Oaks. In 1897, banking on the Salado coalfields and Sacramento Mountain timber, he promised construction of a railroad within ninety days. He built as far as Alamogordo before the Salado coalfields abruptly played out. At this critical juncture, the White Oaks town fathers, sure that there was no route other than one through their town—with big profits to be had—refused to offer any inducements; there would be no compromise. It was a fatal mistake.

His dreams of a White Oaks spur crumbling, Eddy changed course. The El Paso and Northeastern Railroad reached Carrizo "flats" (present-day Carrizozo) twelve miles from the Oaks in 1899 and chugged right on north, bypassing White Oaks. Eddy's turnabout, the greed of the town fathers, and the decline of the Old Abe Mine spelled the town's doom. Mines and businesses closed; people left in droves. White Oaks never recovered.

The stories of White Oaks' above-mentioned founders, inhabitants, and passersby are arranged alphabetically by last name in the chapters that follow.

Main road from White Oaks going west, 1890s. Photographer unknown. Courtesy William B. (hereafter Bill) Mathews.

THE AGUAYOS IN LINCOLN COUNTY
Descendants of the Spanish and Mexican Houses De Aguayo

J. S. Brimberry, an Aguayo descendant with an interest in family bloodlines, writes, "De Aguayo is Spanish, with a disputed origin. Some authorities state that this name means 'Guardian of the Water'—in other words, 'Keeper of a Town Water Supply or Reservoir.' Other sources indicate that 'Aguayo' means simply 'From the Place by the Water.'"[1]

Brimberry claims that the Aguayos were among medieval Spanish nobility based in Castile. Family lore further holds that the Aguayos arrived in the New World in time to participate in the conquest of Mexico in the sixteenth century.

A Mexican Dynasty

The Marquisate of Aguayo came into existence in 1682; the first marquis was Agustín de Échevers. The daughter of this first marquis came to Patos, Coahuila, in New Spain with husband, José Ramón de Azlor y Virto de Vera, in 1711, who became the second Marquis of Aguayo through marriage. The second Marquis astutely managed his wife's vast latifundio (landed estate), cleared the legal titles to her holdings, and greatly extended their boundaries. The Aguayos acquired 2.2 million acres of unused crown lands from 1717 to 1760 through a series of royal deeds.

In 1719, after offering to drive the French out of the area claimed by Spain, this second marquis was appointed governor and captain-general of Coahuila and Texas. He assumed office on October 21. The viceroy of New Spain officially commissioned him in 1720 to reoccupy the East Texas missions and presidios that had been abandoned during the French invasion in 1719. Eventually the marquis established four presidios and ten missions and drew up colonization plans for Texas. The Aguayo expedition solidified the Spanish claim to Texas so that it was never again challenged by the French. The Spanish king rewarded the marquis for his service to the crown by promoting him to field marshal. He died at Parrás, Coahuila, on March 7, 1734.

In 1735, the marquis's elder daughter, now the Marchioness of Aguayo, married José Francisco Valdivielso, owner of vast estates in Nueva Vizcaya. This marriage united two of the largest landholders in northern New Spain. Valdivielso, the third Marquis of Aguayo through his marriage, established a sheep-raising empire that by the 1760s extended over 14,680,000 acres, with headquarters at Hacienda de Santa Catalina del Álamo in Durango.

By the late 1700s, the Aguayo Marquisate was the most important latifundio in the entire Mexican state of Coahuila.[2] Sheep raising was the major business, with flocks estimated to number between 200,000 and 300,000, but the family ran great herds of cattle and horses as well. The Aguayos also grew some 200,000 grapevines and produced wine and brandy for sale in New Spain. Eventually most of the Aguayos became absentee landowners who lived and moved around the elite court in Mexico City. Their interest in the latifundio lay primarily in its wealth, and the only visits to their lands were for annual vacations at Patos.

Apache Raids on Aguayo Haciendas

The large herds of horses and mules on Aguayo lands proved irresistible to marauding Apaches, more so than the latifundios that exclusively raised sheep. To protect their lands from Apache raids in Coahuila, where only three woefully understaffed presidios existed, the Aguayos were forced to maintain a private cavalry. Nonetheless, Apaches ravaged the countryside throughout Coahuila in the 1770s. Relief finally came in 1779, when the governor of Coahuila formed an alliance with the Mescalero Apaches, who agreed to fight as mercenaries against the Lipán Apaches. Peace was short-lived; the Apaches resumed warring in 1781. The Marquis of

Aguayo's Haciendas de Cuatrocienegas and Carmen ranches became primary targets because they were so close to Apache strongholds.

The Aguayos in the Early 1800s

On the eve of independence from Spain, only two families in Coahuila remained serious competitors for landholdings: the Aguayos and the Sánchez Navarro family. The Aguayos, with nearly fifteen million acres, still dominated. Yet there was a problem in the seemingly happy state of Aguayo affairs: though the Aguayos received huge revenues, they overspent at an ever-faster rate. In 1798, to maintain his lavish lifestyle, the Marquis of Aguayo had borrowed sixteen thousand pesos at 5 percent annual interest from his archrival, José Miguel Sánchez Navarro.

By early 1805, José Miguel had become uneasy about the ten thousand pesos Aguayo still owed and directed his business agent to call in the loan. The Marquis protested vigorously and was reluctantly allowed to continue paying interest until the end of 1805 on his assurance of repaying the principal at that time.

The Collapse of the House of Aguayo

The finances of the Aguayo Marquisate had been mismanaged for decades, but the family limped along for a few more years until 1818, when bankruptcy ensued and creditors took over management of the estates. Two years later, the Marquis escaped his creditors by dying. Eldest son José María Valdivielso inherited the title and latifundio and with it enormous debt and legal problems.

By 1823, the Mexican congress had cleared the way for the new creditors to take action. Aguayo was allowed to keep half his properties; the creditors took the other half. The creditors saw this as a way to finally get their half, and the Marquis of Aguayo saw a way to get his remaining half of the latifundio managed without effort so long as the British gave him enough cash to live well in Mexico City. In 1825, the British bought the marquisate, which included sixty-six villages.

In 1834, the state administrators of Coahuila expropriated the entire marquisate, claiming they had been deprived of tax revenues. The Mexican government stepped in and suspended all expropriation proceedings while Congress debated the matter. Congress declared the Coahuila decree unconstitutional in March 1835. The British retained the marquisate but lost the firm's very capable general manager, Dr. James Grant. At this juncture, Carlos Sánchez Navarro offered to buy the entire latifundio. On November 13, 1840, he bought out the Aguayo creditors and the British interests, extending the Sánchez Navarro holdings to twenty-six thousand square miles. The Aguayo landholdings were no more, and the Sánchez Navarros were now first among Mexican landowners.

The Aguayos in Jalisco

By the late 1700s, Manuel María Aguayo had established himself far southwest of Coahuila in the state of Jalisco.[3] Manuel's son was born in 1805. In June 1824, a year and a half after Mexico had won its independence from Spain and the country became more stable, this son, Angél, married Antonia Pérez. The couple made their home near Teocaltiche in Jalisco and in the next few years had four children: Petra, on July 25, 1825; José María, on March 19, 1827; Sacramento in 1830; and Sabina in 1833.

José María Aguayo is Kidnapped by Bandits

When he was only six years old, José was kidnapped by bandits on his way to school. One of his captors, a man on a large black horse, reached down and snatched him before quickly turning his horse toward the mountains and, after several hours of hard riding, entering a forest and riding up a steep trail until finally arriving at the crest of the mountains. The badly frightened boy kept his wits about him and did not try to struggle or to question his captors.

Riding on after reaching the mountain's peak, at about midnight the riders came to a cabin in the timber, dismounted, and entered a small house where several men sat around a table eating and drinking liquor. They gave José food, water, and a rug on the floor for a bed. The next day the kidnappers contacted José's father and grandfather and demanded a large sum of money in exchange for the boy. The money was to be dropped at a secret place and no officers were to be notified. Two days later the money was left as directed and the boy was returned to his father.

José María Enters the Seminary

José was a studious child who loved school and did well in it, finishing at the age of fifteen. He was a well-educated young man at a time when the majority of Mexico's people were illiterate. Around 1843, his father sent him to the Jesuit seminary in Mexico City to study for the Catholic priesthood. After three months José rebelled. He knew it would take another grueling fifteen years to complete his education in the seminary before he could become a priest. The boy was already becoming a freethinker—a characteristic that would define him throughout his life. He ran away from the seminary on Christmas Day and, still only in his mid-teens, struck out on his own.

José's running away from the seminary angered his father so much that he disowned his son. The boy dared to return home only once, arriving in his father's absence to find his mother in the family courtyard sitting at the fountain and combing out her long, red hair. He kissed his mother, said a hurried goodbye, and left with the haunting memory of tears streaming down her face.

Wandering in Mexico

José fell in with a congenial companion his own age. The young men found employment in a nearby city and stayed several months. But they turned to gambling, which led to a drifter's life. As vaqueros outfitted in charro clothing, they drifted from hacienda to hacienda. The boys were finally able to buy mules and supplies to travel the long trails between towns.

One day as they rode with their pack mules through the foothills of a large range of mountains, José María stopped for a few minutes while his partner continued up the trail. When his partner reached a short turn in the path, two bandits jumped out from ambush and grabbed both the bridle of the horse he was riding and the bag of gold he was carrying.

José soon arrived at the scene of the attack, and the bandits attempted to grab his horse's bridle but narrowly missed. Glancing around quickly, José saw his partner stretched lifeless and bloody on the ground, his throat cut and eyes glassy. In a flash, José reined his horse backward, turned, and sped away. He ran his horse until both were exhausted, but the robbers did not follow. Thoroughly lost in prairie country, José finally slowed to get his bearings. He pressed on, tired and hungry, and late that evening he stumbled on an old deserted adobe hut. After unsaddling his horse, he entered the hut, spread a blanket on the floor, and fell into exhausted sleep, ignoring a strong, unpleasant odor that filled the hut.

At dawn José awoke to find a dead man lying in the far corner, a sombrero covering his face and a blanket over his body. The man had been dead for some time. He bolted from the hut, mounted his horse, and rode off again. That evening he came to a well where a donkey was running a jack pump to draw water into a dirt tank. He watered his horse and drank, then lay down to rest. In a few hours, he remounted and rode on over the plains. Spying a distant light after nightfall, he rode toward a fire at a sheep camp. On reaching camp, he was met by an old sheepherder who invited him to stay the night. The old man gave the famished youngster mutton, bread, and coffee. The sheepherder supplied directions to the nearest town, explaining that he had to be off with the sheep by daybreak.

José slept until almost noon the next day and awoke to find over his head a shade of sheepskin and a young dressed mutton hanging on a nearby bush. After downing a hearty breakfast, he saddled his horse and headed toward Guaymas on the coast of the Sea of Cortés in Sonora, hundreds of miles from home. In a short time he was hired as bookkeeper to Don Juan Martínez on a big hacienda. He stayed with this job a year before traveling farther north. He met a man who dealt in livestock and needed a partner, and together they bought mules to ship to California. José intended to keep a handsome little white mule as his saddle mount. The partners loaded their mules on their hired ship and started up the Gulf of California. Before long an El Niño storm came up and threatened to sink the little ship. The ship's captain told José and his partner that he would have to jettison the mules and then threw all the animals overboard despite José's pleas to spare his favorite.

On American Soil

When they were safe in California at last, the partners headed east across Arizona until they reached the Río Grande near El Paso. They followed the river north, arriving in about 1861 at Albuquerque in New Mexico Territory during what José called the "War of Secession."

In Albuquerque, Aguayo settled down for nearly ten years as a lay teacher at San Felipe de Neri Catholic Church in what is now Old Town. He supplemented teaching with a second job as a bookkeeper. Yet he was still restless, so he decided to head south to Tularosa in Lincoln County by horseback. He followed the Río Grande before cutting east to Tularosa and arrived there late in 1870.

In Tularosa, José settled down and married a De La O girl from Doña Ana village. The couple had not been married long when José's wife died. José then met Fran-

Francisca Hill Aguayo and José María Aguayo. Date and photographer unknown. Courtesy John and Wesley Mottinger.

cisca Hill. Born in 1851 in San Pablo, Chihuahua, Francisca was the daughter of James Alexander Hill and Fernanda Quintero.[4] James, an adventurer from Kentucky, migrated to Chihuahua, Mexico, in the late 1830s and settled in San Pablo Moqui, where he became wealthy raising cattle. He married Fernanda, a Mexican girl, and they had at least ten children. Francisca was third from the last. For unknown reasons, the Hills reentered the United States in 1866, leaving their ranch behind and settling in Paso del Norte (present-day El Paso, Texas).

Francisca and José were married in Tularosa on February 15, 1873. He was forty-five; she was twenty-two. The couple would become parents to nine children, seven surviving to adulthood: Aristoteles "Harry," born 1874; Alejandro "Alex," born 1876; Araminta, born 1878; Ester, born 1884; Sara, born 1886; Lucila, born 1889; and Amanda, born 1891.

The José Aguayos in Tularosa

The Aguayos lived near Tularosa for two years.[5] José started a community at Álamo Canyon in partnership with three gringos. The men built homes, planted fruit trees and gardens, ditched the water, and irrigated their plants and trees. In their short two years at Tularosa, the Aguayos accumulated a small herd of cattle, a team of high-spirited horses, and a small, cozy home, and they welcomed their firstborn, Harry, on December 16, 1874. One summer Francisca grew a garden containing vegetables and a riot of pretty flowers. She also planted a packet of watermelon seeds; watermelons were considered a prime delicacy in the semidesert.

The melons ripened, and one day a Mescalero Apache came to the house accompanied by his woman. The navy beans had just been harvested, and Francisca noticed a pleading look in the eyes of the woman as she watched Francisca shuck the beans. So Francisca gave the woman a small bag of beans, a thick slice of sowbelly, and some molasses. The woman appeared so grateful that Francisca added one of her cherished melons to take back to the reservation. It was not long until another Mescalero man came by and asked for another of the prized melons. When Francisca told him none were ripe, the Indian became very angry, shook his fist at her, and muttered as he rode off.

Threatened by Mescaleros in Dark Canyon

Word came from Lincoln in early 1875 that Francisca was needed to take care of her terminally ill sister, Sarah Hill, who had been a schoolteacher there. José had just found work as a hand at a ranch near Tularosa and could not leave since he needed money to add to the couple's herd in the spring. They decided that Francisca would go to Lincoln, so she hired a boy named Bill Brady to drive the team for her and invited a neighbor girl, Sadie Copeland, to ride with them and help take care of the baby. They were not long on their way when the little party observed three Mescalero Apaches trailing them. Francisca recognized one as the man who had demanded a watermelon the previous summer. The Indians began harassing them, riding close and peering under the wagon cover. Before they reached Dark Canyon, Francisca stopped Brady and told him to go no farther for fear of an ambush.

Twilight of the first day found Francisca and her party camped outside a stretch of badland not far from Dark Canyon. The next morning the travelers woke early and headed into the mouth of Dark Canyon. A clear mountain stream flowed through the canyon. At noon they stopped to eat lunch and rest the horses. Francisca had left the baby with Sadie and wandered a short way up the stream when she heard a small rock break away from the canyon wall and splash into the stream. Looking up onto the rimrock, she spotted the three Indians. She sped back to camp and herded the rest of the party into the wagon, took the reins herself, and whipped the team forward. The canyon widened in a few miles, but the walls began to close in again as they approached the narrow end, which was the most likely place for an ambush.

They urged the horses into a run and raced through the canyon to open country. In camp that evening, Francisca spread a cold supper of sausage, jam, and bread. Again everyone bedded down, but no one could sleep. All the adults huddled together, wrapped in blankets. The next morning, two men in a buggy—the Indian agent from the Mescalero Apache reservation and his secretary—drove out of Dark Canyon. Francisca told them of the shadowing Indians, and the agent promised to make sure the Indians returned to the reservation.

In Lincoln

In 1875, with baby Harry four months old, José and Francisca moved to Lincoln, where the family would stay for about fifteen years. They arrived just in time for a front seat at the Lincoln County War. Also at some time during these years, family legend says that José joined in a card game with a Catholic padre and, being a bit *borracho* at the time, lost most of his property and money. After sobering up, as a form of symbolic revenge, José changed the names of his children to Greek from the Spanish—hence "Alexander" for Alejandro, and so on.[6]

José was elected justice of the peace at Lincoln and presided over a hearing in which three Fort Stanton soldiers were accused of lynching a certain William S. Pearl.[7] A drunk Pearl had shot and killed Pvt. John Downey in Fort Stanton after an auction there and was jailed in Lincoln. A party of soldiers battered down the door of the jail on January 23, 1883, and lynched Pearl.

As a lawyer, José defended Billy the Kid in one of his legal troubles.[8] José's eldest son, Harry, said that when his father was justice of the peace, Billy often came to José's office for advice about legal affairs. Billy would sit in a round-backed office chair next to his father's desk and discuss business, and the two became friendly.[9]

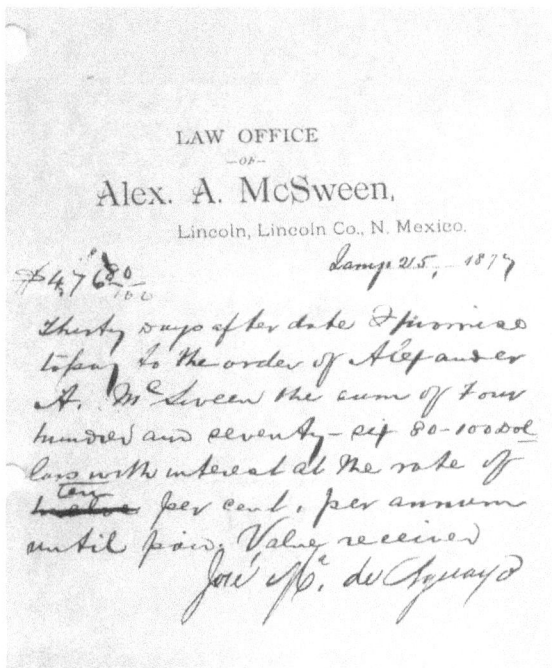

Note for $476.80 from José M. Aguayo to Alex A. McSween, 1879. Courtesy John and Wesley Mottinger.

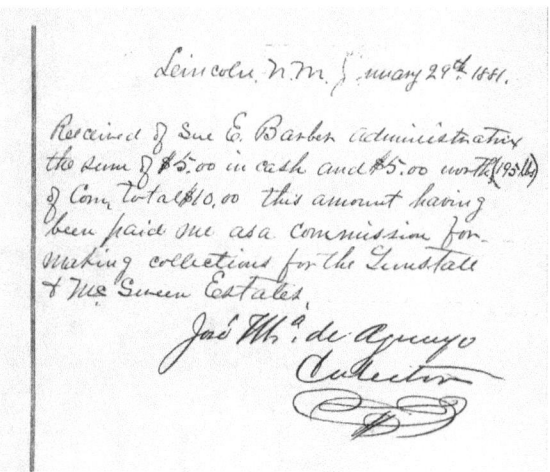

Commission from Sue E. Barber to Collector José M. Aguayo, January 29, 1881. Courtesy John and Wesley Mottinger.

Harry also remembered Billy coming often to their house with his guitar. The Kid loved to play and, according to Harry, sang "like a mockingbird."

José's survival of the Lincoln County War was nothing short of miraculous, especially since he collected debts for *both* the opposing McSween and L. G. Murphy & Co. Mercantile stores. As debt collector, he traveled all

Francisca Hill Aguayo and daughters: *back row right*, Francisca; *back row left*, Lucila; *front row left*, Sara. Date and photographer unknown. Courtesy John and Wesley Mottinger.

over Lincoln County. Farm products were often taken in payment for supplies sold by the mercantile houses. José would return to Lincoln with cattle or horses, loads of corn and wheat, or other farm products. After the "Big Killing" in which Alexander McSween and others died, Susan McSween hired José to collect debts owed her husband's estate. Her report to the court spells his name as "Aguyia."[10] The 1880 census for Lincoln lists José Aguayo as a lawyer, fifty-three years old, with a wife and four children.[11]

The Aguayos Move to Texas Park near White Oaks

In 1890, José moved his family to a small ranch in Texas Park about eight miles out of White Oaks, where they raised goats and lived for over a decade.[12] The Aguayos became famous for their goat cheese, which they sold to the elite of White Oaks.

José, a man of ideas and wide-ranging thoughts, became interested in the events leading up to the Spanish-American War, which included the struggles of the Cuban people for independence from Spain. He was highly critical of the Spaniards and so named his Texas Park ranch Cuba Libre as a show of support for Cuba.[13] José never learned English; nonetheless, he taught school in San Patricio in Lincoln County for almost eleven years at a salary of thirty dollars a month from New Mexico Territory. His children, however, were thoroughly immersed in two cultures, and the Aguayo offspring became proficient in both Spanish and English.

Amanda Aguayo, 1909. Photographer unknown. Courtesy John and Wesley Mottinger.

Sara Aguayo, 1909. Photographer unknown. Courtesy John and Wesley Mottinger.

Five Aguayo relatives. *Left to right*: Sara Aguayo Fisher, Frances Aguayo Vega Alaríd, Humphrey Aguayo, Lucila Aguayo Abeyta, Alex Aguayo. Date and photographer unknown. Courtesy John and Wesley Mottinger.

Eventually José developed a liver ailment and died on November 3, 1901, at the age of seventy-four. Francisca lived on until February 27, 1927, dying in El Paso at a daughter's home. Both are buried in El Paso's Evergreen Cemetery.

Harry Aguayo, Bystander as José Chávez y Chávez Murders a Prisoner

In 1881 when José's son Harry was about six years old, he and his friend Adolpho Romero were playing marbles near Captain Baca's house in Lincoln.[14] A man holding a new rifle walked by, heading toward the nearby makeshift jail where several prisoners were held. The boys were near enough that they could hear the chains rattle as the prisoners moved.

The man said, "Hello, boys, playing marbles, are you?" and walked on. He entered the jail and in a moment the boys heard two quick shots, a man screaming "Oh, God!" and then only silence. The man emerged from the shed, and as he passed by the boys again, he said, "No gringo can take *my* gun away from me!" and continued on to town. The next day José María saw him on his way out of town riding a good horse and leading a pack animal. The man, José Chávez y Chávez, landed in Las Vegas, New Mexico, where he became an outlaw in the infamous White Caps Gang, also known as the Society of Bandits.

In May 1894, Chávez was caught and taken to a Las Vegas jail, brought to trial, and sentenced to hang. Gov. Miguel Otero commuted the sentence to life in prison. A prison riot occurred later in which Chávez assisted the guards; for his actions Gov. George Curry pardoned Chávez on January 11, 1909.[15]

Harry's Take on Who Slipped Billy the Kid His Getaway Pistol

It has never been determined who it was that slipped Billy the Kid the gun he used to kill jailer James Bell on April 28, 1881. Historian Frederick Nolan writes that regular visitor Sam Corbet helped engineer the escape.[16] Corbet was said to have brought the Kid a note informing him that a pistol would be planted in the privy in the back of the courthouse. According to this account, José Aguayo placed the pistol there just before noon on the day the Kid escaped. Another account, by historian Maurice Fulton,[17] has Bell accompanying the Kid downstairs and into the yard to the privy. The Kid retrieved a pistol placed there by a boy who lived nearby and who was an admirer of his. He hid the gun inside his shirt. On the return trip, the Kid preceded Bell up the stairs and, upon reaching the top, drew the gun and whirled on the deputy. Bell turned and ran, but Billy fired, fatally wounding Bell. Fulton names the boy who furnished the gun as José M. Aguayo. The problem with Fulton's story is that José was at this time a respected fifty-four-year-old citizen and his eldest son Harry was a mere child at six years old.

Many years later, Albert Roberts of Carrizozo recalled that Harry Aguayo used to relax in an office chair at Roberts' service station telling tales of yore.[18] Harry had said that as a small boy he overheard his father talking with a young man named Demetrio Perea. Perea liked

Billy and visited him often at the Lincoln jail. On one such trip Perea told Billy that the next day at noon when Deputy Olinger routinely took the prisoners across the street to the cafe, Billy should say he was not hungry.

The following day Billy seized his chance, telling Olinger he was not well and did not want to eat. After Olinger left with the other prisoners, Billy told Bell he needed to visit the privy out back. At the door of the outhouse, Bell unlocked Billy's handcuffs and Billy stepped inside and found the promised pistol hidden on a two-by-four above the door.

Harry Sees a Man Hung in Las Vegas

Harry Aguayo ventured out on his own at about sixteen years of age. He saw a man hung in Las Vegas, New Mexico, about this time when he was working in a livery barn.[19] A freighter with a load of wool had come into town from the north. By a lake, he noticed a lone burro struggling with a load. The load turned out to be two dead sheepherders tied together by the feet and thrown over the burro's back, with their heads dragging on the ground. Upon arriving in Las Vegas, the freighter reported to officers what he had seen. The killer was caught near Los Álamos, brought to Las Vegas, and thrown in jail. Around nine or ten that night, a crowd of masked men arrived with a long rope. They broke the door down, dragged out a man, and took him to a nearby telegraph pole, where a crowd, including Harry, witnessed the hanging. The next morning an inquest was held, but nothing came of it.

Harry, Cowpuncher for the Block Cattle Company

In February 1895, the Block Cattle Company began moving out a herd of two thousand head from the Block ranch to Clayton, New Mexico.[20] Several men from Kansas were to meet the herd at Clayton and take it on to Liberal, Kansas. The sprawling Block ranch, located on the north side of the Capitán Mountains in Lincoln County, was operated by H. K. Thurber of New York City. Tom Pridemore was range boss in charge of some handpicked cowboys, including Harry Aguayo.

This proved to be a hard winter. Riding in bitter cold, the Block hands found the Chisholm Trail near Fort Sumner where they planned to follow it to Clayton. It was tough crossing the Llano Estacado before arriving at Fort Bascom north of present-day Tucumcari.

Early one morning Harry and his partners were riding along in the lead of the herd when they spotted a large yucca. Dry grass was abundant all over the country and piled high around the yucca. They decided to start a fire to warm their frozen feet. They struck a match and soon had a nice fire going. They were enjoying the warmth of the fire when they noticed that another flame had sprung up a few feet away—then another and another. Dozens of fires sprang up before they could be put out. Then the wind began to blow. The cowboys unsaddled their horses and fought the fires with their saddle blankets as hard as they could, but the fires raced out of control to the east. Soon the rest of the Block crew arrived to help, but it was too late—a monster fire roared on toward the Canadian River and followed it far down into Texas. Ironically, not long after the fire, a bad snowstorm blew up. The cowboys had no wood with which to build fires or to cook meals.

The boss sent a wagon into Clayton for coal; there was none. The wagon brought back a load of wooden boxes, but they proved almost useless because they burned so quickly. Pridemore would send out four or five men in the blizzard to hold the herd for fifteen minutes, and then other men would relieve *them* for fifteen minutes. Harry and his partners kept this up through the blizzard. It stormed for three long days; it was a blizzard to remember for many years. Meanwhile, the men from Kansas were holed up in Clayton in warm hotel rooms. They took over the herd when it arrived and moved it on to Kansas while the Block boys returned to the ranch.

In May, Harry helped drive a second herd of 1,800 cattle to Richfield, Kansas. The Block cowboys delivered their herd without incident at Richfield and had a fine time on the way home, thinking up pastimes such as horse racing to break the boredom. Back at the Block, it was business as usual—time for the spring roundup, branding calves, and the customary summer work.

Harry Witnesses the Murder of George Parker by George Musgrave

In October of 1896, riders for the Bloom Cattle Company (made up of the Diamond A and the Circle Diamond) were helping in a cattle roundup on the Río Felix in Cháves County about thirty-seven miles southeast of Roswell. Harry and Dan Welch were riding the range together when they noticed two riders with torn and ragged clothes approaching. As the two riders rode up, Harry recognized one of the men as George Musgrave. Musgrave had worked with them years earlier; the other man was a stranger named Hayes. George called out, "Hello, Harry—Are you still here? You must be a good hand!"

Harry replied, "No; I am still here, though." George continued, "Who is the boss here? George Parker—where is he?" Musgrave then asked whether the cook might give them something to eat as they had not eaten for a while. Musgrave and Hayes rode down to the Circle Diamond chuck wagon, where Sam Butler, the cook, fed them cold meat and bread. Afterward they took out their Bull Durham tobacco, rolled cigarettes, and sat down to smoke.

In the meantime, all the cowboys, including Harry and Dan, were riding into camp for a rest after the herd had settled down. As the cowboys straggled into camp, Musgrave continued to ask questions of the cook: Who is this? Who is that? Butler identified each rider until finally the last three men rode into camp, and again Musgrave asked who they were. Harry answered, "That one is Les Harmon, a good friend of mine and a Block man; Billy Philips, a VV man; and George Parker, our boss."[21]

Musgrave was sitting on one of the cowboy bedrolls scattered around near the water barrel. Parker dismounted and walked toward the water barrel, nodding at Musgrave as he walked but without recognition. Musgrave stood up quickly as if to shake hands but instead pulled out his pistol and shot Parker four times in rapid succession. Parker died instantly as the first shot struck in his left breast at such close range that his clothes caught fire; the second shot lodged in his right shoulder. Musgrave shot Parker again in the top of his head and a fourth time in his back.

As Parker fell, Musgrave's companion drew his gun on the astonished cowboys and warned everyone not to make a move. Musgrave then spoke to the crowd of fifteen or twenty cowboys, claiming that Parker had caused him a lot of trouble and that he was there for revenge. The cowboys watched helplessly; no guns were allowed in camp. Musgrave walked to the remuda and roped Parker's riding horse for himself while his partner roped a horse called Swallow Fork, Parker's cutting horse. Leaving their own worn-out saddles on the ground, Musgrave and Hayes waved goodbye to the cowboys and rode off with new "borrowed" saddles on Parker's horses. Only then did the cowboys notice smoke coming from the burning clothes on Parker's body. The cowboys turned him over and put out the fire.

Three men rode into Roswell to notify the sheriff. Musgrave and Hayes headed west to the C. D. Bonney Ranch southeast of Picacho, intending to kill Bonney as they had Parker, but Bonney was not home. A posse pursued and caught up with the pair on a ranch near the New Mexico–Arizona line, and in the fight that followed Musgrave's partner was killed but Musgrave escaped and rode on to Montana.

A Long-Distance Horseback Ride

Harry Aguayo made several memorable long horseback rides. One was in 1908 from the town of Douglas, Arizona, to White Oaks.[22] In 1903, Harry had married a girl named Edith Sheppard, and by 1908 they had three children. (They would have six sons and three daughters.) Harry moved his family in 1907 to the U.S.-Mexican border town of Douglas, Arizona, a hub for surrounding ranches and smelting operations, where he hoped to find a good job. The Aguayos made the trip to Douglas in two wagons—one of them a new Studebaker wagon. But when Harry saw the slag pouring out of the buckets in the copper smelter, he knew this kind of work was not for him; he had to have the freedom of open spaces. Harry sold his wagon and team to a hardware store in Douglas for $325. Edith bought a train ticket to Carrizozo and from there took the mail hack to White Oaks to await her husband. Harry set off overland for White Oaks on two good saddle horses. He cut across the desert without a map, accompanied only by his bulldog, Bounce, whose job was to guard the staked horses at night.

When the dog's feet got sore, Harry greased them and tied the dog on the packhorse until he could walk again. The two camped out in the desert at night. One day Harry spotted a windmill and headed toward it to water the horses. A huge lynx arrived on the scene and Bounce jumped it. The dog and the cat rolled around scratching and biting until they finally broke apart, giving Harry a chance to shoot and kill the lynx.

Harry arrived on the banks of the Río Grande at Las Palomas, south of present-day Truth or Consequences, where he bought corn for his horses. When the animals later tried to eat the blue and white grain, Harry found that he had been cheated; most of the white grains turned out to be little white rocks.

Las Palomas usually had a ferry boat that could be hired to cross the Río Grande, but this was March and the river was swollen and running swiftly. Even at the best spot the Río Grande ran as wide as a city block, but Harry and his horse plunged in and fought their way

Harry Aguayo and wife Edith in their Nogal orchard. Date and photographer unknown. Courtesy John and Wesley Mottinger.

across. As Harry turned to wave to the men watching on the far bank, he spotted Bounce running back and forth on the other side. He pondered what to do until he saw the dog jump in and disappear up river. Finally the dog's head popped up exactly where Harry was standing watching for him.

By the time Harry reached Engle (a major cattle shipping point on the railroad nine miles east of Elephant Butte), he was broke and hungry. He encountered some Diamond A cowpunchers out gathering steers and headed to their ranch house to ask for breakfast, then rode on to the Diamond A camp to stay the night.

Harry rose early the next day and turned his pony across Lava Gap in the San Andrés Mountains and camped near Jim Gililland's ranch. He rode all the next day to the Bar W Ranch just north of Carrizozo and stayed the night before finally reaching White Oaks. The horseback trip had taken three weeks.

The Harry Aguayo Family Back in Lincoln County

Harry rented a farm below Lincoln and soon learned that he was a better cowboy than farmer. He planted corn, but just before the corn was to be harvested a freeze ruined the entire crop.[23] The next year Harry moved the family to San Patricio on the Río Ruidoso, where they stayed three years and built up a cattle herd. By the time they decided to leave for Tortolita Canyon near Nogal, they had a herd of thirty-five head.

The Aguayos lived in Tortolita Canyon for twenty-three years and retired in the railroad town of Belén south of Albuquerque. Harry died on April 20, 1960, and is buried in Carrizozo's Evergreen Cemetery. Edith lived only three months longer; she died August 6, 1960, and is buried beside Harry.

NOTES

1. J. S. Brimberry, notes on the origin of the Aguayo name, n.d. Written on his letterhead titled "Oilfield Transportation Co., Brimberry Interests, Odessa, Texas," copy in author's collection.

2. Harris, *A Mexican Family Empire*, 6–9, 27, 35–37, 69, 106, 113, 162–66. Harris focuses primarily on the rise of the Sánchez Navarro family of landholders in the Mexican state of Coahuila. For decades that family jockeyed for preeminence as ranchers in Coahuila and Zacatecas with the Aguayos. Harris mentions the Aguayos in his book, and it is primarily from his research that the Mexican history of the Aguayo latifundios is taken. See also Thrapp, "Aguayo, Marquis de," 10; Buckley, "The Aguayo Expedition," 15; and Hackett, "The Marquis of San Miguel de Aguayo," 49.

3. The Lincoln County, New Mexico, branch of the Aguayos wrote a family biography, recounting José M. Aguayo's beginnings in the Mexican state of Jalisco and following him through his travels north to California and, eventually, to New Mexico Territory and a life in old Lincoln County. The Aguayo story continues with the memoir of José's oldest son, Aristotle, or Harry. These stories—resulting from a cooperative effort among Harry Aguayo, his wife, and his sisters Araminta, Sara, and Amanda—were compiled by Harry's daughter Lorraine Aguayo Brimberry of Ruidoso. The papers, undated and not paginated, were furnished to the author by Brimberry. Author's collection. Hereafter cited as "Aguayo family biography." John Mottinger of Santa Teresa, New Mexico, also furnished material about the family.

4. Benjamin Hill, "The Descendants of James Alexander Hill," n.d., copy in author's collection.

5. Aguayo family biography.

6. L. A. Brimberry, telephone conversation with author, January 26, 2007.

7. Notes on the lynching of William S. Pearl by Fort Stanton soldiers, January 23, 1883, Robert N. Mullin Collection (hereafter, "Mullin Collection").

8. A. H. Aguayo obituary, *Lincoln County News*, April 29, 1960.

9. Ibid.

10. Notes on the lynching of William S. Pearl.

11. U.S. Bureau of the Census, *United States Census of 1880*, Lincoln, Lincoln County, New Mexico.

12. Aguayo family biography.

13. Brimberry, telephone conversation with author, January 26, 2007.

14. Aguayo family biography.

15. Metz, *Encyclopedia of Lawmen*, 43. Emphasis added.

16. Nolan, *Lincoln County War*, 418.

17. Fulton, *History of the Lincoln County War*, 394–95.

18. Roberts to Stearns, November 15, 1978, copy in author's collection.

19. Aguayo family biography.

20. Ibid.

21. Ibid. See also Tanner and Tanner, *Last of the Old-Time Outlaws*, 69–75, 190–95.

22. Aguayo family biography.

23. A. H. Aguayo obituary.

JOHN AH NUE, A.K.A. "JOHN CHINAMAN"
Celestial Extraordinaire

By 1852, a mere three years after the California gold rush and only five years after a grand total of three Chinese immigrants first sailed for San Francisco, some twenty-seven thousand Chinese people could be found in California.[1] More and more Chinese were fleeing the extreme poverty, famine, and chaos in their homeland in search of what they called the great "Gold Mountain." Although they were welcomed initally to do work that whites would not, the Chinese came to represent unwanted immigration. Whites derisively dubbed them "Celestials" or "John Chinaman," and mutterings of violence began. The Chinese retaliated by calling the whites "barbarians."

Chinese men, with their traditional queues or pigtails symbolizing allegiance to the Chinese emperor, at first did a lot of what whites thought of as women's work: cooking, laundry, and other domestic jobs. Eventually they began branching out into working in the goldfields and mines.

In 1870, Congress granted the Chinese civil rights but at the same time refused to grant them citizenship. Although Chinese immigrants had been in the country for nearly two decades, xenophobia was picking up steam. By 1873, the Chinese began to be demonized. Denis Kearney of the Workingman's Party sought their expulsion from California. After Kearney became mayor of San Francisco, he took his campaign nationwide. Life soon became very dangerous for Chinese workers.

In 1881, eleven bills calling for Chinese exclusion were submitted in Congress. On May 6, 1882, the Chinese Exclusion Act became law. The Chinese could no longer enter the United States unless they fit into certain strictly defined categories, such as merchant or teacher. It was the first anti-immigrant law in U.S. history and remained in effect until 1943, when China became the United States' ally in World War II.

Ah Nue Sails Into San Francisco

Sometime in the early 1870s, a Chinese man in his twenties named Ah Nue arrived in San Francisco on a sailboat.[2] (In Chinese naming traditions, "Ah" would denote the family surname and "Nue" the given name. Although both Nue himself—as well as legal documents—spelled his name as "Nue," most everyone else referred to him incorrectly as "Nu.") He was born in Canton, China, probably in June or July of 1847. In 1900, Ah Nue told the census taker he was fifty-five, three years older than his real age, and by 1910 he had added another five years to his age, saying he was seventy.[3] Because age was venerated in China, he may have exaggerated his life span to gain respect.

For several years Ah Nue worked in San Francisco. He remembered when the railroad was built into San Francisco from San José.[4] "Go West Young Man" turned into "Go East" for Ah Nue. Cooking his way across the mining camps of Arizona, he arrived in Lincoln County in the mid- to late 1870s. He spent a few years as a cook on the Bar W Ranch just north of present-day Carrizozo, keeping the Bar W cowboys well fed. He knew of Billy the Kid, who sometimes rode by the Bar W or stopped in for a meal, but he was just another cowboy to Ah Nue, who said, "Plenty boys shoot in those days." Ah Nue also told a newspaperman that "Albuquerque big town now, I hear. I remember Albuquerque and Las Vegas I visit there 56 years ago. Albuquerque not much big then. Las Vegas better town then."[5]

After a time, Ah Nue felt the lure of White Oaks near to the Bar W and was himself bitten by the gold bug.

The Partners Ah Nue and Wah Sing Arrive in White Oaks

Ah Nue moved the short distance to White Oaks in 1884 when it was still thriving and was the liveliest city in the southeast part of the territory. He brought with

Ah Nue's log house today. The cabin sits on what once was Willow Street. Photograph by Bob Workhoven. Author's collection.

him Wah Sing, his friend, companion, and business partner. Wah Sing was five years younger than Ah Nue, about thirty-two years old. Everyone in White Oaks called Ah Nue "John Chinaman," and he even received much of his mail addressed this way.[6] The partners built a small cabin and operated a laundry out of the log house where they also lived; this structure can still be seen on the east end of White Oaks. Behind the cabin they planted a few apple trees.

Soon after arriving, the thrifty pair put their accumulated savings into a restaurant located in the block next to the Brown building.[7] Then they began investing in White Oaks real estate. An April 1887 *Lincoln County Leader* reported, "Our Chinese laundryman, not satisfied with the lots he already had, this week purchased the house in which Mrs. Frain teaches school and the lot on which it stands. John sees the coming boom as well as those whose eyes are not almond colored."[8]

Ah Nue cooked in the restaurant; Wah Sing operated the laundry. The two men continued to operate their laundry and to do odd jobs, always saving their money. On November 1, 1888, Ah Nue applied for a license as a hotelkeeper.[9] The "hotel" was actually a boarding house that advertised itself as offering meals at all hours. As a sideline Ah Nue sold teas. The *Lincoln County Leader* of April 1887 carried what it called a "Chinese Notice":

Ah Nue's ad for a restaurant and boarding hotel from the *Interpreter*, January 17, 1890. Author's collection.

Ah Nue's ad for a restaurant and boarding hotel from the *Interpreter*, March 23, 1888. Author's collection.

Ah Nue's license to operate a hotel, issued July 22, 1889, by George Curry, clerk of the Board of County Commissioners. Courtesy Chris and Jack Harkey.

"The subscriber is in receipt of a fine assortment, direct from China, of all kinds of Chinese Fancy Goods, to which he invites attention, and will sell low."[10]

A favorite hobby among Chinese men was collecting songbirds, sometimes rare and costly birds. Often they would take the birds, still in their cages, for walks to keep up their avian spirits. The *Golden Era* newspaper reported that "Ah Nue, the good natured Chinese launderer, has a number of canary birds and spends his leisure time caring for his little pets. Several of them are setting, one having successfully hatched three birds. 'John' takes pride in showing his feathered family to callers."[11]

The Brutal Murder of Wah Sing

On the night of August 8, 1893, Wah Sing died in a vicious attack that today would be called a hate crime.[12] In the still of the night, terrible screams from Ah Nue and Wah Sing in their log shack awoke the sleeping town. John Owens, a constable of Lincoln County who lived in White Oaks, immediately went to investigate. He found Wah Sing hacked to death and Ah Nue injured, both lying in pools of blood. Wah Sing was only forty-one years old.

Though badly hurt, Ah Nue survived, but the murderer had chopped off his queue in the attack. Ah Nue had lost his honor as a Chinese man; he could never return to his homeland, never send for a Chinese bride. Eventually he regrew his queue and proudly wore it long and white through old age.

Although no one was ever brought to justice for this terrible crime, it is suspected that a disgruntled gambler had left the Little Casino Saloon where he had lost all his money gambling. He then went on a rampage, looking to replace his lost money by robbing the Chinese partners who he was certain had money hidden in their log house. Town gossip had the Chinese partners keeping their money hidden at home in coffee cans.

Ah Nue in Later Years

Eventually Ah Nue became a handyman for Judge John Y. Hewitt, the wealthiest and most powerful man in White Oaks, until Hewitt's death in 1932.[13] During these later White Oaks years, newspapermen would interview Ah Nue from time to time. Ah Nue would say such things to them as "People too smart these days. God will punish people for being too smart" and "It will get hotter and drier. People don't believe it, but it will get hot and dry like it was 240 to 400 years ago. I talk to God and he tell me. Used to rain more here, you see. Pretty soon burn everything up. Maybe this year rain pretty good,

Ah Nue and friends in front of his log house, 1925. Photographer unknown. Courtesy White Oaks Historical Association.

but after that it be hot and dry."[14] He was also recorded as saying, "Lots more gold here. But people can't find it. Must go down 100 or 150 feet to get it. Nobody got enough money to do that. I know where is plenty gold; I talk to God and he tell me where it is, but these other people don't know."[15]

A boy in the 1920s, Robert Leslie remembers that after he got his first .22 rifle and began hunting, Ah Nue would buy cottontail rabbits from him. The old man gave him fifteen cents for a rabbit, which he would then boil and feed to his chickens, whose eggs he also sold. On holidays, Robert's grandmother filled plates of food and took them to the old man, who would promise her, "Grandmother, when I hit it rich, you're gonna be the first one I think of."[16]

Ah Nue kept digging in his gold hole, hoping to strike it rich. After Robert grew up, the young man would ride his horse to Ah Nue's hole to find him down underground filling his bucket.[17] He would climb out, hoist up the bucket, and descend into the hole again.

Ah Nue and the Big City

In 1913 or 1914, Ah Nue became seriously ill and his neighbors, the Kastler couple, became concerned.[18] Ah Nue believed he would get well if only he could consult a Chinese doctor. The Kastlers' daughter took him to a Chinese doctor in El Paso. His friends found comfortable quarters for him in a rooming house and left him in the care of a Chinese physician. The following day Ah Nue arrived by train in Carrizozo and was taken by friends back to White Oaks. When his White Oaks friends asked him why he had not stayed in El Paso, Ah Nue replied, "Oh, it worse than San Francisco. They just go, come, go, come, all night along. It drive a man clazy [sic]."

End of the Road

After losing patron Judge Hewitt, Ah Nue eked out an existence for another four years on a county pension of six dollars a month and the charity of neighbors who liked him and had also stayed on in White Oaks after the "bust."[19] In spite of their long acquaintance, Ah Nue and the other citizens of White Oaks continued to live differently. Robert Leslie did not think Ah Nue ever rode in any kind of automobile.[20] After Robert acquired a 1928 Model A, he would find the old man weaving up the street to the store. Robert would stop, and Ah Nue would look over Robert's Model A, stand there in the road for a moment, and without a word walk on. He never got in.

Ah Nue primarily ate chicken and salmon and bought a little rice, and he stripped boards off people's fences to burn for fuel. When he was almost ninety years of age, on an early Friday morning in January 1937, he was found on the floor of his log shack nearly frozen to death and unable to move hand or foot. People in White Oaks tried to give him all possible assistance, but he died about noon that day.

The Funeral

Ah Nue's funeral was a community affair. All the men of the town banded together and gathered at the Little Casino Saloon. There, under the gaze of all the pictures still hanging in the long-abandoned saloon, they took razor blades and cut off the green felt from the pool tables.[21] Then they built a wooden coffin and thumbtacked that green felt as a lining on the inside of the coffin. They laid out Ah Nue themselves and dug the grave. Dave Jackson's old Model T, nicknamed "Teddy," carried the coffin at a snail's crawl toward Cedarvale Cemetery. The townspeople slowly walked behind the car the long mile to the cemetery to bury their friend.

It has been said that if a Chinese should die on Gold Mountain, a proper funeral must be held and the body shipped back to China. For as all good Chinese sons know, if he is not buried in his home soil, his spirit will wander the earth forever. But if the spirit of Ah Nue wanders the streets of White Oaks, as some people believe, he is more likely to be looking for "lots more gold" instead of a return ticket to Canton.

Ah Nue's headstone in Cedarvale Cemetery. Ah Nue's year of death is given erroneously as 1936. It should be 1937. Photograph by Bob Workhoven. Author's collection.

NOTES

1. Hall, "A Walk with My Great-grandfather," 54–72. Hall's depiction of the early Chinese in America details their political and legal battles, as well as their cultural practices and heritage.

2. "Ah Nu [sic] remembered much history during his 40 years at White Oaks," *Lincoln County News*, March 6, 1969, reprinting an unsigned article sent to the *Albuquerque Tribune*, January 16, 1937. This news article was intended as both obituary and tribute.

3. U.S. Bureau of the Census, *United States Census of 1900 and United States Census of 1910*, White Oaks, Lincoln County, New Mexico.

4. "Ah Nu [sic] remembered."

5. Ibid.

6. Chris Harkey, undated communication with Haldane. Zane Leslie told Chris how the mail was addressed.

7. "Ah Nu [sic] remembered."

8. *Lincoln County Leader*, April 30, 1887.

9. License issued July 22, 1889, by George Curry, clerk of the Board of County Commissioners, copy in author's collection.

10. *Lincoln County Leader*, April 30, 1887.

11. *White Oaks Golden Era*, May 8, 1884.

12. As recalled by Brown in "Major step for White Oaks," *Lincoln County News*, June 12, 1969.

13. Dan Burroughs, "Plenty More Gold," from an article he wrote ca. 1930 and later gave to the Glencoe Women's Club. Reprinted by the *Lincoln County News*, April 11, 1958.

14. "Ah Nu [sic] remembered."

15. Burroughs, "Plenty More Gold."

16. Robert Leslie, taped interview by Reily-Branum, August 8, 1995, author's collection.

17. Ibid.

18. Ted Rayner, "Ah Nue and the Big City," The Folklore Corner, *Lincoln County News and Carrizozo Outlook*, February 17, 1956.

19. Burroughs, "Plenty More Gold."

20. Robert and Dorothy Leslie, interview by Haldane, July 22, 1999.

21. Ibid.

JAMES A. ALCOCK
Feisty Little Irishman and Manager of the Carrizozo Cattle Ranche Company

James Alcock was a young Irishman,[1] small of stature, likable, and outgoing. He first set foot in the northern Tularosa Basin in October 1882. As an agent for an English syndicate, he negotiated the purchase of the Lawrence Murphy ranch twelve miles southwest of White Oaks, intending to manage it himself.

Murphy (of the Lincoln County War's House of Murphy) had advertised his "Carisosa" Ranch for sale in February 1878 in the *Grant County Herald* because of poor health: "Desiring to leave Lincoln County, I offer for sale, cheap for cash, all my right, title and interest in the best Cattle Ranch in New Mexico, known as the 'Carisosa.' It secures twenty miles square of first class pasturage, and abundant water for ten thousand head of stock. There is also about 600 head of cattle of different ages, but mostly composed of young cows, with 14 young Durham bulls. Also, a residence, corrales, outhouses, library, poultry, choice hogs, horses, wagons, young orchard and many other advantages and inducements which can only be appreciated when seen."[2]

Murphy then left Lincoln in May 1878 for Santa Fe and died there on October 20. Before his death he renewed a note to Thomas B. Catron of Santa Fe, pledging the Carrizozo ranch as security. After Murphy's death, Catron took possession of the ranch and eventually sold it in 1882 to an English syndicate fronted by Alcock. Alcock also bought three other ranches from Catron, with the total paid for all four ranches amounting to $225,000.[3] Alcock was part owner of the Carrizozo Cattle Ranche Company (which in time became the Bar W) as well as its manager. Fall roundup had already been completed with more than 1,400 head of cattle branded by the time "little Jimmy" arrived. The *Lincoln County Leader* welcomed Alcock with this hopeful assessment: "The young Englishman [Irishman], Mr. Alcock, who negotiated the purchase of the ranch, did well, but anyone who asks him to sit down or take another ride just now, is liable to get hurt. Alcock stood muster first rate as a beginner. He will soon be a full-fledged cowboy."[4]

Most people around Lincoln County thought Alcock a "fine chap," according to cowboy Jack Thorp.[5] But he was sadly out of place running a cattle ranch and eventually almost drove the company on the rocks. The ranch as it was when Alcock arrived covered about forty miles by fifteen, with buildings of the usual low, single-story adobe type and the main house a model of comfort. Even though Alcock had not yet married, the house was furnished luxuriously and elegantly throughout, complete with a Steinway piano and fine engravings.

As his cook on the ranch, Alcock hired Gottfried Georg (Godfrey) Gauss, a German who had immigrated to the United States in 1853. Gauss was the same man who in 1879 witnessed Billy the Kid's escape from the courthouse in Lincoln and the murder of Deputy Sheriffs Olinger and Bell.

Alcock Helps Open the Salado Coalfield

In 1884, coal was discovered about twenty miles from White Oaks near present-day Capitán, and Alcock with others did considerable work in opening up this coalfield, known as the Salado.[6] At first the coal was hauled to El Paso or San Antonio in wagon trains pulled by horses or mules. Later on, Charles B. Eddy's El Paso and Northeastern Railroad extended its rails to Coalora (as the coalfield town came to be called), and for a short time about twenty cars of coal were shipped daily to El Paso. The coal played out, however, in only a matter of two or three years.

The Sheep Killing Cases

A growing problem throughout New Mexico in the 1880s concerned the rights to public domain. Along with the recent influx of foreign capital into Lincoln County came sheepmen who believed they too had rights. Within months of his arrival, Alcock—who ran only cattle—was in open conflict with the sheepmen in a battle the newspapers dubbed the "Sheep Killing Cases."

In 1883, Alcock bought Dionicio "Dio" Chávez y Sánchez's permanent water and grazing rights for $3,800 in order to get sheep out of his Three Rivers ranges.[7] However, Chávez had other water in the same country and continued to run hundreds of sheep. Before long, Chávez was accused of setting fire to the grass on Alcock's range and was brought in to court on December 26, 1883. This case was continued, and the defendant posted bail in the amount of $100. The case came to trial on January 14, 1884, and an all-Anglo jury found the defendant not guilty.[8]

Meanwhile, Arcadio Sais continued to graze a large number of sheep on a *partido* ("in equal shares") contract with Dio near the sheep ranch of J. E. Sligh, a sheepman who was also a circuit preacher at his place near White Oaks, on the outskirts of the Carrizozo Cattle Ranche Company's range. In June 1885, the *Golden Era* newspaper of White Oaks reported the slaughter of hundreds of sheep on the Carrizozo range: "Seven hundred sheep were killed one night last week on the Carrizozo range. The sheep belong to Arcadio Sais of this place, but Dionisio [*sic*] Chávez has been overseeing the herd. Some time ago J. A. Alcock, manager of the Carrizozo company, paid Chávez $3,800 for his ranch, with the understanding that Chávez was to move his sheep out of the country. It seems that Chávez still owns water close to the ranch sold to Alcock and that instead of driving his sheep out of the country, he has repeatedly drove [*sic*] them on his former range."[9]

Tongue in cheek, the article concluded, "It is barely possible anyone will ever know who killed those sheep." The *Golden Era* continued with this pronouncement: "Sheep and cattle were not made to run together. It is an indisputable fact that sheepmen have a lawful right to go on government land in the center of vast cattle ranges, and take up a small piece of land, and turn loose thousands of sheep if they can find sufficient water . . . but would it be right between men? It always has the appearance of blackmail. . . . It would make trouble every time. Let the cattlemen get to themselves and let the sheepmen get to themselves."[10]

Parson Sligh also entered the fray about this time. It is thought that Alcock and Sligh originally had bad blood between them because Alcock had refused to buy Sligh's mining claim. In July 1885, Sligh started a newspaper he called the *Interpreter*. The newspaper soon was dubbed the "Interrupter" because it did nothing but stir up trouble. Its first issues attacked prominent citizens of Lincoln County on the sole grounds that they were cattle owners. The principal target of Sligh's wrath was Alcock.[11]

On August 24, 1885, one of the young men in White Oaks boasted after a few too many drinks that he knew all about who had killed Dio Chávez's sheep. The informant went on to make a statement accusing Jimmy Alcock and four other men who were of previously good standing in the country. Hearing this drunken talk, Sligh rushed to the nearest magistrate and swore out three separate complaints. In the first complaint, Sligh stated that he had good reason to believe "that on the 12th day of June, 1885, the said parties did willfully and

Caricature of Parson Sligh as "Holy Jim," from the *Golden Era*, November 26, 1885. Author's collection.

maliciously shoot, kill and maim 500 sheep, property of Dionicio Chávez." The second complaint charged that the same men had "assembled with the intent, and with force and violence, unlawfully to maim, disfigure, disturb, interfere and drive away from the place where the same were being herded and pastured [illegible] thousand sheep, the property of Dionicio Chávez, and did then . . . proceed to execute and carry out said intent."[12] The third complaint charged that the same men unlawfully drove and interfered with a herd of sheep belonging to Dionicio Chávez.

Alcock and the other men named in the complaints were all arrested. In October 1885, cases against the "Sheep Raiders" were called for a third time before Justice Collier's court in White Oaks. All charges were dismissed on October 30, 1885. Sligh's principal witness was present in court to have his say, but he was never called.[13] The stockmen had the final word in the *Golden Era* of November 1885, in an editorial titled "HOLY JIM: The Sly Hypocrite Takes Shelter behind His Religion." In response to Parson Sligh's claims that his enemies were trying to destroy his reputation and character, the editorial countered, "The Parson is mistaken . . . We can't destroy his reputation and character as a minister of the Gospel, for the reason that he has none. We can show him up as a traducer of private character, as one who foments discord among neighbors, or as the friend, associate and confederate of cattle thieves—but his character as a preacher is something too intangible and impalpable." In another column Sligh insisted, "The beginning of the end is near at hand. The sheep kukluxers will be punished, and crime and criminals will yet be made odious in Lincoln county."[14]

Hospitality at the Carrizozo Ranch

Meanwhile, the Carrizozo ranch was becoming widely known for its hospitality to travelers. An unidentified wayfarer signing himself only as "H. S." describes his reception there in the *El Paso Daily Times*:

> East of this [the lava] beds is the Carrizozo cattle ranch . . . with the popular and efficient young manager, J. A. Alcock, whose residence is palatial and the surroundings more like Kentucky than New Mexico. Mr. Alcock came here from his native Ireland in 1882, and purchased the property from Tom Catron, Edward Walz and James Scott. It comprises a stretch of country 45 miles by 15 and is watered by thirteen mountain springs, one of which is near Mr. Alcock's residence and from which thousands of cattle obtain water. . . . The ranch numbers 15,000 head of cattle and rarely if ever have I seen the number look so fat and sleek at this season of the year. The capital of the Carrizozo is 80,000 pounds, all held abroad. The water when piped to different points on the ranch . . . will afford an ample supply for 100,000 cattle.[15]

Will Hudgens Challenges Alcock to a Pistol Duel

Besides the Sheep Killing Cases, Alcock became embroiled in at least one other serious argument with a White Oaks citizen, though this argument almost certainly arose over the Sheep Killing Cases. On December 9, 1884, George Ulrick for the territory of New Mexico brought suit against William Hudgens, owner of the Pioneer Saloon in White Oaks, " charging that on the 7th day of December 1884 in the town of White Oaks . . . appeared one William H. Hudgens [who] did deliver a challenge to fight a duel with deadly weapons to one James A. Alcock."[16]

On the same day, George Ulrick for New Mexico Territory brought the exact same charges against James A. Alcock for *accepting* Hudgens's challenge to fight a duel with deadly weapons.[17] Hudgens posted a $4,000 bond. Nothing further was heard of these charges.

Alcock Departs the Carrizozo Ranch

Over time the English syndicate owners became dissatisfied with financial losses under Alcock's management. Waiting in the wings to take over the business was a clever young White Oaks lawyer named William C. McDonald.[18] Around 1890, the English owners of the Carrizozo ranch, for whom McDonald had made several surveys and maps, appointed him as manager. Their first object was to find a purchaser. After a couple of years, and now more knowledgeable of the business, McDonald himself purchased the ranch and became sole owner. He cut down the number of cattle the ranch stocked, and, in later years of good rain, liquidated the ranch's debts accumulated by Alcock.

Alcock left the Carrizozo ranch permanently sometime in the late 1880s, accompanied by his new wife, Lulu. In 1889, the couple joined in Colorado's Cripple Creek gold rush (the national census for 1890 shows them in Teller County, Colorado), but they returned to New Mexico by 1894.

A few years later, writer Emerson Hough gives us our final glimpse of Alcock. Reminiscing about his return to White Oaks in 1904 after an absence of two decades, Hough wrote, "I don't find my old friend Jimmie Alcock, of the Carrizozo stock ranch. Poor Jimmie; he left White Oaks, went to Colorado, lost all his money, and had both legs broken in an accident. He was plucky, however, and is now over in Arizona, foreman in a mine at Bisbee, instead of foreman of a cattleranch."[19]

Lulu Alcock and child. Date and photographer unknown. Herman B. Weisner Papers, Archives and Special Collections, Ms249, New Mexico State University Library.

NOTES

1. Some historians, notably William Keleher, maintain that Alcock was English. However, the *United States Census of 1900* for Teller County, Colorado, lists "James A. Allcock [*sic*]" as age thirty-eight, having been born July 1861 in Ireland and married for thirteen years to wife Lulu. Lulu lists herself as born November 1868 in Okanogan, Washington, of Canadian parents. Alcock lists his occupation in the 1900 census as gold miner and noted that he had first immigrated to the United States in 1882 (he must have traveled directly to New Mexico Territory). Alcock returned to his homeland at least once. In 1888, the New York passenger list for the ship *Aurania* lists him as a rancher. In 1888, Alcock would have been twenty-seven years old, and his new wife, Lulu, twenty years old.

As further confirmation of Alcock's Irish origin, in 1887 a traveler identifying himself only by the initials H. S. wrote a long account of his trip from El Paso to White Oaks, "On to White Oaks. The Trip from El Paso to White Oaks," which appeared in the *El Paso Daily Times*, April 17, 1887. Crossing the malpais, H. S. arrived at Alcock's Carrizozo ranch and stopped over. He says that "Mr. Alcock came here from his native *Ireland* [italics added] in 1882." A copy of the *El Paso Daily Times* article of April 17, 1887, is on file in the archives of the Lincoln County Historical Society.

2. *Grant County Herald*, February 16, 1878.

3. Keleher, *Fabulous Frontier*, 114; notes on Murphy's Fairview Ranch, n.d., Mullin Collection.

4. *Lincoln County Leader*, November 18, 1892.

5. Thorp, *Along the Rio Grande*, 47.

6. Dorothy Guck, "Coalora," undated note, copy in author's collection.

7. *White Oaks Golden Era*, June 18, 1885.

8. *White Oaks Golden Era*, January 10, 1884; Justice of the Peace Court Records for White Oaks, 1884, Criminal Action No. 23, Lincoln County Courthouse, Carrizozo.

9. *White Oaks Golden Era*, June 18, 1885.

10. Ibid.

11. *White Oaks Golden Era*, November 5, 1885.

12. Ibid. A detailed recounting of the three separate complaints is also contained in this article.

13. The *White Oaks Golden Era* of September 3, 1885, says that W. H. Hudgens, Sligh's principal witness, had "taken a walk."

14. *White Oaks Golden Era*, November 26, 1885.

15. H. S., "On to White Oaks."

16. Justice of the Peace Court Records for White Oaks, 1884, Criminal Action No. 42, *The Territory of New Mexico, Plaintiff, vs. W. H. Hudgens, Defendant*, Lincoln County Courthouse, Carrizozo.

17. Justice of the Peace Court Records for White Oaks, 1884, Criminal Action No. 41, Lincoln County Courthouse, Carrizozo.

18. Morris Brown Parker, "The Parker Family: Ancestors, Relatives, Descendants," 1940, 54. Morris Parker wrote a fifty-nine-page account of the Parker and Wells families, including relevant dates, places, events, and business affairs. (William McDonald was Parker's father-in-law.) Lincoln County Historical Society.

19. Hough, "West after Twenty Years."

SUSAN McSWEEN BARBER
Cattle Queen of New Mexico

Susan McSween Barber was one of the most elusive, mysterious, and fascinating characters who ever lived in old White Oaks. As a key figure in the infamous Lincoln County War and an impressive survivor of the climactic Five-Day Battle of July 15–19, 1878, she proved herself to be a woman with an abundance of "sand" even as her husband was shot and her house burned to the ground.

Susan's arsenal of weapons during the war included crying, swooning, begging—and threatening. At one point, mistakenly believing that Sheriff Peppin's posse had killed her husband over on the Pecos, she grabbed a shotgun and marched in a rage to the Murphy-Dolan store to confront Jimmy Dolan.[1] While the informant does not record how or by whom this situation was defused, Susan was finally calmed down and bloodshed was averted.

Both Susan's early life and her involvement in the Lincoln County War have been documented by many historians—among them Frederick Nolan, Robert N. Mullin, and John P. Wilson—and will not be repeated here except for a brief summary. This section focuses primarily on shedding more light on what happened to Susan after that war—the years in which she married and divorced second husband, George B. Barber, and built a stunning career on her own as one of New Mexico's most successful women ranchers.

In Susan's last decades, she moved to White Oaks and to the substantial house from which she reigned for a time as the town's leading social doyenne. Then, incredibly, that house also burned. She struggled to maintain an upbeat outlook in spite of having to live like a pauper in what was little more than a shack throughout her final years at White Oaks. At the age of seventy-eight, she wrote a friend, "I am now living in this place and am very old, but supple."[2]

People constantly sought Susan out in White Oaks to hear about Billy the Kid but apparently failed to ask her about her own life. Among the visitors were former New Mexico governor Miguel A. Otero, writers Emerson Hough and Walter Noble Burns, and historians J. Evetts Haley and Maurice G. Fulton. Burns interviewed Susan for his book *The Saga of Billy the Kid* but earned her lasting displeasure when she read it after publication. Susan also carried on an extensive correspondence with Professor Fulton in which she engaged in lengthy and bitter diatribes. In reaction to reading a copy of Pat Garrett's recently revised (1927) book *The Authentic Life of Billy the Kid*, sent to her by Fulton, Susan wrote, "Received your letter and the Book I'll say thanks but it sent a dagger to my heart to learn that you had gotten down to a level with such people to believe what they have had to say . . . in defence of their hideous atrocious work or doings from the beginning to end. . . . I only read one half of your book 'The Authentic Life of Billy the Kid' I could not have borne up under any more of it."[3]

The Girl from Gettysburg

Susan began life as Susanna Ellen Hummer, and at various times also went by Sue Ellen Homer and Sue E. Homer. She was born on December 30, 1845, on the Hummer family farm in Tyrone Township in Adams County a few miles north of the southern Pennsylvania town of Gettysburg.[4] She was one of eight children (seven girls and one boy) of Peter and Elizabeth Stauffer Hummer and must have found life in the Hummer household both busy and noisy.

After her mother's death when Susan was only six years old, her father married his deceased wife's younger sister Lydia, and eight more children were born. For someone with as well-developed a sense of self and entitlement as Susan, competing among this welter of siblings could not have been ideal. There was another problem: the Hummers were Dunkards, a German American Baptist sect that migrated from Germany to Pennsylvania in 1719. A plain people, Dunkards

Susanna Hummer as a young woman. Date and photographer unknown. Courtesy Robert N. Mullin Collection, negative no. 150.2, Haley Memorial Library and History Center, Midland, Texas.

Believed to be Susan McSween's wedding photo, 1873 (?). This pose prominently displays a gold wedding band on her left hand, perhaps indicating that it was taken at the time of her marriage to Alexander McSween, when she was twenty-seven. Photographer unknown. Maurice G. Fulton Papers, Box 19, Folder 5, Special Collections, University of Arizona Library.

believed in simple, unadorned dress and were opposed to alcohol, tobacco, the taking of legal oaths—and war. Susan had no quarrel with the prohibition against alcohol and tobacco, or even the views about oaths and war, but objected to plain clothing. Throughout her life she adored fashionable dress, hairstyles, makeup, expensive jewelry, and social position.

At some time afterward (she does not appear in the 1870 census in the Hummer home), Susan rebelled against the restrictions of that household and clambered out the second-story window of the family home to strike out on her own. The next seven or eight years are difficult to account for. Susan recalled meeting her first husband, the asthmatic Presbyterian minister Alexander McSween (whose last name she often misspelled as "McSwain") in Pekin, Illinois, in 1870.[5] The two were drawn to each other right away. Soon after the couple met, McSween left for Washington University at Saint Louis and a short stint at law school. On April 13, 1873, Susan registered with the First Presbyterian Church in Atchison, Kansas, probably in preparation for marrying McSween. The couple was married in August 1873 by the pastor of the First Presbyterian Church of Atchison. They departed for Eureka, Kansas, where Alexander began practicing law.

The summer of 1874 saw an extreme drought that ruined crops; then the Eureka bank failed. By fall of the same year the McSweens, ruined financially, began looking farther west to a new beginning.

Life in Lincoln

Leaving behind debts destined never to be paid, the McSweens set out from Eureka with their few belongings in a wagon pulled by a team of horses. They traveled through Dodge City on to Colorado, where they fell in with a Mexican freighter. The trio headed to Punta de Agua in the territory of New Mexico. At that critical juncture the couple bumped into young Miguel A. Otero, who advised them to try their fortunes in the southern New Mexico town of Lincoln.[6] (In later years

Otero would become a governor of the New Mexico Territory.) The McSweens arrived in Lincoln on March 3, 1875. They were only the second Anglo family in town, along with the Boltons. There was no attorney, and it seemed an ideal place to begin anew. Susan's arrival caused quite a stir; she was a slim, attractive woman who always wore elegant dresses. Local Hispanic women disliked her from the start and thought she had too high an opinion of herself.

Alex set out to build a practice and to make himself rich by expanding into the business world. In 1876, he teamed up with a young English newcomer named John Tunstall and a Pecos rancher named John Chisum. Susan curbed her tongue somewhat during this time and (in public) took a backseat to her husband, but Alex's partner Tunstall observed that she had a shrewd business mind. Long afterward Susan also claimed to have forewarned Alex that he was headed for big trouble if he went up against the Murphy-Dolan powerhouse that controlled business in the area.

By 1877, the McSweens had finished building a large and imposing home that also housed a law office for Alex and David Shield, husband of Susan's sister Elizabeth. Susan made a showplace of the house, furnishing it with a piano, lace curtains, Brussels carpets, and fine wall hangings. Inevitably, McSween went afoul of the established House of Murphy. When Tunstall was murdered in February 1878, tensions escalated quickly, culminating in the Five-Day Battle of July 15–19 in which Alexander McSween and four other men were killed. Susan watched her dream house torched by the sheriff's men and destroyed under the noses of a complicit U.S. Army commanded by Colonel Nathan Dudley. Her husband was killed as he tried to run for cover out of his burning house on the evening of July 19. During the battle on that day, Susan left the shelter of her home to march in plain sight to Dudley's quarters to demand help in saving her home and her husband's life. She only enraged the colonel, who began a vicious campaign to brand her as "profane and unreliable," no better than a common prostitute.[7] Not one to retreat from a fight, Susan stayed on in Lincoln after Alex's murder, finding shelter in the Bolton home.

Picking Up the Pieces

Four days after the battle, Susan emerged wearing the only dress she now owned—and sooty at that—to pick through the ashes of her home for anything salvageable. When some of Dolan's men appeared on the scene to threaten her, Susan stood her ground and defied them to kill her. The men were shamed into leaving. In one sense, this, her lowest moment, was also the moment when Susan's story really began.

Susan thought it wise to join her sister Elizabeth, who had fled with all the Shield family to Las Vegas after the Five Day Battle. Following a brief stay in Las Vegas, Susan returned to Lincoln on November 23. By January 1879, Susan had got herself appointed administrator of her husband's estate and the estates of John Tunstall and Dick Brewer (another victim of the war) and set about collecting debts. She found little legal justice for herself. She swore out a libel suit against Colonel Dudley that went nowhere; a military court of inquiry exonerated him. A civilian court also acquitted him of the arson of her house.[8] From the estates of her husband and the unmarried Tunstall and Brewer, Susan gained a great deal, aided considerably by the proceeds of Alex's $10,000 life insurance policy. Soon she made her home at the former Baca house, built a rock wall and plank fence around it, and mapped out her future with the help of lawyer Ira Leonard.[9]

The 1880 census of Lincoln County lists forty-five Hispanic widows as heads of household (with children living at home); twenty-one widows lived in other households.[10] This unusually large number of widows testifies to the aftermath of the Lincoln County War, a telling statistic setting Lincoln County apart from other New Mexican counties of the time.

In the same census, Susan lied and shaved three years off her age, saying she was thirty-two; she was actually thirty-five. (This habit of lying about her age continued until her death; in the 1900 census, she gave her age as forty-seven when her true age was fifty-five.) Susan continued to live in Lincoln, where she managed a 160-acre homestead with the help of two black hired hands, raised corn on fifty acres of the homestead, and kept farm animals.[11]

Second Marriage and the Three Rivers Ranch

On June 20, 1880, two years after the death of her first husband, Susan married George Barber, another Lincoln attorney and a man five months younger. Susan now owned several properties scattered throughout southeast New Mexico, from south near Roswell to Gray in the north (later called Capitán), and to Nogal and Tularosa. As a widow, she entered a claim under the

Timber Culture Act to 160 acres near the South Spring River near Roswell.[12] This parcel adjoined the property of Pitser M. Chisum, John Chisum's brother. In February 1881, she paid Pitser for one-fourth of the water running in the Chisum Acéquia to use in irrigating her own crops.[13]

Sometime later in the same area Susan entered a claim to (and patented) 320 acres under the Desert Land Act. From Sebrian Bates, a former black employee in Lincoln, she bought another 160 acres that Bates had acquired under the Homestead Act.[14] Within ten years she gave up the Timber Culture claim and sold 480 acres of her other claims for a profit. In 1883, the Barbers bought a large parcel of grazing land at Three Rivers (Tres Ríos), the confluence of three streams about sixty miles southwest of Lincoln. In a biography that George Barber himself penned for inclusion in the 1895 *Illustrated History of New Mexico*, he said that he subsequently organized the Three Rivers Land and Cattle Company (the Tres Ríos ranch) in 1885.[15]

The Barbers' marriage seemed to have begun traditionally enough as a partnership, though the two always maintained separate brands for their horses and cattle. Susan registered her own two brands as "SB," for "Susan E. Barber," and "SUE" for "Susan Barber."[16] Susan ran the ranch while George manned law offices in Lincoln and (later) the gold-mining boomtown of White Oaks thirty-five miles to the northwest. Apparently the couple also maintained separate homes during much of their marriage, leading one to speculate that they did not actually live together for long.

Boss Lady

For almost two decades, from 1883 to 1902, Susan reigned at Three Rivers as one of New Mexico's most successful women ranchers. She constructed a dam of sorts across Three Rivers (actually a stream, not a river) and a half-mile ditch to carry the water to her property.[17] She was said to have acted as foreman over her own cowboys. Others claim that she was assisted in ranch work by a black foreman named Pompey. In either case, she ran some five thousand head of cattle and presided over 1,158 acres of Three Rivers land.

Susan's Mescalero Apache neighbors called her "Jefa Poquita" (Little Chief).[18] The Mescaleros also said that she rode sidesaddle on cattle drives outfitted in chaps and could be spotted from miles away riding at the head of her cowboys in a bright fuchsia riding habit. (Others said that they had "never seen her on a horse; she was a lady and rode in a buggy.")[19]

In September 1884, the *Golden Era* newspaper at White Oaks reported Susan as driving 125 head of stock from John Chisum's Spring River Ranch in southeast New Mexico through Nogal bound for the Three Rivers ranch.[20] Amelia Bolton Church maintained in a 1951 interview with Eve Ball that these cattle were the last of a starter herd that was originally paid as part of a claim owed Alex McSween, sold to her by Chisum, or given her outright by Chisum.[21] More likely, however, Susan's starter herd came from a February 1879 sweep by Captain Henry Carroll and his posse after they were dispatched to gather stolen livestock at Seven Rivers on the Pecos.[22] Captain Carroll returned with 294 head to Roswell, where various owners sorted out their own stock, of which 140 were claimed for the Tunstall estate by his legal representative Susan McSween.

Sometime in the mid-1880s Susan built a stone house with three-foot-thick walls on the ranch.[23] In 1888, Susan planned and supervised the location of the fences, corrals, and all such works on the ranch. She had crews establishing grain fields, vegetable gardens, fruit orchards, and berry patches. She ran a tight ship, making her cowboys wash and comb their hair and doff their hats in the house.[24] They always respectfully called her "Mrs. Barber."

Sometime in 1888 or 1889, a pair of young Texans, Lin Branum and Os Hyde, were mavericking on Susan's land. This was a discouraged but legal practice of catching unbranded cattle and branding them with one's own brand. Branum and Hyde managed to build up a fairly large herd of cattle in this manner. In time this infringement on Susan's territory came to her notice, and she dispatched her foreman, Pompey, to run off the pair. When the foreman confronted Lin Branum, Branum shot him in the head.[25] The foreman did not die, but Branum thought that he had killed him and so ran for Oklahoma Territory.

In those days there was always a threat of trouble. One of Susan's cowhands, a boy named Severo Padilla from Capitán country, killed an older cowboy on a cattle drive to Kansas City.[26] The older cowboy had taken a dislike for Padilla and had tormented him at every opportunity without the least provocation. The timid boy finally turned on the bully after a severe horsewhipping as he lay in bed. Severo jumped from his bed, gun in hand, and pumped a bullet through the man's heart. Susan, her foreman, and the other cowboys testified on the boy's behalf in court, and he was acquitted of murder.

The Barbers Divorce

The Barbers traveled some together during their marriage. Newspapers reported the Barbers returning from New Orleans in early 1885 and mentioned one of Susan's visits to Baltimore.[27] She also traveled from Three Rivers to Las Vegas, New Mexico; Saint Louis; and New York. At one time she visited former New Mexican friends Stephen and Hallie Elkins at their summer home in Maryland, where she was said to have danced with former New Mexican governor Lew Wallace.[28]

In January 1887, the Barbers sold one-half interest in their Three Rivers property to John Rugee and Emil Durr of Milwaukee for an undisclosed sum. The new partnership retained the name "Three Rivers Land and Cattle Company" and employed Susan as general manager.[29]

By 1889, the Barbers' marriage had soured. George sold out his interest in the Three Rivers Land and Cattle Company to engage in the cattle business on his own on the west side of the Sacramento Mountains.[30] He pursued his political ambitions to become representative to the Republican Territorial Committee in 1890 and district attorney of Lincoln, Cháves, and Eddy Counties in February 1892.[31]

On September 15, 1891, Susan filed for divorce in the district court of Lincoln County. She claimed that George had never supported her during their marriage and had abandoned her on March 1, 1891. The divorce was granted on October 16, 1891.

Susan Becomes a White Oaks Fixture

During the next phase of Susan's life, local papers took notice of her varied activities, making her a social fixture of the area. Building continued apace at the ranch. In May 1891, according to the *Lincoln County Leader*, White Oaks carpenter Charley Littell went to Three Rivers to build a barn for Susan.[32] Susan bought many lots at the Oaks in her own name—probably believing that a railroad would arrive and the town would mushroom. She also grubstaked many miners in hopes of sharing a big strike, never pressing for repayment.

Susan often visited White Oaks, which was by then a thriving business and social center. She brought along (and gave away) much of the fruit and produce raised on her ranch. In April 1891, the *New Mexico Interpreter* reported, "Mrs. S. E. Barber, who is in from Three Rivers, states that the last cold spell injured the fruit in that vicinity but slightly. On her ranch, peach, apricot and plum trees are in bloom, and apple, cherry and pear trees in bud."[33] And in August 1891, the editor of the *Lincoln County Leader* opined, "We are deeply indebted to Mrs. Barber for a generous supply of blackberries . . . very toothsome to ye editor."[34]

Living full time in White Oaks began to figure more prominently in Susan's thoughts. In March 1892, the *Lincoln County Leader* reported that "Mrs. Sun. [the reporter's abbreviation for "Susan"] E. Barber is preparing to build an elegant residence on her property on the south side of town adjoining Judge John H. Hewitt's residence."[35] In April 1892, the *White Oaks Old Abe Eagle* reported Susan driving seven hundred to eight hundred cattle from her ranch to Engle down south—then the most accessible railroad point in southeast New Mexico. From Engle the cattle were shipped in "38 foot New England cars" to the Jones and Nelson feedlots in Grand Summit and Strong City, Kansas.[36]

In July 1892, the *Lincoln County Leader* reported that "Mrs. Barber has completed her neat and commodious cottage [in White Oaks] and is now grading and otherwise beautifying the grounds surrounding. Mrs. B. is a successful florist. We predict that next year she will have the prettiest flower garden in town."[37] Back East Susan also warranted notice. The *White Oaks Old Abe Eagle*, quoting from an article in the *New York Commercial Advertiser*, reported:

> Near the town of White Oaks, N.M., [meaning the Three Rivers ranch] lives one of the most remarkable women of this remarkable age, at the present time a visitor in this city [New York]. The house in which she lives, a low, white-walled adobe building, is covered with green vines and fitted out with rich carpets, artistic hangings, books and pictures, exquisite china and silver, and all the dainty belongings with which a refined woman loves to surround herself. The house was built with her own hands. The huge ranch on which it is located with its 8,000 cattle, is managed entirely by her. It is she who buys or takes up the land, selects and controls the men, buys, sells and transfers the cattle. She is also a skilled and intelligent prospector, and found the valuable silver mine on her territory in which she now owns a half interest. She sings charmingly, accompanying herself as dexterously as

she uses an adze or jack plane. She entertains delightfully at her home, whist [card game] parties, little dances, and even an occasional german [dance]. Her name is Mrs. Susan E. Barber. A woman who can run a ranch, build a house, manage a mine and engineer a successful german, deserves a prominent place in the ranks of women of genius.[38]

Susan must have basked in all this praise. It was short-lived, however, as news of her ex-husband's new marriage was reported in the *Lincoln Republican* in October 1892: "Saturday, October 1st, at the residence of Sheriff D. W. Roberts, the hearts of Mr. George B. Barber and Miss Eugenia Roberts were made to 'beat as one.'"[39]

Later on Susan was to maintain that Eugenia had "stolen her man."[40] To escape it all, the new divorcée left on a protracted trip to the East, also revisiting her childhood home in Pennsylvania for the first time since climbing out that window more than twenty years ago.[41] She must have caused a sensation among her Dunkard relatives, sweeping about in silks, taffetas, diamonds, and gold jewelry. She returned to Lincoln County in December 1892.

Surviving ledgers of the Paul Mayer livery stable show extensive business transactions having to do with the Three Rivers ranch. Entries from June through December of 1898 alone attest to the purchase of a staggering 8,300 pounds of flour—presumably converted to mounds of biscuits, pancakes, and bread for hired hands and cowboys. Also listed in these accounts is gear for farm work and for horses—saddles, bridle reins, and plows, among other items.

Susan may have hired a significant number of Hispanic cowboys and laborers for her ranch, because in a later interview by the Lincoln County Historical Society, Urbano Carrillo and his wife maintain that she spoke Spanish "better than anybody in the country." They also said that she was a "very fine lady . . . [who] associated well with people."[42]

By now Susan apparently had come to believe that, while ranches and cattle were all well and good, diamonds were a girl's best friend. She liked keeping her jewelry close, carrying around a box filled with diamonds, pearls, and other gems, and wearing a solid gold chain as a necklace.

In 1901, Susan bought back the half-interest of Rugee and Durr in the Three Rivers Land and Cattle Company.[43] In July 1902, Susan sold most of her Three Rivers acreage, what then remained of the cattle (some three hundred head), and the home ranch to Monroe Harper for a sum variously reported to be $32,000 or $35,000.[44] She retained some land and a ranch house.

Susan Barber's account at Paul Mayer's livery stable. The ledger lists Susan's accounts from mid-1898 through September 1904 and reflects the purchase of astonishing amounts of flour, not to mention chains, bridle bits, riding bridles, and saddles—attesting to a large ongoing ranching and farming operation at Three Rivers. Courtesy John and Wesley Mottinger.

Parties at Three Rivers and White Oaks

Outwardly Susan recovered from her divorce to continue an active social life. The September 28, 1899, *White Oaks Eagle* reported that "Mrs. Susan (McSween) Barber gave a delightful house party at her ranch at Three Rivers last week, beginning Thursday and lasting through Saturday. The guests were royally entertained."[45] Likewise, the November 1, 1899, issue had her traveling from Three Rivers to attend a masked ball at the Gumm House in White Oaks.[46]

J. Denton Simms, a young boy in a Texas family who had become one of her neighbors at Three Rivers, remembered Susan during this period well:

> One of our more interesting neighbors was Mrs. Susan Barber, the elderly widow of Alexander McSween. . . . Mrs. Barber, as she was now called because of a subsequent marriage . . . lived near us in a historic landmark known locally as "The Rock House." She was thought to be very wealthy and . . . enjoyed being called the "Cattle Queen of New Mexico." Wealthy or not, she still owned and wore an impressive display of diamonds and visited around the community in a fine buggy driven by a man she called Pompey. Mother recalled that when Mrs. Barber came to call on us she was dressed in taffeta, wearing a wide hat decorated with ostrich plumes. . . . Mrs. Barber loved to go to the dances that were held at various ranches. My brother Donald, who was forced to dance with her on one of these occasions, complained about how stiff her legs were. Mother wondered who got those jewels when she died, since we knew of no heirs.[47]

Susan in the Early Years of the Twentieth Century

The first years of the new century went by pleasantly enough. The *Carrizozo Outlook* reported on Susan's activities:

> Mrs. S. E. Barber, probably the best known and most energetic business woman in the Southwest, came up from her Three Rivers ranch last Monday to do a little shopping and look after her properties here. She reports everything flourishing down in that locality with prospects of an abundant fruit crop. . . . Several years ago she sold her home ranch to Monroe Harper, realizing therefor a sum ranging near the $35,000 mark. Since that time she has been improving another place not far away which bids fair to rival the one

Susan Barber at seventy-four years of age. (Are those large diamonds in her pierced ears?) Date: 1917. Photograph by Susan's nephew by marriage James Kirkness, Baltimore Photo Co., Baltimore. Courtesy John and Wesley Mottinger.

> disposed of. She has added over 1,000 grape vines and several hundred apple and other fruit trees to her already large orchard during the present season.[48]

Susan now began to spend most of her time in White Oaks. Always vain, she continued to dye her hair to maintain a youthful appearance. Larue Wetzel recounts, "Aunt Pansy, as a young new bride, went to call on Mrs. Barber. It was the custom of newly married young brides to call on the older ladies in White Oaks. . . . When Aunt Pansy got to the door of Mrs. Barber's house and knocked, Mrs. Barber had been dyeing her hair with henna. She was a bit of a vain woman and liked to keep her hair colored red. She answered Aunt Pansy's knock with a towel on her head and the red henna running down her face and neck. It scared Pansy to death because it looked so much like blood."[49]

Fire Destroys the White Oaks Home

In the summer of 1923, Susan's house, for the second time in her life, burned to the ground. Forced to jump from an upstairs window to save her life, she described her escape in a letter to Paul Mayer in July 1923: "When I jumped I took hold of a wire to lessen the jump but

Susan Barber with her nephew and niece. Susan stands next to her nephew Mr. Hummer from Pennsylvania and her favorite niece Minnie Shield Zimmerman. Date unknown, probably early 1920s. Photograph by Maurice G. Fulton. Courtesy Robert N. Mullin Collection, negative no. 128, Haley Memorial Library and History Center, Midland, Texas.

Susan Barber in her later years, standing with unidentified relatives in California, mid-1920s. Photographer unknown. Courtesy John and Wesley Mottinger.

after I jumped and tore the flesh of hands from the bone in three different places the wire broke loose then I fell with my side unto a brick wall and my arm and shoulder was already burned to a crisp in some places."[50]

For a time Susan had to use a walker and a wheelchair. Salvaging a few scorched items from the ashes, she moved into a decrepit house owned by friend and neighbor Laura Stewart Leighner. The house was located near the intersection of Jicarilla and Carrizo Streets.

All Those Diamonds

In October 1980, Barbara Branum of La Luz, New Mexico, interviewed Truman Spencer about his early days in Carrizozo and White Oaks. Spencer was the son-in-law of William McDonald, the first governor of the state of New Mexico, and had married Frances McDonald, McDonald's only child. After his father-in-law's death in 1918, Spencer assumed control of El Capitán Livestock Company and the Carrizozo Cattle Ranche Company. Over the years he had come to know Mrs. Barber well and was able to answer questions about her well-remembered diamonds. Truman Spencer was ninety years of age at the time of this interview with Branum. Mary and Jack Krattiger also participated in the discussion.

TRUMAN: She's buried up in White Oaks, you know. Hell, I knew her well, talked to her. You may remember old George Barber, the lawyer at Carrizozo? . . . She quit him and he quit her. . . . She was at Three Rivers and had cattle down there and finally sold out . . . and then moved to White Oaks. She lived there quite a while. She used to come down to Carrizozo in a Ford automobile, had a driver. She was sittin' in the back enjoyin' the scenery.

MARY KRATTINGER: Nobody knew that she didn't have any money, did they, Truman?

TRUMAN: Not for a long time.

MARY KRATTINGER: Did he ever tell you [Barbara] that she told him she was going to give him her diamond rings?

TRUMAN: She was gonna give me two diamond rings when she died. . . . She had me up there [after her house burned in 1923] helpin' her sift the ashes to find these diamonds. And the old fella that was drivin' her says to me, "Well don't be lookin' any more. She doesn't have any diamonds. She's got rid of all of 'em.

White Oaks Women's Club, 1921. The women pose in front of the Gumm House at White Oaks. *Left to right*: Mary Rustin, Ollie Harman, Mrs. Bob Forsythe, Bertha Crenshaw, Edna Burnett, Susan McSween Barber, Myrtle Smith, Edith Van Schoyck, Kate Grumbles, Mrs. H. S. Hanner, Edna Cleghorn, Sally Owens. The two little girls in front are (*left to right*) Opal Crenshaw and Dixie Harman. Photographer unknown. Courtesy John and Wesley Mottinger.

That's what she's been livin' on." And I traced it down and she had. She sold the diamonds in El Paso. She didn't have any diamonds left.[51]

A Remarkable Life Draws to a Close

Visitors continued to flock to Susan's door, even at her advanced age. A group of Boy Scouts came from El Paso, Texas, all eager to hear about Billy the Kid. Howard Newton, one of the scouts, recounted,

> Mrs. Barber, a lovely lady, invited all of us boys into her house. We sat everywhere as the great lady told us about the Battle of Lincoln. . . . After giving us Scouts the story of the Battle of Lincoln, this great lady answered questions. What was Billy like? The Boy Scouts of Yucca Council . . . took 16 mm movies of George Coe and Mrs. Barber. We put it in a good movie to show about summer camp. . . . Several years ago I went to the Boy Scout office in El Paso . . . and inquired about [the movie]. A young Scout executive thought it was taking up space and burned the film.[52]

Robert Leslie, who grew up in White Oaks, recalled, "After she moved in with Laura Leighner a boy by the name of Ben Cochran and I did chores for them, at which time she was as straight as a pine and moved about with ease. The two of us had been doing chores for her before her house burned, stocking in wood, pumping water, running errands, etc. She had a cistern in the lower room with a pitcher pump. The cistern was always full; you could almost dip it out. I remember once she payed [*sic*] us each off with a silver dollar."[53]

According to Johnson Stearns of Carrizozo, who worked in Albert Roberts's gas station where Susan bought gas for her car in which she was always chauffeured about, "I began working for Albert . . . in 1929. I would have been 12. . . . We had one gas pump and only white gasoline at 19 cents per gallon. She usually bought just four or five gallons of gas for a total of 95 cents."[54]

Dorothy Leslie, Robert's wife, who lived in White Oaks as a girl, often visited Susan in the Leighner house. She recalled, "She gave piano concerts after school for the local youngsters. Susan was always pleasant and happy, and customarily invited the neighborhood children into her home where she usually gave them cookies and talked and laughed with them all. . . . The White Oaks kids would come through her yard. She played the Ouija board a lot. She would invite us in to watch the Ouija board and tell us our fortunes with it. Then she'd . . . laugh. It was all a game. She was always happy, smiling, loved kids to come in. And had none of her own." Dorothy also remembered Susan as a woman who cared deeply about appearances: "Her hair was . . . arranged in small curls over her forehead, and she was trim and tidy. She loved to tell us about her reigning as Cattle Queen of New Mexico and we loved to listen."[55]

Niece Minnie Shield Zimmerman from California (born ca. 1868) visited often; so did nephew Edgar Shield (born ca. 1878). Friends from White Oaks moved on as the town shriveled and died. Susan wrote one of her special friends, Bertha Mayer Hunter (now living in California), several letters that revealed a wicked humor. For instance, in a letter dated March 23, 1926, Susan wrote,

> Everybody is gone and nothing doing in the mines Jackson and Lane are the only person working and that is for Hodspeth "Wild Cat" work. "kin sabe" if they will strike anything or not we shall hope they will.... Vanschoycks [Van Schoycks] moved down to the Hotel and are fixed up nicely but they have only about four customers, so now they are talking about going to the Rio Doso as everybody has gone wild over that place for a summer home.... Mrs Brown is still in business but she has to divide up with the Van Schoycocks trade but she talks brave. I sit here with the blues listening to their complaints but try to look happy & brave when there whining and tell them as usual never mind the change will come now soon. I dart into Van Schoyechk' door sometimes and exclaim aloud say the RailRoad is comming for a moment they think of jumping up to see. This is sure a graveyard Bertha, and yet I have not selected a lot.[56]

In July of 1926 former governor Miguel Otero, accompanied by Marshall Bond of California, drove to White Oaks to query old friend Susan about her memories of Billy the Kid. Fifty years after their first meeting, he found in her "aged face and form unmistakable traces of former beauty."[57] Susan kept in touch with Otero and wrote to him in June 1930: "I have been very unfortunate in having a dreadful fall. My left leg was so lacerated, ligaments so badly torn, it had to be carried on a pillow for three months to the person to handle me for the same length of time, and can't walk yet, and happened one year ago next Sunday."[58]

When writer Walter Noble Burns visited Susan in White Oaks to gather information for his book about Billy the Kid, she informed him, "I came to White Oaks because it is so peaceful here. My life has been... strenuous. I have seen so much fighting and killing during my life, that in my old age I want peace."[59]

The End of the Road

In the winter of 1930, Susan caught the flu and could not shake it off. Flu turned into pneumonia. She died on January 3, 1931. The townspeople sat up with the body until the funeral. Dorothy Leslie recounts, "The turnout was not too large, but you can be sure all of us kids that she had treated so royally were there. We were sad and missed her a lot in days to come. She was truly a great little lady and her memory lives on."[60]

Susan McSween Barber in doorway of clapboard house, July 1926. Standing to her left: former New Mexico Governor Miguel A. Otero and Marshall Bond, Jr. Seated on the step is Marshall Bond. The Bond-Otero party drove to Lincoln County to interview survivors of the Lincoln County War. Susan's White Oaks home had burned down in the summer of 1923, forcing her to move into the home of her friend Laura Leighner. Susan, trim and tidy as always, musters a game smile. Photographer unknown. Archives and Special Collections, RG85-88, M/1/11, Ms 249, New Mexico State University Library.

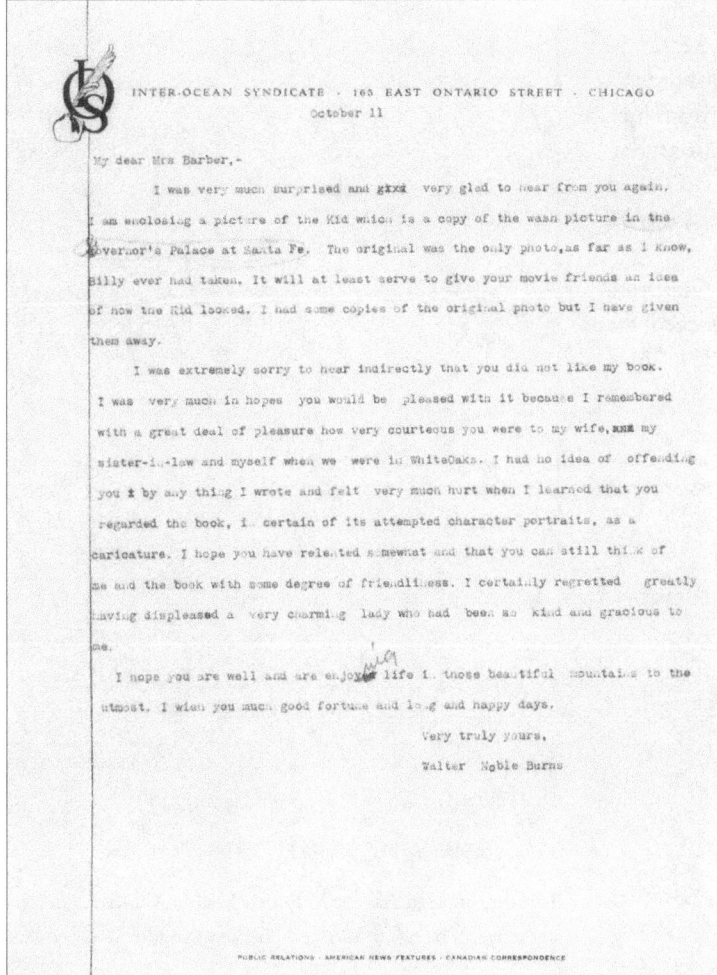

Burned cloth from Mrs. Barber's house fire. In the summer of 1923, Susan's house in White Oaks burned to the ground, an eerie second such occurrence, as she herself escaped by jumping thirteen feet from an upper window. (Her house in Lincoln had been torched—and her husband killed—during the "Big Killing" on July 19, 1878.) This embroidered cloth with its highly visible burned area was salvaged from the second fire. It probably had covered the top of a piano or organ. Photograph by Bob Workhoven. Courtesy John and Wesley Mottinger.

Letter from Walter Noble Burns to Mrs. Barber, October 11 (1926?). Burns apologizes for any offense she took from his account of the Lincoln County War in his book *The Saga of Billy the Kid*. Courtesy John and Wesley Mottinger.

Her funeral was held in the old Congregational rock church in White Oaks. The *Carrizozo Outlook* obituary reported, "The funeral was well attended by the many old friends of the deceased who gathered to pay their last respects to the memory of this estimable lady and to take a last look on the remains of one whom they so loved, after which the cortege moved to the cemetery where the remains were interred, just as the sun was sinking behind the mountains."[61]

Truman Spencer had paid for Susan's tombstone as part of a package when he bought a tombstone for former governor William McDonald. Truman asked the tombstone maker if he would consider making a stone for Susan. The man furnished the stone without charge, saying that Truman had paid for Susan's memorial along with the big stone bought for the governor.

No doubt Susan would have relished finagling a free gravestone on the coattails of Governor McDonald's final marker; she might have relished it as much as her private joke of watching the foremost rancher in southeast New Mexico scrabble through the ashes of her White Oaks house in search of phantom diamonds.

Probate

Sallie Sweet, administrator of Susan's estate, found that her worth—including real estate and personal effects—amounted to only $400 in her final report of April 19, 1932. Assets were offset by claims amounting to $2,340.90.[62] The real estate included nineteen now-worthless White Oaks lots and half-interest in seven other lots, all bought by Susan in the glory days of White Oaks.

Part of the claims against the estate included $267.92 owed to nephew E. P. Shield for undertaking charges and the cost of hiring an auto to transport the body from Carrizozo to White Oaks. A larger amount, $2,019.10 owed niece Minnie Shield Zimmerman, was for services rendered to Susan as nurse and housekeeper from June 1929 to June 1930. Another debt was owed Montgomery Ward & Co. for "Budget Plan Merchandise" on which Susan had made four partial payments, leaving a balance of $23.45.

Old friend and former Regulator Frank Coe from Lincoln County War days of long ago penciled the following entry in his diary on January 4, 1931: "Mrs. Susan McSwain Died in White Oaks at 8 oc. She was 91 years old. Poor old Girl out of H Misery."[63]

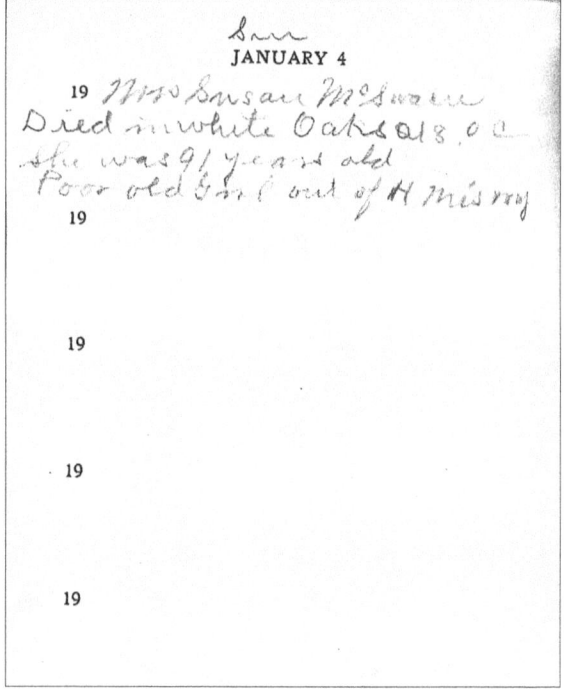

Diary entry by Frank Coe upon Mrs. Barber's death, January 4, 1931. Courtesy Eleanor B. Shockey.

NOTES

1. Wilson, *Merchants, Guns, & Money*, 94.

2. Miller, "The Women of Lincoln County," 183.

3. Susan E. McSween Barber to Major Maurice G. Fulton, March 10, 1928, Mullin Collection. Transcribed by Lynn Koenig, November 19, 1988, and copy furnished to author by Nora Henn of Lincoln, New Mexico.

4. Nolan, *Lincoln County War*, 15.

5. Ibid., 14.

6. Chamberlain, "Billy the Kid, Susan McSween," 202; Chamberlain, "In the Shadow of Billy the Kid," 39.

7. Chamberlain, "Billy the Kid, Susan McSween," 202.

8. Miller, "The Women of Lincoln County," 179; Chamberlain, "Billy the Kid, Susan McSween," 211.

9. Wilson, *Merchants, Guns, & Money*, 129.

10. Ibid., 131–32; Miller, "The Women of Lincoln County," 183.

11. Miller, "The Women of Lincoln County," 188.

12. Ibid., 189.

13. Lincoln County Deed Book C, 175–76, transaction dated February 15, 1881, Lincoln County Courthouse, Carrizozo.

14. Miller, "The Women of Lincoln County," 189.

15. "George B. Barber," 582–83.

16. From undated drawing by Robert Leslie of Carrizozo, depicting early-day brands of White Oaks, copy in author's collection.

17. Miller, "The Women of Lincoln County," 189.

18. From information furnished by Nisha Hoffman of Ruidoso, storyteller who plays the part of Susan McSween Barber in reenactments and/or Chautauqua presentations.

19. Urbano Carrillo and Rosa Montaño Carrillo, interview by Lincoln County Historical Society, November 7, 1977, recording TR771107-B, copy in author's collection.

20. *White Oaks Golden Era*, September 18, 1884.

21. Amelia Bolton Church, interview by Eve Ball, December 3, 1951, Roswell, New Mexico, copy in author's collection.

22. Nolan, *Lincoln County War*, 374.

23. Chamberlain, "Billy the Kid, Susan McSween," 212; Chamberlain, "In the Shadow of Billy the Kid," 50; Fulton, *History of the Lincoln County War*, 420. A note from the Mullin Collection mentions a "stone house, with walls three feet thick." More probably, the house was of adobe bricks.

24. Chamberlain, "Billy the Kid, Susan McSween," 212.

25. Reily-Branum and Haldane, *Corralled in Old Lincoln County*, 21.

26. J. S. Chávez, "Cattle Queen of the Territory," *Lincoln County News*, August 12, 1960.

27. *White Oaks Golden Era*, March 12, 1885.

28. Chamberlain, "In the Shadow of Billy the Kid," 50.

29. Chamberlain, "Billy the Kid, Susan McSween," 212; Miller, "The Women of Lincoln County," 189–90.

30. "George B. Barber," 583.

31. *Lincoln Independent*, February 26, 1892; Curry, *George Curry*, 236.

32. *Lincoln County Leader*, May 30, 1891.

33. *New Mexico Interpreter*, April 10, 1891.

34. *Lincoln County Leader*, August 15, 1891.

35. *Lincoln County Leader*, March 12, 1892.

36. *White Oaks Old Abe Eagle*, April 21, 1892.

37. *Lincoln County Leader*, July 22, 1892.

38. *White Oaks Old Abe Eagle*, September 1, 1892. The newspaper quotes from an article in the *New York Commercial Advertiser*.

39. *Lincoln Republican*, October 7, 1892.

40. Mrs. Lula Sisson (George Barber's niece) to Cora Barber Dutton, March 12, 1948, Nogal, New Mexico, copy in author's collection. (Cora Dutton's first husband was George Barber's son, Ralph.) Sisson corresponded with Susan for years after George and Susan Barber divorced and said that Susan "declared she loved my Uncle to the last, and she accused 'some other woman.'"

41. *Lincoln Republican*, December 16, 1892.

42. Carrillo and Carrillo interview.

43. Miller, "The Women of Lincoln County," 190.

44. Ibid; *Carrizozo Outlook*, May 25, 1905; Chamberlain, "In the Shadow of Billy the Kid," 51. Fall tendered two notes totaling $3,250 on March 29, 1917, for the rest of the Tres Ríos property. The first was payable June 1, 1918, and the second a year later.

45. *White Oaks Eagle*, September 28, 1899.

46. *White Oaks Eagle*, November 1, 1899.

47. Simms, *Cowboys, Indians and Pulpits*, 24–26.

48. *Carrizozo Outlook*, May 25, 1905.

49. Larue Lane Wetzel, interview by Roberta Haldane, May 29, 2001, White Oaks, New Mexico.

50. Barber to Mayer, July 18, 1928, copy in author's collection.

51. Haldane, "What Became of Susan McSween Barber's Diamonds?" 39–41.

52. Howard Newton, "Memories of a Boy Scout Adventure in Lincoln," n.d., copy in author's collection.

53. Robert Leslie to Haldane, February 22, 1999.

54. Stearns to Haldane, October 17, 2007.

55. Dorothy Leslie to Haldane, undated.

56. Several letters between Susan McSween Barber and Bertha Mayer Hunter were generously furnished to the writer by Ina Hunter Dow of Salinas, California. Ina's mother (Bertha Mayer Hunter) and grandmother (Ina Wauchope Mayer) had been good friends of Susan Barber during their years in White Oaks.

57. Otero, *Real Billy the Kid*, 112–23; Bond, *Gold Hunter*, 169–72.

58. Barber to Otero, June 12, 1930. Copy furnished to author by Nora Henn, Lincoln, New Mexico.

59. From the obituary article, "Funeral of Mrs. S. E. Barber Held At White Oaks Wednesday," *Carrizozo Outlook*, January 9, 1931.

60. Dorothy Leslie to Haldane, undated.

61. "Funeral of Mrs. S. E. Barber."

62. "In the Matter of the Estate of Susan E. Barber, Deceased," final report and account of Sallie Sweet, Administratrix, April 19, 1932, Susan E. Barber Probate File, Probate Records, Lincoln County, New Mexico, Lincoln County Courthouse, Carrizozo.

63. Copy of Frank Coe's diary entry on January 4, 1931, generously furnished by Eleanor Shockey of Ruidoso. Eleanor, Frank Coe's granddaughter, has the original diary in her possession.

JAMES W. BELL
Deputy Sheriff Shot Dead by Billy the Kid during His Escape from Lincoln

James W. Bell (also referred to as "Long" Bell, "Lone" Bell, and even in one instance as "Jim Long")[1] was shot and killed by Billy the Kid in Lincoln during the Kid's escape from the Lincoln County jail on April 28, 1881. Bell was only twenty-seven or twenty-eight years old at the time.

Where Was He from, and What Did He Look Like?

In the 1880 national census for White Oaks in Lincoln County, Bell gave his age as twenty-seven, his place of birth as Georgia (with both father and mother having been born in Virginia), and his occupation as a miner.[2] Other than this information, not much more about him is known with any certainty. (Some historians claim that he was born in 1842 in Maryland.[3])

Bell has been alternately described as a "tall, dark man with a livid knife-scar across his left cheek," a "rather quiet, slender man, not much over thirty years of age, whose left cheek was disfigured by a knife slash from ear to mouth," and a man "of medium height, soft-spoken, with a knife scar across his left cheek from mouth to ear." The latter and probably most credible description comes from O. W. Williams, who organized a mining party from Dallas, Texas, to New Mexico that included Bell.[4]

Bell may have had relatives in White Oaks; several Bells are listed in the 1880 census. When Bell's estate was being settled after his death, I. Bell (presumably Isaac Bell, also a resident of White Oaks) claimed before justice of the peace Frank H. Lea that he was owed the amount of $7.75, which he had lent James before his death. This may be coincidence, however, and not proof of kinship.

Was Bell a Texas Ranger?

Several historians have also claimed that J. W. Bell was a private in Company D, Frontier Battalion, of Capt. Dan W. Roberts's Texas Rangers.[5] This association with Bell probably arose from J. B. Gillett's memoirs *Six Years with the Texas Rangers*, in which Gillett lists the names of Rangers to be retained in service in the fall of 1875 after the Texas state legislature failed to vote the necessary appropriation for their annual maintenance:

> As the day for reduction arrived there were some anxious moments among the men of Company D as no one knew just who was to be retained in the service.... On December first Captain Roberts formed the command in line and explained it was his sad duty to reduce the company to twenty men, and announced that the orderly sergeant would read the names of those to be retained.... The sergeant then stepped forward and began to read. First Sergeant Plunk Murray ... First Corporal Lam Seiker ... and Privates Charles Nevill ... J. W. Bell ... were the lucky ones to be retained in the command.[6]

This J. W. Bell, however, is *Joseph* William, born May 17, 1849, in Ohio, according to researcher and writer Bill Reynolds.[7] Reynolds reports that records of the Department of the Interior dated March 4, 1918, also show a J. W. Bell who served under Captain Roberts attempting to obtain a pension as an Indian War veteran. This Bell cannot be the White Oaks Bell, of course, who died on April 28, 1881.

Records of the Texas State Library contain a book by Sammy Tise, *Texas County Sheriffs*, which lists another J. W. Bell as a private in G. W. Campbell's Montague County Rangers.[8] This Bell enlisted on December 13, 1873, and was discharged on February 13, 1874. This may be James Bell of Lincoln County fame; there are no records of any others. No conclusive evidence exists to prove James W. Bell of White Oaks was ever a Texas Ranger. However, records for the Rangers are sometimes incomplete, and Oscar W. Williams, Bell's friend and mining companion in 1879, should have known and had insisted that he was a Ranger.[9]

Mining Ventures in Colorado and New Mexico

Williams formed a party of men in the spring of 1879 for an overland trip from Dallas to Leadville, Colorado, to take part in mining ventures there. Williams wrote, "J. W. Bell, an ex–Texas Ranger, joined us. But for him the fates began here to spin a thread of life with an evil ending. He was killed by Billy the Kid at the Lincoln County Court House two years later."[10]

The Williams party traveled northwest across the Staked Plains, arriving at Tascosa, a small settlement in the Texas Panhandle, on July 5, 1879.[11] At that time Tascosa consisted only of a post office, two stores, and the house and corrals of a Mexican freighter named Casimero Romero. The men had hardly arrived in Tascosa when they learned of a new strike at Carbonateville some twenty-four miles south of Santa Fe near present-day Cerrillos in New Mexico. Changing their plans, they headed for the town.

After arriving at Carbonateville, the party disbanded. Bell and Williams set up camp together, staked out some mining properties, and made forays into the surrounding hills. They stayed in Carbonateville the rest of the summer until word arrived that White Oaks to the south offered better promise. Bell then drifted alone on to White Oaks, arriving there some time in 1880. He actively engaged in mining activities in both the Nogal and the Jicarilla Mountains, as shown by Lincoln County records in the county courthouse.[12]

Newspaper accounts support the records of the Lincoln County Courthouse. The *Golden Era* recorded, "Messrs. Bell and Lypard have struck it big in the Nogals. It is galena ore in an immense body. No assays yet. Further mention will be made as soon as we receive particulars. . . . Deputy Sheriff Bell and R. D. Lypard have a fine prospect in the Jicarilla mountains."[13]

Bell Was a Lincoln County Deputy Sheriff, but Was He Also a U.S. Deputy Marshal?

In 1881 Bell, already one of Sheriff Pat Garrett's friends, became one of Garrett's deputy sheriffs, according to the *White Oaks Golden Era*: "J. W. Bell has received his appointment as deputy sheriff of Lincoln county."[14] Garrett himself confirms that "J. W. Bell, afterwards my deputy, was known by Mason to be a friend of mine."[15]

Whether he was indeed a U.S. Deputy Marshal cannot be confirmed, though historian William Keleher accords him that title.[16] So does historian Donald Lavash in his book *Wilson & the Kid*. Discussing U.S. Secret Service special agent Azariah F. Wild, assigned to New Mexico to locate a ring of counterfeiters, Lavash states that Wild needed help in catching a counterfeiting ring and contacted "Deputy Marshal John D. Bell who had become familiar with most of the criminals. Bell maintained his residence at White Oaks where some of the outlaws occasionally imbibed the liquor and cavorted with the women. Wild asked Bell to make a deal with [suspects] Cooper and West."[17] The *Las Vegas Daily Optic* twice called Bell "United States Deputy Marshal Bell" in 1881. A January 21 article spoke of the high esteem in which White Oaks citizens held Bell: "United States Deputy Marshal Bell came in yesterday bringing a 'coon' [sic] who some time ago stole a horse here and decamped for pastures new. Mr. Bell is a very cool and daring man. The citizens of White Oaks have full confidence in him and believe that he will conscientiously discharge his duties at any cost."[18]

Against this claim, a letter dated June 15, 1989, from David C. Paynter of the Civil Reference branch of the National Archives to Reynolds says that searches of the archives reveal no confirmation that Bell was ever a U.S. Deputy Marshal: "We have searched the records of the United States Marshal's office, 1789–1960, and found no mention of Mr. Bell. He may have been employed on an 'as-needed' basis by the United States Marshals' service. In any case we have no record of his employment or appointment in our files."[19]

Albert Regensberg, an archivist with the New Mexico State Records Center and Archives, wrote to Reynolds that he had "searched our records hoping to find that either James W. Bell or Robert Olinger . . . were deputy United States Marshals, but the results are negative."[20]

Bell Rides with White Oaks Posses in Pursuit of Billy the Kid

On the evening of November 22, 1880, someone tried to steal horses belonging to a J. B. Bell, who lived in the southwest part of White Oaks. On November 23, 1880 (dates vary with different accounts), the Bell of this chapter rode with a White Oaks posse headed by Deputy Sheriff Will Hudgens to the gang's camp at Blake's sawmill near town only to find the camp deserted.

The posse trailed the outlaws to Coyote Springs,

where a gunfight erupted in which several horses but no men were killed. The rustlers escaped, and the posse returned to White Oaks.[21] That night, some of the outlaws brazenly rode into White Oaks, and one of them shot at Jim Redman in front of Hudgens's saloon. Garrett says that the outlaws then rode out of town and on the outskirts came upon "J. N. [sic] Bell" and his friend Jimmy Carlyle, whom Hudgens had left on guard there. The two men fired at the fleeing outlaws but missed.[22]

Another posse was immediately formed, including Bell and Carlyle, and it trailed the Kid, Billy Wilson, and Dave Rudabaugh to the Greathouse-Kuch ranch. (Jim Greathouse and his partner Fred Kuch had a ranch, trading post, and rooming house about forty miles north of White Oaks. Locals generally referred to this place as the Greathouse ranch or tavern.) Carlyle was killed in a shootout at this location.

Bell Helps Sheriff Pat Garrett Deliver the Kid and His Gang to the Santa Fe Prison

On December 15, Pat Garrett and his posse again set out in pursuit of the Kid, this time heading for Puerto de Luna.[23] After leaving the ranch of Alexander Grzelachowski, the men headed downriver in a raging snowstorm for Los Ojitos, where Garrett was tipped off that the gang had moved on to Fort Sumner. Sneaking into Fort Sumner, Garrett learned that the gang was at the Wilcox and Brazil ranch and roadhouse twenty miles east of town.

Garrett then hired a runner to go to the Kid's hideout on the edge of the Llano Estacado and tell him it was safe to come back to Fort Sumner. Late that night the gang showed up in Sumner, and in a gunfight amidst a thick fog, Tom O'Folliard was shot and killed. The rest of the gang headed back to the Wilcox and Brazil roadhouse.

The posse pursued them beyond the roadhouse to a rock hut at Stinking Springs about four miles east of the Wilcox and Brazil place, where the Kid, Dave Rudabaugh, and Billy Wilson finally surrendered the next morning. With the prisoners securely shackled, the lawmen set off toward Las Vegas, arriving on December 26 to a large waiting crowd.

On the next morning, Deputy James Bell joined the Garrett contingent as one of Garrett's guards. (It is not known whether he was sent for or had been part of Garrett's posse all along.) He helped load the prisoners aboard a wagon, and everyone headed for the railroad.

While the party waited aboard the train to start the short trip to Santa Fe and the territorial prison there, an angry crowd began to gather, intending to snatch Dave Rudabaugh. Rudabaugh had earlier killed a Las Vegas jailer during his escape from jail there. In a determined show of force, Garrett backed off the mob until the train could pull out.

The *Las Vegas Daily Optic* gave an account of Bell's meeting with Billy Wilson in the railroad cars at Vegas. Wilson's first words were, "Bell, help me out of this scrape!" Bell replied, "That is a hard thing to ask of me after you killed Carlyle in cold blood, as you did." Wilson hung his head and reportedly responded, "I didn't shoot at him and tried to keep the others from doing so." At this point, Rudabaugh, overhearing this last remark, chimed in, "You are a damned liar. We all three shot at him. You and I fired one shot apiece and Kid twice."[24]

Bell Is Killed by the Kid

The Kid was subsequently sentenced at Old Mesilla to hang for the murder of Sheriff William Brady and was brought back to Lincoln. James Bell and Robert Olinger were assigned to guard him until the day of the hanging. Folklore says that Bell was mild-mannered, polite, and friendly toward the Kid and casts Olinger as the overbearing bully. It is hard to believe, however, that Bell could be described as friendly since he held the Kid responsible for the recent murder of his good friend Jimmy Carlyle. The truth probably lies somewhere in between, with Bell most likely an unsympathetic but fair jailer.

At any rate, on April 28, 1881, Bell walked the handcuffed Kid downstairs from the second floor of the courthouse to the privy. The prisoner either found a previously hidden six-shooter under the toilet seat and hid it on his person, or he wrested Bell's from him as they started back up the stairs. Either way, the Kid shot Bell inside the courthouse. One shot went wild; the other killed Bell.

Gottfried Georg (Godfrey) Gauss, caretaker of the courthouse at the time of Billy's escape, gives his version of events: "That memorable day I came out of my room . . . and was crossing the yard behind the court house, when I heard a shot fired, a tussle up stairs in the court house, somebody hurrying down stairs, and deputy sheriff Bell emerging from the door running toward me. He ran right into my arms, expired the same moment, and I

James W. Bell's memorial in Cedarvale Cemetery. (Note: No one alive today knows where Bell's body is buried; this is a headstone erected to his memory.) Photograph by Bob Workhoven. Author's collection.

laid him down, dead.... When Billy went down stairs at last, on passing the body of Bell, he said, 'I'm sorry I had to kill him but couldn't help it.'"[25]

In a document dated April 29, 1881, the Lincoln County constable formally requested Jesús Sueras, justice of the peace of the First Precinct of Lincoln County to hold an inquest over the dead bodies of Olinger and Bell, saying that they were "now lying in a room in the correll of the Court House in the Town of Lincoln."[26] George Washington Mitchell was appointed administrator of Bell's estate. The property and effects were pitifully small. Debts consisted of $7.75 owed to I. N. Bell and $2.20 to J. Tomlinson.

Location of Bell's Grave

As with so many aspects of Bell's life and death, it is not certain where he is buried. Records of Lincoln County cemeteries have been examined; none lists a burial for Bell. According to Lily Klasner, "We reached the town [of Lincoln] late in the day, and found that the reports we had heard were only too true. Bob Olinger and J. W. Bell had been killed by the Kid.... The commanding officer at Fort Stanton had sent down and had the bodies of Olinger and Bell, the other guard killed by the Kid, taken to Fort Stanton. The next day Bob Olinger, being a U.S. Deputy Marshal, was buried in the cemetery there, but Bell's body was sent over to White Oaks where his family lived."[27]

An article in the *White Oaks Golden Era* contradicts Lily's report:

> The city of White Oaks was startled on last Friday by the intelligence that Billy Kid had killed two of his guards and made good his escape.... It seems that while Mr. R. Olinger was at supper on last Thursday evening, as usual Deputy Sheriff J. W. Bell was guarding the prisoner, when it is supposed that Kid, watching his opportunity, slipped his handcuffs and struck Bell over the head and then grabbed the revolver from Bell's scabbard while he was in a dazed or stunned condition and shot him through the heart, killing him instantly.... J. W. Bell was made deputy sheriff last January and filled the office, becoming an officer of the law and had won the esteem of all law abiding citizens. His loss is greatly mourned by all the citizens of White Oaks and surrounding camps.... Mr. Olinger was buried at Fort Stanton and Deputy Sheriff Bell was buried at Lincoln.[28]

A Fitting Monument to
"The Other Guard Killed by the Kid"

In 2003, the White Oaks Cemetery Association headed by Bessie Leslie spearheaded the idea of a monument to Bell, and the New Mexico Sheriffs' and Police Association cooperated in the project.[29] The two groups chose a site in the White Oaks cemetery that oral tradition of the Leslie family of White Oaks maintains marks the spot of Bell's burial. A handsome monument to Bell's memory was erected on July 19, 2003, 122 years after the deputy sheriff was violently cut down in the Lincoln County Courthouse.

NOTES

1. Reynolds, *Trouble in New Mexico*, 153. Much of the background information in this chapter relies on Reynolds's thorough research. Reynolds has uncovered further convincing facts that discount Bell's service as a Texas Ranger with Captain Roberts. The Bell who served under Roberts was described as being 5'8" tall, having fair skin, blue eyes, and dark hair, and hailing from Green, Ohio. See also Keleher, *Violence in Lincoln County*, 332; Bartholomew, *The Biographical Album of Western Gunfighters*; and Metz, *Encyclopedia of Lawmen*, 20.

2. U.S. Bureau of the Census, *United States Census of 1880*, White Oaks, Lincoln County, New Mexico.

3. Thrapp, "Bell, James W."; Keleher, *Violence in Lincoln County*, 332; Metz, *Encyclopedia of Lawmen*, 20.

4. Reynolds, *Trouble in New Mexico*, 153; Thrapp, "Bell, James W.," 91. Thrapp quotes his source for Bell's description as simply "Oscar Williams."

5. Bartholomew, Thrapp, and Metz are among these historians.

6. Gillett, *Six Years with the Texas Rangers*, 53, 56–57.

7. Reynolds, *Trouble in New Mexico*, 158.

8. Ibid., 159.

9. Williams, *Pioneer Surveyor*, 89.

10. Ibid.

11. Reynolds, *Trouble in New Mexico*, 154.

12. Ibid., 160 (deed dated September 25, 1880, executed by J. W. Bell and J. B. Bell [not a relative] to A. W. Jones of several mining claims in the Nogal Mining District, recorded November 1, 1880, Book B, p 230; and deed dated March 11, 1881, executed by J. W. Bell to Pat F. Garrett of a one-half interest in the Georgia Placer Mine, Jicarilla Mountains, White Oaks Mining District, recorded March 11, 1881, Book B, p 433).

13. *White Oaks Golden Era*, March 10, 1881, from a reprint of "Happenings in White Oaks Twenty Years Ago," *White Oaks Eagle*, January 24, 1901.

14. *White Oaks Golden Era*, February 3, 1881, from a reprint of "Happenings in White Oaks Twenty Years Ago," *White Oaks Eagle*, January 10, 1901.

15. Nolan, *Pat F. Garrett's*, 128.

16. Keleher, *Violence in Lincoln County*, 296.

17. Lavash, *Wilson & The Kid*, 60.

18. *Las Vegas Daily Optic*, January 21, 1881, and May 3, 1881.

19. Reynolds, *Trouble in New Mexico*, 161.

20. Ibid., 160.

21. Nolan, *West of Billy the Kid*, 233–35.

22. Nolan, *Pat F. Garrett's*, 115–17.

23. Nolan, *West of Billy the Kid*, 242–55. This is a complete account of Garrett's chase of the Kid from Puerto de Luna through Los Ojitos, Fort Sumner, and finally to Stinking Springs. After the Kid's capture at the Springs, Garrett took his prisoners to Las Vegas to catch the train to Santa Fe.

24. *Las Vegas Daily Optic*, January 21, 1881.

25. Lincoln County Historical Society, "From Mr. G. Gauss," 1–5.

26. Request to Jesús Lueras from the Constable of Precinct No. 1, filed April 29, 1881, by Ben H. Ellis, Probate Clerk.

27. Klasner, *My Girlhood among Outlaws*, 189.

28. *White Oaks Golden Era*, May 5, 1881, from a reprint of "Happenings in White Oaks Twenty Years Ago," *White Oaks Eagle*, February 14, 1901.

29. Karen Mills, Lincoln County Historical Records Clerk, Carrizozo, telephone conversations with author, January 30, 2006, and February 9, 2006. Descendants of Robert Leslie (born May 17, 1853; died February 1932) say that Leslie showed them the burial site of James Bell in Cedarvale Cemetery, White Oaks. Robert arrived from Texas at the foot of the Tucson Mountains in the spring of 1884 and, after several years at his homestead there, bought a place in White Oaks.

EDWIN R. BONNELL
Builder of the Bonnell Opera House

Ed Bonnell, the second son of James Bonnell and Rachel Buffington, was born on September 15, 1848, in Franklin Township, Iowa.[1] The Bonnell ancestors were French Huguenots who fled southwestern France to England because of persecution by French Catholics about the time of the Edict of Nantes. The French ancestors spelled their name "Boinnelle."

Civil War Drummer Boy

On October 12, 1863, fifteen-year-old Ed Bonnell enlisted in the Civil War as a drummer boy in Company E, Twenty-fifth Iowa Volunteer Infantry.[2] Besides falling victim to malaria, he also suffered some kind of injury or rupture (not a battle wound) that never healed and would plague him throughout his life. This problem caused his discharge by reason of disability in January 1864. Undeterred, he re-upped with Company E of the Forty-fifth Iowa Volunteer Infantry in May 1864. He served until September 1864, when he was again discharged because his term of service had expired. A deep attachment to his Civil War service would last the rest of Ed Bonnell's life.

Marriage

Ed married Clotilda Ainsworth (born May 14, 1850) in Franklin Township, Iowa, on November 1, 1868.[3] The young couple became parents to five boys in rapid succession: Edwin Allen, born September 10, 1869 (and died at White Oaks April 10, 1888); Levi Irwin "Erva," born February 18, 1871; Harvey Lewis, born February 24, 1873; Bertram "Bert" Jay, born March 24, 1876; and Daniel Nelson, born February 22, 1878.

Over the Santa Fe Trail to Larned, Kansas

On August 25, 1873, Ed and brother Frank (seven years younger) arrived at Larned, Kansas, a small town on the Arkansas River adjacent to Fort Larned on the Santa Fe Trail. In September, Ed filed on land south of the Arkansas River.[4] The following March when he arrived to take possession of his claim, he found the river at flood level. In an attempt to cross, he hopelessly mired his wagon and was forced to unhitch the team, take the wagon apart, and carry it out of the bog piece by piece. He abandoned this claim and later filed on a piece of land on Ash Creek.

Ed and Frank Bonnell were among the first settlers at Larned from the "Iowa colony." According to a Pawnee history written by Capt. Henry Booth, "The Settlers of '73 were not numerous. Only 132 heads of families are entitled to the proud distinction, but like most pioneers whose primary object in following the Star of Empire westward is to seek amplitude of space, their progeny was numerous, and sons of settlers, re-enforced by their sisters, now swell the ranks and support the honor of Israel. He [Ed] returned to Iowa several times and kept in touch with friends and relatives there, influencing them to come to Kansas."[5] Clotilda died in Larned on September 28, 1878, and is buried in the Larned cemetery.[6] Ed struggled to care for their five boys, one a mere baby.

Sometime during his Larned years, Ed filed an application for membership in the B. F. Larned Post No. 8, Department of Kansas Grand Army of the Republic (GAR).[7]

The First Tree Claim to Be Proved in the United States

Ed Bonnell held the distinction of proving up on the very first tree claim to be filed in the United States.[8] The origination of tree claims is found in the Timber Culture Act of March 3, 1873. At the time of the act's passage, the public referred to this act as the "Tree Claim Act." The act was passed to encourage the growth of timber on western prairies: "Be it enacted by the Senate and House of Representatives . . . That any person who shall plant, protect, and keep in a healthy, growing condition for ten years forty acres of timber, the trees thereon not being more than twelve feet apart each way on any quarter-section of any of the public lands of the United States shall be entitled to a patent for the whole of said quarter-section at the expiration of said ten years, on making proof of such fact by not less than two credible witnesses; *Provided*, That only one quarter in any section shall be thus granted."[9] The Timber Culture Act was formally repealed by an act of Congress on March 3, 1891.

To White Oaks in 1880

Two years after his wife died, Bonnell arrived in White Oaks to set up a lumber, mercantile, and hardware business.[10] Four of his young sons accompanied him. Bonnell had not been in White Oaks long before he was drafted as a member of a posse organized by Constable "Pinto Tom" Longworth to pursue Billy the Kid and his gang to Greathouse-Kuch ranch some forty miles north of White Oaks.[11] The posse rode through heavy snow to the ranch, reaching it after dark the evening of November 27. It was the second posse formed at White Oaks to chase the Kid.

Bonnell's business supplied building materials (lumber, shingles, doors, windows, etc.) for the town.[12] In time he dabbled as well in real estate, stocks, and mines and was a notary public with an office on White Oaks Avenue. His business letterhead included the name "Pat F. Garrett, Ex-Sheriff of Lincoln County" and "All the Business Men of White Oaks, New Mexico."[13] He also busied himself establishing a ranch near McPherson and Biggs's sawmill. As noted in the *White Oaks Golden Era*, "Ed R. Bonnell and F. O. Blood have been putting in troughs at the former's ranch."[14]

Bonnell was a staunch believer in the future of White Oaks; he bought town lots as he was able.[15] Account ledger sheets from 1888 show sales from his lumber/mercantile/hardware store to many Lincoln County notables: George Barber, George Coe, J. N. Coe, Judge

Edwin Bonnell sometime in the 1880s. Photographer unknown. Archives and Special Collections, Ms04320002, New Mexico State University Library.

John Y. Hewitt, Marcus Whiteman, Rolla Wells, E. W. Parker, Pinkerton detective Charles Siringo, and even former Sheriff Pat F. Garrett.[16] Garrett bought five 12 ×14 windows, three doors, four hundred feet of lumber, and seven 1 × 10 inch pieces of lumber 16 feet long. His total bill as of December 28, 1888, came to $80.08. According to a Bonnell descendant, the invoice was never paid.[17]

A Second Family with Viola Albright

In 1884, Ed married a beautiful young music teacher from Kansas named Viola J. Albright and began a second family that came to include four children, all born at White Oaks: Minnie, born August 27, 1885; Charles, born September 13, 1887; Edna, born September 1, 1889; and Archie, born July 5, 1892.

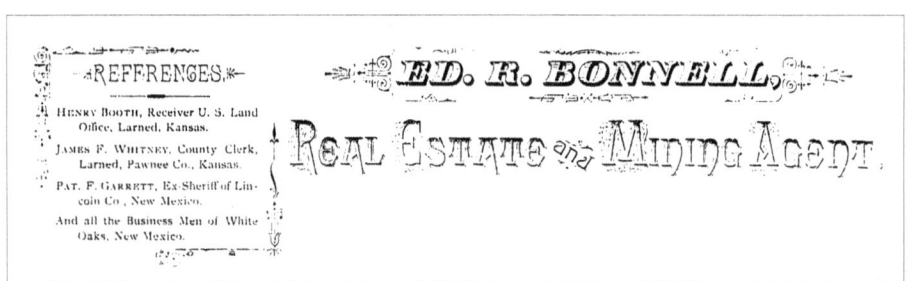

Ed Bonnell's letterhead from 1886. Note Pat Garrett's name as one of Bonnell's references. Author's collection.

Viola Albright Bonnell, second wife of Ed Bonnell. Viola was reared in the Good Intent community near Atchison, Kansas, and was a music teacher before marrying Edwin. Photographer unknown. Courtesy Timothy Bonnell.

The new Mrs. Bonnell kept busy with her family in the same manner as most women of her time. Her life was full with her family and such tasks as sewing for her children.[18] The couple kept in close touch with their Iowa relatives. Son Erva (fourteen years old) wrote his aunt in 1886:

> Dear Aunt,
>
> I received a letter from you and was glad to hear from you. My little sister can stand up now and soon will walk. Her name is Minnie. The little boys are as proud as they can be of her. They play with her all the time when they are not in school.
>
> I have been going to school until here lately. I have been sick for about two weeks. . . . Bert and Nel are having big times at school and are learning well. We have a good teacher now.
>
> Your nephew,
> Erva Bonnell[19]

Ed Builds an Opera House

Nearly every western town of any size in the late 1800s boasted an opera house. Activities at the opera houses

Viola Bonnell's account with the Ozannes. Author's collection.

were as popular in daily life in newly settled areas like White Oaks as they were in the more established regions "Back East."[20] In post–Civil War years, the public clamored for entertainment as well as culture, and so there was an increasing number of traveling shows, minstrel shows, circuses, medicine shows, lectures, and even Shakespearean plays. Musical entertainment of all sorts was extremely popular. Traveling theatrical companies spread popular culture by sharing jokes and songs with their audiences. A bevy of pretty young ladies added to the fun. While operas were performed in the opera houses, musical comedies and operettas were much more common on small-town circuits.

A public town hall had been built early on in White Oaks. It was a 24 × 48 foot structure with walls of 4 × 6 planks laid flat and spiked, covered by a corrugated sheet-iron roof, built with mostly donated labor and materials furnished at cost. This first town hall quickly became the center of such public gatherings as Sunday and day schools, dances, political meetings, and local theatricals in the early 1880s.

Ed built a second town hall for use primarily as an athletic club, where boxing, wrestling, and gymnastics held sway.[21] Later, in 1892, he added other activities—dancing, musicals, and entertainments such as the play *The Flowing Bowl* staged by locals—and named his hall the Bonnell Opera House. The *Lincoln Independent* reported in January 1892 that "Ed Bonnell has put a stage in his hall and has employed Mr. Frank Robinson to paint some scenery for same. He intends to fix the hall up for theatrical purposes."[22]

The End of a Too-Short Life

In September of the same year, the Civil War veteran made a sentimental journey to Washington, D.C., with two comrades. The *Lincoln County Leader* reported, "On Thursday, E. Bonnell, W. S. Redding and Stanley Rudison started for Washington DC there to take in the GAR Encampment."[23] Before the advent of automobiles, the logistics of getting to Washington and back must have been daunting. The group of three probably caught a train east out of Las Vegas.

By early 1893, Ed's health was deteriorating rapidly, as he wrote to friend Charlie Siringo, "I am still confined to my bed. Have not been out of the house for over a month past. I will have Charlie Bull look up the numbers of your lots today and I will send them to you tomorrow. White Oaks very dull, weather disagreeable and windy."[24]

The pain from his old malady increased, causing Ed great suffering. White Oaks surgeon Doctor Paden performed an operation attended by a crowd of onlookers; it was unsuccessful. After a second operation, Ed died en route from White Oaks to Las Vegas on September 28, 1893. His memorial in Cedarvale Cemetery is a simple white marble tombstone that reads, "EDWIN BONNELL, CO. E, 45TH IOWA INF." Not long after Edwin died, Viola left White Oaks and returned with her young children to Atchison, Kansas, to be near her parents.

In 1900 John Hewitt, administrator of the Bonnell estate, brought suit against the widow, Ed's five sons by his first marriage, and Viola as guardian of the four

Ad from *Old Abe Eagle* for play at the Bonnell Opera House, February 23, 1893. Author's collection.

children of the second marriage. The suit was filed in the district court of the Fifth Judicial District of New Mexico Territory "for the purpose of having an account taken of all the real estate of which the said Edwin R. Bonnell died seized and possessed; to have said real estate . . . sold for the payment of the debts of the said decedent's estate and for such other relief as equity may require and said court may deem just."[25] Viola never remarried and died in Atchison on August 10, 1935; she is buried at the Alderson Cemetery west of Atchison near her parents.

Bert Bonnell Marries Sydney Coe, Daughter of Frank Coe

In 1892, the year before their father died, sons Bert and Nelson had been sent back to Pomona, Kansas, to stay with relatives and attend school.[26] From there Bert went to the Ozarks in Missouri and worked on a farm for three years. Then he returned to Lincoln County and found his first job punching cows on the VV Ranch for Mrs. Cree.

On September 9, 1899, Bert Bonnell went to work on the Ruidoso ranch of Frank Coe of Lincoln County War fame.[27] After the cattle roundups were over, Bert stayed on to help supervise while brother Nelson, who had also returned to Lincoln County, did the hauling and freighting. Every member of Frank Coe's family played some kind of instrument, with young people riding horseback as far as twenty-five miles to attend the Coe dances. With the Coe daughters and son Wilbur, there was always music to dance by.[28]

Frank Coe had a pretty twenty-two-year-old daughter named Sydney. It was not long before Bert noticed the girl. However, a young man named Irvin Lesnet had been "keeping company" with Sydney, and they were planning to be married in spite of Frank Coe's disapproval. Irvin wrote her a note telling her what night he would be at her place. They were to ride off to Lincoln and be married.[29]

Lesnet made the mistake of giving the note to Ross Coe, son of Jap Coe and cousin of Frank Coe, to deliver; Ross opened the note and read it. No one now knows just what it said, but Ross promptly took it to Sydney's father. On the night that Irvin rode up leading a horse for Sydney to ride during their elopement, Frank Coe was waiting with a shotgun. He stepped out on the porch and killed Irvin with one blast. He was never brought to trial for the shooting. The way was now clear for Bert to court Sydney; the pair was married at the Frank Coe ranch in Glencoe on December 18, 1900. Bert and Sydney began their marriage by moving to the so-called goat ranch where Frank Coe grazed five hundred goats.[30] Bert continued to work for Frank for a time.

The old one-room schoolhouse built by the Coes to educate their children gave way to the start of the Bonnell ranch house after Bert and Sydney bought it from J. Landly Pool. They gradually expanded the house until it became an eighteen-room residence.

Bert Bonnell family, ca. 1914. Sydney Coe Bonnell (*left*), her father Frank Coe (*middle*), Bert Bonnell (*right*), with Sydney and Bert's two sons, Frank and Ralph Alfred. Photographer unknown. Courtesy Haley Memorial Library and History Center, Midland, Texas.

The Famed "Bonnell Ranch on the Ruidoso"

The location of the Bonnell Ranch at the confluence of the Ruidoso River and Eagle Creek, and at a road junction connecting Tularosa, Lincoln, Fort Stanton, and Capitán, made it ideal for a dude ranch. At the two extremes of the road lay Roswell and El Paso.[31] The Pickwick Corporation operated a bus (stage) through southern New Mexico that regularly stopped at the Bonnell Ranch. In 1922, Bert bought stock in the Pickwick Corporation, but it later filed for bankruptcy.[32]

Bert recalled his days on the ranch:

> I came to the F. B. Coe Ranch on the ninth day of September 1899. At that time the mail came from Fort Stanton, if some one in the neighborhood was over that way. They would bring the neighborhood mail to what was at that time the Jap Coe Ranch, now the Bonnell Ranch.... On Sundays the young folks in the neighborhood would get together, sometimes a dozen or more—the boys mounted on broncs and the girls would be on day herd. We would bring the mail to the Jap Coe Ranch and there we were met by most of the people who got their mail through Fort Stanton.... I decided we should have a better mail service so I got in touch with Postmaster General Burleson of Texas. ... He told me if I would carry the mail 90 days free of charge to the government and if someone would take care of the office for 90 days, also with no charge ... it could be worked out. Frank Coe was named for the job, with Mrs. Bonnell assistant.... Some wanted to call it Coeville, others suggested ... Coe. That didn't quite fit, so Mrs. Frank Coe, mother of Mrs. B. J. Bonnell, suggested we call it Glencoe, New Mexico.[33]

During the summer, the Coe orchestra played every Saturday night for the community and guests of the Bonnell Ranch. Over the years people enjoyed some of the top performers of the day, including Louise Massey of "My Adobe Hacienda" renown.[34]

The friendly and outgoing Bert was a fine square-dance caller and floor manager. After the musicians began playing, he would make his way to the center of the floor and call for the couples to "wash their faces and take their places and get set for an old-time square." Sydney beat out the melody and rhythms at the piano, while Bert and Sydney's son, Ralph, and Harold and Joe Coe picked the guitar. Wilbur, Frank Coe's son, fiddled away on his violin. They played as long as anyone cared to dance.

The fame of Bonnell's ranch spread far and wide through ads Bert placed in newspapers in Chicago and elsewhere offering home-cooked food and ranch-style lodging. In distant Honduras, Englishman Rowland Cowper picked up a copy of a magazine featuring one such ad and immediately caught a freighter to New Orleans, then a bus to the Bonnell Ranch. He stayed for two years, leaving in 1939 only after WWII had started and he needed to get back to England.

Later on, Sydney and Bert Bonnell also donated land for Saint Anne's Chapel across the highway from Bonnell Ranch headquarters. Bert Bonnell died at Glencoe on November 23, 1951.

NOTES

1. "Descendants of Nathaniel Bonnel," 1–3, family genealogy in the collection of Eleanor Bonnell Shockey.

2. Ed. R. Bonnell, application for membership in B. F. Larned Post No. 8, Department of Kansas Grand Army of the Republic, n.d., copy in author's collection.

3. "Descendants of Nathaniel Bonnel," 3.

4. Isabelle Worrell Ball, "Early Pawnee History," *Larned Eagle-Optic*, October 27, 1892, read before the Pawnee County Old Settlers' Association, fall 1892, Early Pawnee County Collection.

5. Capt. Henry Booth, "Pawnee County History," read at the Kansas centennial celebration, July 4, 1876, and printed by the *Larned Press*, July 7, 1876, Early Pawnee County Collection.

6. Larned Cemetery Book, Early Pawnee County Collection.

7. Bonnell, application.

8. Crawford-Gore to Haldane, July 9, 2005.

9. Stephen Beyerlein, U.S. Department of the Interior, Bureau of Land Management, Santa Fe, to Haldane, n.d. (reference No. NM 952 [9600]). As an enclosure to this letter, Beyerlein sent a copy of the Timber Culture Act of 1873, enacted by the Forty-second Congress, March 3, 1873.

10. Shockey to Haldane, June 21, 2003.

11. Nolan, *West of Billy the Kid*, 234.

12. Advertisement in the *New Mexico Interpreter*, June 6, 1890.

13. Copy of letterhead furnished by Eleanor Bonnell Shockey.

14. *White Oaks Golden Era*, July 31, 1884.

15. Jones Taliaferro to Ed R. Bonnell, June 15, 1888, Records of the Probate Clerk, Lincoln County Courthouse, Carrizozo.

16. Account ledgers for business of Ed Bonnell, collection of Eleanor Bonnell Shockey. Purchases by Pat Garrett are from the ledger dated December 28, 1888.

17. Eleanor Bonnell Shockey, interview by Haldane, May 31, 2005, Ruidoso.

18. Originals of the Ed Bonnell account with Ozanne & Armstrong Dry Goods and Clothing, White Oaks, in account ledgers for business of Ed Bonnell, collection of Eleanor Bonnell Shockey.

19. Erva Bonnell, to "Dear Aunt," November 11, 1886, White Oaks, copy in author's collection.

20. Information about opera houses in the late 1800s came from Booker, "Did They Really Sing," 132–53.

21. Parker, *White Oaks*, 70–71.

22. *Lincoln Independent*, January 21, 1892.

23. *Lincoln County Leader*, September 10, 1892.

24. Ed Bonnell to Charles Siringo, February 27, 1893, copy in author's collection.

25. Notice of Suit, *White Oaks Eagle*, October 4, 1900.

26. Eleanor Bonnell Shockey, interview by Haldane, May 31, 2003.

27. B. J. Bonnell, from an article written for the Glencoe Woman's Club in the 1930s and reprinted in Paul Baker, "B. J. Bonnell," Ramblin' Around Lincoln County, *Lincoln County News*. Undated copy in author's collection.

28. Coe, *Ranch on the Ruidoso*, 257–61.

29. Avant, "Bundy Avant Story," 14.

30. Coe, *Ranch on the Ruidoso*, 257–61.

31. Ibid., 257.

32. Shockey interview, May 31, 2003.

33. Bonnell, article.

34. Coe, *Ranch on the Ruidoso*, 259–61.

JAMES CARLYLE, ALIAS BERMUDA CARLISLE
Gunned Down in the Greathouse-Kuch Ranch Shootout

Jimmy Carlyle (also Carlisle) was a popular young White Oaks blacksmith who was killed in the Greathouse-Kuch ranch shootout north of White Oaks in late November 1880. Carlyle was one of a trio of "Unlucky Jims," a group of young men in their twenties who showed considerable promise on their arrival in the area in the late 1870s but who would not live long enough to fulfill expectations. Lincoln County was a raw place, filled with liquor, guns, and gambling. Often young men became edgy and dangerous after a few drinks or after real or imagined insults at the gaming tables and the *bailes*.

Early Life

James Carlyle is listed in the June 1880 census as twenty-six years old, born in Ohio of Scots parents.[1] Historian Philip Rasch believes that Carlyle was born in Trumbull County, Ohio, about 1854.[2] Rasch also thinks that Carlyle might have been a boyhood schoolmate of Billy Wilson, a member of Billy the Kid's gang at the Greathouse ranch.[3] Another story has Wilson and Carlyle first meeting at Dodge City, Kansas, where many restless young adventurers spent time before moving on to Texas to hunt buffalo.[4]

Buffalo Hunter and Buyer of Hides on the Texas Range

In the 1870s, Carlyle operated on the Texas buffalo range as one of many buffalo hunters, skinners, and hide buyers. An article concerning the Greathouse-Kuch ranch shootout sheds further light on Carlyle's background:

> What bestows [on Greathouse] this standing [of conditional immortality] is that brief moment in Jim's anonymous passage through his years when his name became linked forever with that of . . . Bonney. This occurred during the attempt to capture Bonney at Greathouse's road house on November 27, 1880, in which Jimmy Carlyle was killed by Dave Rudabaugh in our opinion. . . . This Jimmy Carlyle is not to be confused with that Jim Carlyle who was a gunslinger in and around Dodge City during its trail days, and who had been a hostler in a Denver livery stable during the 1860s. . . . our Jimmy Carlyle was one and the same with that "Burmude" [*sic*] Carlyle who was in Hanrahan's Saloon at the 'Dobe Walls with Bat Masterson and others.[5]

Historian Rasch names Carlyle as one of the participants in the Adobe Walls battle in Texas in June 1874.[6] By the following year, he was in charge of a hide yard for Rath, Reynolds & Company. He continued to buy hides at Rath City as late as March 1878, before coming to Lincoln County.

In "Jim Greathouse or Whiskey Jim," fellow buffalo hunter Frank Collinson reminisces about his days as a buffalo hide buyer on the Texas plains:

> This man, Jim Carlyle, I knew well. He had come from Kansas with Lee Reynolds and Rath in 1875 and had charge of the hide yard. It took several men all this time to load and unload hides, and stack them up; all the hides had to be weighted down to keep them from blowing away or getting badly scattered. Carlyle was also hide buyer and classer for Reynolds and Rath. Nearly all winter hides were classed. The best hides of cows and young bulls went for robes. In March 1878 I sold over two thousand hides to Carlyle. He was a good hide man. A fine, quiet man, not the least inclined to be a gunman. I had not heard what had become of him until I heard the Kid had killed him. . . . I always figured the killing of Carlyle was the most uncalled for of any of the Kid's killings, although he laid all the blame on the White Oaks party.[7]

Later, in January 1938, Collinson wrote a letter to Bruce Gerdes from his home in El Paso, Texas, in which he again touched on his knowledge of Carlyle and repeated much of the same information in the above paper.[8] Another buffalo hunter by the name of W. S. Glenn, recounted one of Carlyle's misadventures in the spring of 1877:

> I decided to move back into the buffalo [range], being then camped on Croton Creek.... The camps were not far apart. One of the Moody brothers stayed at the old camp and the other went to the new one ... leaving at the old camp a fellow by the name of Jim Bermude [*sic*] or Jim Carlyle, afterwards killed by "Billy the Kid." Jim was the hunter, and a man by the name of George Jilds [Giles?] was doing the skinning. Bermuda made a killing and was helping Jilds skin.... Jilds, thinking that he could finish the pile by night, told him [Carlyle] to go on and make another killing. ... Something had gone wrong with Jim's gun and he had taken it to pieces. They were camped in a gulch for protection from the wind. [Jilds] went out of the gulch and down to where the horses were hobbled. As he looked up there sat an Indian, another unhobbling Jilds' horse and one already on his [Carlyle's]. So [Jilds] went back, hollering to Jim that the Indians had their horses [and] getting his [Carlyle's] six-shooter, Jim hurriedly put his gun together, and when he went to shoot found that the breech block was gone and ran after it.... The Indians rode off and by the time Jim came back with his gun were out of sight ... Jilds going with him they soon came up [upon the Indians], one Indian was down, wearing a linen duster, cinching up his saddle. Jim fired three or four shots, but without effect, and all they could do was to walk back to camp.[9]

It is clear from the context of both the Collinson and Glenn writings that Carlyle knew "Whiskey Jim" Greathouse and many other buffalo hunters in the Texas buffalo range. He would encounter some of them again, including Pat Garrett, after moving to New Mexico.

Prior to settling in White Oaks to join the gold rush, Carlyle also rode briefly with the Murphy-Dolan faction of the Lincoln County War. Carlyle had lived in White Oaks a little more than a year before he met his death at the Greathouse-Kuch ranch shootout.

The Greathouse-Kuch Ranch Shootout

After receiving word that Billy the Kid and his gang were nearby with stolen horses, a White Oaks posse that included Carlyle trailed the thieves to Coyote Springs. A gun battle ensued, and after the Kid and Billy Wilson had their horses shot from under them, they escaped on foot into the hills. The posse returned to White Oaks but not before Jimmy Carlyle is said to have appropriated the Kid's prized new pair of gloves from the gang's campsite.

The Kid's gang slipped into White Oaks that night and put up at the West and Dedrick stable. The following evening they hurrahed the town on fresh mounts, exchanging shots with James Bell, Jim Redman, and Carlyle. Other accounts say that only Redman was actually fired at and that Bell and Carlyle returned fire as the gang galloped out of town. In either case, after the exchange the Kid's gang headed forty miles north to the Greathouse-Kuch ranch.

A second posse formed by Constable Thomas "Pinto Tom" Longworth tracked the fugitives through heavy snow and bitter cold to arrive after dark at the Greathouse-Kuch ranch and tavern, known locally as the Robbers' Roost. The posse surrounded the tavern as Constable Longworth returned to White Oaks for reinforcements and supplies, leaving Carlyle in charge to negotiate terms of surrender.

The parley continued through the next morning with the tavern's recently hired German cook and handyman, Joe Steck, traipsing back and forth between the posse and the gang to deliver messages. Finally, Wilson demanded that Carlyle come inside the tavern to talk face to face. Carlyle strongly resisted but agreed after Greathouse offered himself up as hostage for Carlyle's safety.

Precisely what occurred within the tavern is still unclear. One story has the Kid losing his temper when he discovered that Carlyle was wearing his prized, missing gloves. Whether or not that is true, it is known that the parley continued amid many drinks of the plentiful liquor on the premises, until suddenly a random shot was fired. Carlyle—befuddled by too much whiskey and believing that Greathouse had been killed—leaped out a window in a desperate attempt to escape. Amid a wild barrage of bullets and the crash of glass, he was shot and killed.

Road sign near site of Greathouse-Kuch ranch shootout. This official scenic historic marker was placed by the State of New Mexico on U.S. Highway 54 a few miles south of Corona, New Mexico. Photograph by Bob Workhoven. Author's collection.

Steck, who had been busy rustling up a meal for the men inside the tavern, dropped everything and ran for cover outside along with Greathouse's partner Fred Kuch. Steck, Kuch, and Greathouse, who had managed to flee from the posse, caught some horses and headed for the nearby Abel's ranch while the Kid's gang jumped on their horses and headed for the William Spence ranch. The posse also scattered, riding for Jerry Hockradle's place roughly twenty-five miles south in the Jicarillas, leaving Carlyle's body where it lay in the snow.

Who Killed Jimmy Carlyle?

There has been lingering speculation as to who actually killed Carlyle. Was it someone in the Kid's gang or could it have been someone in the White Oaks posse? In the December 3, 1880, issue of the *Las Vegas Gazette*, the Kid sent a letter to Gov. Lewis Wallace denying any responsibility for Carlyle's death:

> The house was Surrounded by an outfit led by one Carlyle, Who came into the house and demanded a Surrender I asked for their Papers and they had none. So . . . told Carlyle that he would have to Stay in the house and lead the way out that night. Soon after a note was brought in Stating that if Carlyle did not come out inside of five minutes they would Kill the Station Keeper [Greathouse] . . . in a Short time a Shot was fired on the outside and Carlyle thinking Greathouse was Killed Jumped through the window, breaking the Sash as he went and was Killed by his own Party thinking it was me trying to make my Escape.[10]

The *Las Vegas Daily Optic* in January 1881 reported Billy Wilson, as he was being transported to jail in Santa Fe, pleading with Deputy Sheriff James Bell to help him out of his predicament and claiming no responsibility for Carlyle's death. In early March 1881, Rudabaugh was arrested by Deputy Sheriff Barney Mason as an accessory to Carlyle's murder. He was given a preliminary trial before Justice Lea and placed under a $3,000 bond to appear at the next term of the district court.

Apparently a Lincoln County grand jury also thought that the Kid's gang was responsible. On August 8, 1881, a grand jury returned an indictment naming William Wilson, Rudabaugh, and Jose Sánches as the killers: "With the leaden bullets aforesaid out of the guns aforesaid . . . discharged and shot . . . the said James Carlisle [*sic*] upon the head, breast, and body . . . [and did] effect the death of him the said James Carlisle."[11]

Random chance picked Carlyle as the first of the White Oaks Unlucky Jims to die. Next was James Bell, shot dead at age twenty-seven by Billy the Kid during the Kid's escape from the Lincoln County Courthouse in April 1881. The last was "Whiskey Jim" Greathouse, who, some said, was turning his life around after a trou-

James Carlyle's unmarked grave. Date and photographer unknown. Author's collection.

bled start as whiskey supplier to Comanches in Texas. Greathouse never realized his dream of establishing a freighting empire in New Mexico; he was bushwhacked in his late twenties by Joel Fowler in December 1881.

Carlyle's senseless death helped galvanize citizens into a resolute push for law and order. Newly elected sheriff Pat Garrett would soon swing into action, but it would be another five or six years before law would be firmly planted in Lincoln County.

Where Is Carlyle Buried?

The morning after the Greathouse-Kuch ranch shootout, Steck and Kuch returned to find Carlyle frozen stiff; they tied a blanket around the body and buried it in a shallow grave. A second White Oaks posse arrived shortly after to disinter Carlyle, rebury the body in a wooden coffin, and burn the roadhouse to the ground. The burial was not in White Oaks—no listing appears in undertaker John Slack's meticulous records. Eyewitness accounts agree that Carlyle was buried near where he died by the posse who returned to the Greathouse-Kuch ranch the day after the shooting. Robert Leslie, White Oaks old-timer born in 1912, also maintains that "Carlyle was killed and buried there [the Greathouse-Kuch ranch]. His grave could still be seen there in the '30s."[12]

NOTES

1. U.S. Bureau of the Census, *United States Census of 1880*, White Oaks.

2. Rasch, "Alias 'Whiskey Jim,'" 97.

3. Rasch, "Billy Wilson," 70.

4. Ibid.

5. Hutchinson and Mullin, *Whiskey Jim*, 3, 17.

6. Rasch, "Alias 'Whiskey Jim,'" 97.

7. Collinson, Frank, "Jim Greathouse or Whiskey Jim," n.d., article on file with Panhandle-Plains Historical Museum, Canyon, Texas.

8. Frank Collinson to Bruce Gerdes, January 4, 1938, on file with Panhandle-Plains Historical Museum.

9. Glenn, "The Hunters' War."

10. *Las Vegas Gazette*, December 3, 1880.

11. Grand Jury Indictment of William Wilson, ____ Rudebaugh [*sic*], and José Sánches, Territory of New Mexico at the County of Lincoln, signed by Simon Newcomb, District Attorney, Third Judicial District, August 8, 1881, copy in author's collection.

12. Robert Leslie to Haldane, September 16, 1999.

THE DE GUEVARA, LALONE, AND LACEY FAMILIES
Mexico and Texas Meet Canada in White Oaks

The de Guevaras of White Oaks originally bore the old Spanish name "Niño Ladrón de Guevara," meaning "Little Boy Thief de Guevara."[1] "Niño Ladrón" was dropped from the name in New Mexico in the mid-1800s, with most of this family thereafter calling themselves de Guevara, de Guebara, or, in modern times, Guevara. A prominent Ladrón de Guevara family lived in Guadalupe del Paso (now El Paso/Juárez) as early as the 1600s.[2]

Present-day de Guevaras name Don Juan Joseph Niño Ladrón de Guevara as their first ancestor of record.[3] The title "Don" indicates an upper-class family. Don Juan's grandson, Don José María, married Doña Guadalupe de La Riva at San Pedro de Gallo, Mexico, in 1800. This couple's second son, José Antonio Pelagio Ygnacio Niño Ladrón de Guevara, was born March 23, 1803, in Villa de los Cinco Señores (now Nazas), Durango. It is this Ygnacio who figures in White Oaks history.

The region around present-day Nazas is located within sight of central Mexico's rugged Sierra Madre Occidental near the road from Mexico City to New Mexico. New Mexico was then under Mexican rule and was known as part of New Spain. The Mexican state of Durango was an important silver-mining center, and the upper-class de Guevaras would have been greatly interested in mining affairs. Ygnacio Niño Ladrón de Guevara, patriarch of the White Oaks de Guevaras, was likely connected to the de Guevaras of Guadalupe del Paso and as a young man would have traveled through that place on his way from Durango to New Spain.

Ygnacio Niño Ladrón de Guevara

A year or so before setting out for New Spain, Ygnacio married Dolores Cayales. Son Nasario was born around 1830. In 1833, at age thirty, Ygnacio and two partners ventured north over many hundreds of dangerous miles from Durango to the Ortiz Mountains of New Mexico.

The partners filed a claim on a gold mine at Sierra del Oro ("Hill of Gold"). Two months later, Ygnacio sold his share for the tidy profit of three hundred silver pesos.

A second son, Plácido, was born about 1834. Dolores died within a few years of the couple's marriage, and in 1844 Ygnacio married Anamaría Tórres at San Felipe, New Mexico. Anamaría came from Manzano, a village on the old trail to Santa Fe by way of the Ortiz Mountains. Ygnacio moved south to Manzano and farmed there. In Manzano, in addition to Ygnacio's two sons from his first marriage, Ygnacio and Anamaría raised an orphaned niece named María Estanislada "Lada" Padilla, born in the banner year of 1846 in which New Mexico was annexed as part of the United States. Lada would in turn care for her foster parents in their old age.

Around 1866 or 1867, the de Guevara clan left Manzano for Lincoln County and a new little settlement called Plaza de Missouri, located eighteen miles upriver from the confluence of the Hondo and the Pecos Rivers. By 1870, they were at Picacho, a few miles southeast of Lincoln. Ygnacio's name does not appear on the 1870 census, but he resurfaces in the 1880 census as living in Lincoln (First Precinct) in a house next door to son Plácido.

Ygnacio was to live through, and survive, a rich but tumultuous era of the Old West—Indian attacks, the Lincoln County War, and assorted gunfights. At one time he served on a Lincoln grand jury, indicating that he was respected in that town. The de Guevara patriarch lived until he was at least eighty-two with Teófilo and Lada Padilla LaLone at their ranch on Lincoln County's Magado Creek at Nogal. He is listed with this couple in the 1885 New Mexico state census, and he died shortly afterward.

Plácido de Guevara

The census of 1860 places Ygnacio's younger son Plácido and Plácido's wife, María Sánches, as living next door to Ygnacio in Manzano. Nine children were born to Plácido and his wife while Ygnacio had followed work in several New Mexico villages including Magdalena, Manzano, Lincoln, and, finally, White Oaks.

At least four of Plácido's children with María were Lincoln County citizens: Maximiano "Max," born August 19, 1856, in Magdalena; Manuel, born in 1870 in Lincoln; Lorenzo, born in 1873 in Lincoln; and Francisco "Frank," born 1878 in Lincoln. Manuel became a coal miner at White Oaks and a sheepherder; Lorenzo became a farmer, miner, and teamster. Frank was sent to a Catholic seminary but would not stay. Plácido's relationship with Tecla Sisneros also produced a son, Felís (anglicized as "Felix"), born in 1859.

María de Guevara died February 13, 1884, in White Oaks. According to the *White Oaks Golden Era*, "A beautiful tribute was paid the deceased Sen. María Sánches De Guebara, by the Americans of White Oaks. Mr. M. Branham read the funeral service, and the last sad rites were all performed by Americans."[4]

In 1892, Plácido married second wife Librada Barela. According to the 1900 census, five de Guevara families lived in White Oaks, with Plácido, Lorenzo, and Max in adjacent homes, and Manuel and Felix next to each other. Some de Guevaras continued to live in White Oaks for thirty years or more. Plácido died May 31, 1904, in White Oaks.

Maximiano "Max" de Guevara

Max de Guevara lived in Lincoln during the worst of the Lincoln County War. He made his first recorded appearance there on July 23, 1869, as the godfather at a baptism. He grew up to be widely regarded as an honorable, upright citizen and married Felicita Montoya, a native of the village.

As a youthful twenty-one-year-old and Lincoln's new town constable in October 1877, Max became the jailer of Jesse Evans and his gang of "Boys" of Lincoln County War notoriety. (These troublemakers were allied with Jimmy Dolan in the feud heating up at the time.) The Boys—Jesse Evans, Frank Baker, and others in Jesse's gang of rustlers—had been captured, shackled, and confined in the newly completed county jail. The jail, built of adobe by George Peppin, consisted of an aboveground house for the jailer and two underground rooms built of logs for the prisoners.

By November 7, the Boys had managed to file through their shackles and cut holes in the jail's underground log walls and were on the verge of escaping. All kinds of augers and files had been packed in goods (which turned out to be rocks) ordered from Jimmy Dolan's store by the prisoners and delivered to them at the jail. On hearing the news, Sheriff William Brady—a man also in Dolan's camp—ambled over to assess the situation.

Young Max de Guevara suggested to Sheriff Brady that it might be wise to obtain new shackles, but after a sharp exchange with the sheriff, he resigned and walked away from the jail.[5] (Curiously, not only were the prisoners left unguarded the night Max resigned, but no one even bothered to lock the jail doors after Brady's visit.) A new jailer was found, but it was the following day before Max turned over his keys to Sheriff Brady and the new jailer could assume his duties. It was time enough; by morning a man at the jail door peered in and reported the jail empty, the Boys long gone. Sheriff Brady declined to pursue the escapees.

By mid-December 1878, the de Guevara house in the lower part of Lincoln village had become a gathering place for Billy the Kid and some of the Regulators, including Doc Scurlock. One day Billy had ridden back to town hoping he might benefit from a general amnesty that territorial governor Lewis "Lew" Wallace had issued for some participants in the Lincoln County War. Soon after, 2nd Lt. James H. French, newly arrived in Fort Stanton and a man prone to "ill health" (he was an alcoholic), secured an on-the-spot appointment from Sheriff Peppin as deputy sheriff to go after some of the Regulators at the lower end of town. Peppin issued warrants to French for Doc Scurlock and others and suggested that French might find them at the home of Max de Guevara.[6]

French arrived with three men early on the morning of December 13 at the former jailer's home. After bursting into the house by smashing a board that Max had propped against his door to keep out intruders, the very drunk Lieutenant French started waving a pistol around and threatening the inhabitants. Finally, he sat down on a bench near the fire, shoving paper after paper in Max's face and demanding to know the whereabouts of Doc Scurlock. When Max maintained that he did not know, French called him a liar. Max stood his ground as French cocked the gun and pointed it at him. Then Max's wife, Felicita, started screaming. Fortunately, the lieutenant gathered up his papers and lurched on, first to the Copeland house and finally to the old Baca house occupied by Susan McSween and her lawyer Huston Chapman. There French again waved his pistol, then removed his overcoat and tunic and challenged Chapman to a fight before dramatically donning his clothing again and sweeping out.

As a result of these incidents, de Guevara, along with Susan McSween and her lawyer, swore out warrants against Lieutenant French, charging him with felonious entry and assault with intent to kill.[7] French was arrested, and Fort Stanton's commanding officer, Colonel Dudley, ordered a board of inquiry. Max, along with Susan McSween, Juan Patrón, George Barber, and others, made sworn statements about the incident before the court of inquiry. On December 21, the board recommended no action be taken.

Having somehow survived the worst of the Lincoln County War, in February 1885 Max moved his family from Lincoln to a house he had bought in the nearby mining town of White Oaks. Later, in the early 1900s, he would become a mercantilist at Ancho and Jicarilla. Max died November 22, 1915, at Carrizozo, New Mexico; Felicita died on October 21, 1930.

Felix de Guevara

Felix worked for Paul Mayer at White Oaks as a young man. Mayer owned the livery stable but had a dangerous sideline: he couriered gold from the White Oaks mines to the railroad stop at San Antonio. As part of his job, Felix gathered information about strangers in town and relayed what he learned to Paul; sometimes he also accompanied Paul on his surreptitious gold runs.

In 1883, Felix married Carmelita Tórres and built a house in White Oaks. The couple prospered, making it possible for Felix to branch out into raising stock. He bought a ranch five miles east of White Oaks near Patos Mountain from a man named Farr.

Felix and his wife did not have the first of their four children until 1902. Felix was an accomplished fiddle player; he taught his three sons the fiddle, and they all played at dances. Felix died in 1930, and Carmen died four years later. They are buried in Cedarvale Cemetery.

The LaLone Family

The census of 1870 places some of the de Guevaras living in Picacho next to a farmer named "Teofil [*sic*] LaLonde." Theophilus LaLonde, a French Canadian, had married into the de Guevara family. His wife, Estanislada "Lada" Padilla, was the four-year-old girl shown in the census of 1850 as being raised in Manzano by Ygnacio de Guevara and his wife. Somewhere along the line, Theophilus changed his name to Teófilo LaLone, which sounds closer to the French pronunciation of "LaLonde."

The LaLonde family traces its roots to the Canada of 1665 and an ancestor named Jean de LaLonde who hailed from Havre de Grace, France. Part of the family later crossed the Saint Lawrence River from Ontario to settle on the opposite side of the river in New York State, and it was there that Theophilus was born June 4, 1837.

He left home at age fourteen to work on the Saint Lawrence River—first as a pilot, then ship's mate, and finally captain. In 1858, he departed Kingston, Canada, for Shorter County, Nebraska. By about 1862, he had arrived in New Mexico, probably by the Santa Fe Trail. Along the line he met the de Guevara family and Ygnacio de Guevara's niece Lada.

Theophilus and Lada were married in 1867 or 1868. Lily Casey Klasner in her memoirs[8] describes meeting a Frenchman in 1868 in the small adobe hamlet of Missouri Plaza whose wife was a Spanish American woman named Lada. Lily said that her father, Robert Casey, bought a ranch on the banks of the Hondo that had first been settled on in the early 1860s by a Frenchman named LaLonde.

Teófilo, as he was now known, moved to farm at a place called Picacho only a few miles from Lincoln. While living there he made many friends, including Capt. Saturnino Baca, the "father of Lincoln County."

Teófilo and Lada LaLone. Date unknown. Photograph by Max H. Koch. Courtesy Richard (hereafter Rich) Eastwood.

The LaLone family. Standing, *left to right*: Fanny, Carrie, Fred, Louie, Addie. Sitting, *left to right*: Lada, Teófilo, Becky. Date unknown. Photographer Max H. Koch. Courtesy Rich Eastwood.

Eventually the LaLones moved from Picacho to the lower Bonito River valley and later still to Magado Creek at Nogal. Teófilo farmed and sold hay. He also supplied milk and butter to Fort Stanton and vegetables to White Oaks. According to a descendant, Teófilo was not much of a businessman and drank too much; as a result, his family often went hungry.[9]

The 1890, 1900, and 1910 censuses all show the Teófilo LaLone family in White Oaks, lured there by jobs and the town's relative prosperity. Lada's cousins, the de Guevaras, lived there as well and worked in the White Oaks mines.

The LaLone children—ten in number born over almost two decades (only six survived to adulthood)—grew up in the Oaks: Louis "Louie," Rebecca "Becky," Epifania "Fanny," Federico "Fred," Adelaida "Addie," and Caroline "Carrie." Louie was the only child who kept the "LaLonde" spelling of the family name.

Doc and Fanny LaLone Lacey

Daughter Fanny eventually married Doc Lacey from a nearby ranch. Peter Elijah "Doc" Lacey was born to Lewis Madison Lacey and Margaret Lodriski White in the Locke Hill community about twelve miles from downtown San Antonio, Texas. Margaret's father, Peter, had been a Stephen Austin colonist. Doc attended Boerne Academy until he was seventeen, receiving a good education for the times. In the late 1880s, Doc and his brothers rode range, caring for cattle, and drove some herds of cattle up into Kansas and Wyoming.

In 1890, the Laceys moved into San Antonio. Sometime between 1894 and 1897, Doc made his way to White Oaks in Lincoln County and worked in the Old Abe Mine. He met pretty Fanny LaLone, then employed in a wealthy household. On November 7, 1898, Doc and Fanny were married in White Oaks.

J. C. Lacey dugout home at the ranch of Doc Lacey's father three miles north of White Oaks on the west side of the road. Doc lived here as a young man. Date and photographer unknown. Courtesy Rich Eastwood.

After a brief stint back in Doc's hometown of San Antonio, the couple returned to Lincoln County and settled in Carrizozo, where Doc worked for the railroad. The year 1907 was hard; five of the couple's eight children came down with scarlet fever. Son Louis died on June 9, and when Doc found no cemetery in the new railroad town of Carrizozo, he donated a parcel of land to start Evergreen Cemetery.[10]

The following year Doc went to work for the Parsons Mining Company on the Bonita. In 1909, he designed and built a fairground for Carrizozo, fencing forty acres of ground and building a half-mile track and baseball diamond. He collected 40 percent of the gate money for all games, and when he had enough money, he built a grandstand on the grounds.[11]

In 1918 or 1919, the Lacey family moved to El Paso with Doc continuing to work for the railroad. In the 1920s, most of the LaLone family members relocated to Sierra Madre, California. In the mid-1920s, one of Doc's cousins talked him into moving to Encinitas, California, to grow potatoes. With the failure of the much-vaunted potato scheme, Fanny's brothers and sisters then convinced Doc and Fanny to join them in Sierra Madre, California, where Doc became caretaker of the Mia Italia estate.

Doc died January 20, 1937, a few months shy of his sixty-seventh birthday, and is buried in the Sierra Madre Cemetery. On November 7, 1978, Fanny died and was buried next to her husband.

Wedding portrait, Fannie LaLone and Doc Lacey. Date and photographer unknown. Courtesy Rich Eastwood.

Doc Lacey cooking outdoors over a campfire. Date and photographer unknown. Courtesy Rich Eastwood.

Doc Lacey at a roundup on the Block ranch. Doc Lacey (check mark) near the chuck wagons east of White Oaks. Date and photographer unknown. Courtesy Rich Eastwood.

NOTES

1. One version of the origin of the "Niño Ladrón de Guevara" title appears on the Ortiz family website http://perso.wanadoo.fr/rancho.pancho/Ortiz.htm. According to Garcia Carraffa and Garcia Carraffa, *Diccionaria Heráldico*, the founder of the family in Spain—Sir Sancho Guillermo, Duke of Brittany of the royal house of France—fought the Moors in Navarre in 716. In 885, his descendant, Sancho de Guevara, also fought the Moors and killed King Garcia Iñiguez and his wife Doña Urraca in the battle of Aibar or Baldillón. When Guevara found the bodies of the pair, the pregnant queen had been cut up and the arm of a baby boy protruded from her body. Guevara rescued the baby and raised the boy in the mountains, then had him named King of Navarre. The boy was given the name "Niño Ladrón de Guevara."

2. Chávez, *Origins of New Mexico Families*, 202–203.

3. The author is indebted to three descendants of the families discussed in this chapter for their extensive genealogical and historical research. Rich Eastwood of Mexico (near Enseñada), grandson of Doc Lacey, has spent much time digging into the Lacey-LaLonde history and has also accumulated a large collection of photographs—many taken by White Oaks' premier photographer Max Koch—which he has generously shared.

Charles Hayes of Morro Bay, California, was married to Rebecca "Margaret" Martínez, whose mother was a Guebara and daughter of Lorenzo Guevara—mine laborer, teamster, and clay miner at Coyote, New Mexico. Margaret died before Charles; they had been married for thirty-six years. After her death Charles realized how little he knew about his wife's family and set about searching for information. He plowed his way through many U.S. censuses and the 1885 New Mexico state census. He also traveled extensively to retrace the steps of the Guevaras from the time of Ygnacio Guevara's entry into New Mexico to the present.

Arturo "Bud" Guevara and his wife, Pat, of Payson, Arizona, shared their painstaking research into the Guevara family tree, all the way back to Don Juan Joseph Niño Ladrón de Guevara, whom they name as their first New Mexico ancestor of record. Arturo is a fifth-generation descendant of Felis (Felix) and Carmelita Guebara of White Oaks.

These individuals obtained much of their genealogical information from the U.S. censuses of 1860, 1870, 1880, 1900, and 1910 and the New Mexico census of 1885. Charles Hayes also enlisted the help of family history libraries at various Churches of Jesus Christ of Latter-day Saints.

4. *White Oaks Golden Era*, February 21, 1884.

5. Nolan, *Lincoln County War*, 171, 362–63.

6. Rasch, "Murder of Huston I. Chapman," 155, 244.

7. Nolan, *Lincoln County War*, 368–69.

8. Klasner, *My Girlhood Among Outlaws*, 38. Klasner refers to Lada LaLone as a "Spanish-American" woman and not Mexican, perhaps denoting higher status in the community and in the estimation of the Caseys.

9. Eastwood, "Theophilus LaLone [Lalonde] and Estanislada Padilla," in Genealogy of the Lacey and LaLonde families, 3, copy in author's collection.

10. Eastwood, "Doc and Fanny, A Chronology," in Genealogy, 3.

11. *Southwestern Outlook*, June 4, 1909.

SCOTSWOMAN JANE MALCOLM GALLACHER AND HER THREE SONS
Founders of the J + H Ranch

Jane Malcolm was born into her Presbyterian family on July 26, 1856, in the small village of Kilbirnie near the Firth of Clyde in Scotland. Jane's father, John, died when she was in her early teens, and her mother, Margaret Mackie Malcolm, became determined to seek a better life for herself and her seven offspring in the United States.[1] The Malcolms set sail across the Atlantic in 1871, with thrifty Margaret Malcolm shipping all the family goods in a false-bottomed trunk to avoid paying import duty taxes.[2] Margaret was a strong-minded Scotswoman, ambitious and hardworking, and would prove equal to the daunting task of guiding seven fatherless children to success in their adopted country.

Getting Started in a New Land

The Malcolms landed in Chicago in early October 1871 just in time to see most of that city burn in the Great Chicago Fire that allegedly began when "Old Lady O'Leary's cow kicked over a lantern."[3] The fire killed 250, destroyed 17,450 buildings, and left nearly 100,000 people homeless.[4] If Margaret Malcolm was ever tempted to return to Scotland, it was then. Fortunately for Lincoln County in the territory of New Mexico, she did not. Margaret and her children headed for Braidwood, Illinois, a town a scant ten or twelve miles south of Joliet. The Gallacher family received U.S. citizenship in 1871.

Jane Malcolm Gallacher. Date and photographer unknown. Courtesy Jane Gallacher Shafer.

Mother Malcolm and her daughters. *Left to right*: Jane Malcolm Gallacher, Margaret Mackie Malcolm, Jeannette Malcolm Barr. Date and photographer unknown. Courtesy Jane Gallacher Shafer.

William Gallacher, Sr., in Masonic regalia. Date and photographer unknown. Courtesy Jane Gallacher Shafer.

William and Jane Gallacher's children. *Left to right*: William, Harry, John, Margaret. Date and photographer unknown. Courtesy White Oaks Historical Association.

It was in Braidwood that Jane Malcolm and William "Bill" Gallacher (also a recent emigrant to the United States) first laid eyes on each other.[5] A ramrod-erect, slim girl who loved to dance, Jane was still in her teens when she met the charming twenty-seven-year-old Bill. On June 2, 1848, William Gallacher (originally pronounced "Gallaher," but now "Gallager") was born the eldest of nine children in Kelso, Roxburgh, Scotland. Family records show that William's parents moved the family to Hartlepool, England, on the North Sea in 1850.[6] They then moved back to Scotland to Pollock's Haws in 1852, Barrsmills Beith in 1854, and finally to Hurlford on the west side of Scotland ten miles from the Firth of Clyde in 1856. There the last five Gallacher children were born and William grew to manhood.

Bill came from a polished family, had a good singing voice, and loved to socialize. He promptly set out to win Jane for his wife. The young couple married in Joliet, Illinois, on July 20, 1875, just before Jane's nineteenth birthday. By 1879, the Gallachers were in Diamond, Illinois, where first son, Harry Alexander, was born on August 11.[7] Son John Malcolm followed on May 11, 1882.

To the Coal and Gold Mines in New Mexico

In 1885, Bill and Jane, their two small sons, and Jane's mother, Margaret Malcolm, all moved west to Carthage, New Mexico. Bill was the first man to discover rich deposits of coal in the Socorro country and was instrumental in opening the mines there.[8] He quickly worked his way up to become a coal mine superintendent.

A third son, William Wilson Gallacher ("Willie" or "Will"), was born in Carthage on April 1, 1887. About the time Will was born, the coal at Carthage gave out while mining in nearby White Oaks picked up. The Gallachers traveled to White Oaks by stagecoach in August 1887.

While Bill worked in the White Oaks mines, Jane and Margaret took over operation of the existing Ozanne Hotel. By February 1888, Jane and her mother had opened their own boarding house opposite Whiteman's store and advertised a "Good Table...Reasonable Prices."[9]

Carizo Hotel.

Wm. Gallacher, Prop.

WHITE OAKS, - - - NEW MEXICO.

This HOTEL is a new BRICK structure and is furnished through out with new funiture. Sleeping rooms are well supplied with clean and comfortable beds, and provided with light and ventilation.

Table supplied with the best the MARKET Affords.

Every care taken of and attention paid to the wants of Transient Guests.
CHARGES REASONABLE.

William Gallacher's ad for Carizo [sic] Hotel, *New Mexico Interpreter*, June 6, 1890. Author's collection.

The Carrizo Hotel

Bill and Jane furthered their ambitions in the Oaks by building a new brick hotel. By June 1890, the hotel was completed and outfitted throughout with new furniture.[10] The hotel opened almost at the same time as the Gallachers celebrated the arrival of their last child, Margaret Mackie, on June 2, 1890.

Jane later said that she could always locate her small daughter in any crowd because she could read "The Pride of Denver" slogan emblazoned across the seat of Maggie's bloomers.[11] In those days, flour sacks were often used to make the underwear of small children.

Bill Gallacher Dies, Leaving Jane to Hold Her Family Together

On November 19, 1893, Bill Gallacher was killed when he fell 390 feet down a mine shaft. White Oaks oldtimer Robert Leslie relates one version of events:

> It was in 1930 that we climbed down the manway to the bottom of the South Homestake shaft. We followed the tunnel to the shaft on the blacksmith level. It was there where Bill Gallacher fell down the shaft to his death. There was a chain across the opening and it was not to be unfastened until the bucket was lowered down and stopped for the men to get on to go to the lower levels. . . . Gallacher was in the lead, he unsnapped the chain, no bucket was there, he stepped out in the open shaft and yelled: "Watch out below, boys!" This was in 1894 [1893] November 19. He was 44 [45] years of age.[12]

After the death of her husband, Jane and Mrs. Malcolm stayed on in White Oaks to operate their hotel/boarding house. As always, Jane's mother helped with the business. During these years, it was Mrs. Malcolm's custom to go to her friends' houses at Christmas and bake her famous Scottish shortbread for them.[13] Together these two women cooked for more than seventy men who worked around White Oaks during that time. At the same time they also did the laundry for some fifteen families. Life was grindingly hard for these frontier women.

When Jane and her mother were doing the laundry for so many families, White Oaks wells had only "hard" water. Soft water was very scarce unless it could be caught in cisterns after rains. To get soft water for the laundry, Jane would wake up her boys about two o'clock each morning, saying, "Time to go!"[14] The boys got out of bed, dressed, and made the rounds of their neighbors' cisterns with a barrel mounted on wheels to "borrow" a little water from each cistern. The two older boys also helped their mother as soon as they were old enough to work outside the home. Harry, the eldest, worked for Price and Walker Grocery, and John worked at the post office. Later, as adults, John taught school and Harry mined.[15]

Jane stored the money everyone earned in a gallon lard bucket. Each Saturday night the boys would turn their wages over to Jane, who then gave each boy an allowance that had to last for the next week.[16] Jane continued running her boarding house in White Oaks until 1914.

A Game of Fox and Hounds

The youngest son, Will, and his friends used to play the game Foxes and Hounds in the Oaks.[17] On one such occasion he and his friend Ralph Treat were the foxes. They took gunnysacks, filled them with torn up bits of newspapers, and ran off, dropping clues in the form of newspaper scraps every so often so the hounds could trail them. They ran until, exhausted, they were forced to stop and rest at the White Oaks mill. As Will sat down, he happened to lay his hand in a corner and touch a partially buried gunnysack.

Pulling out the sack, he and Ralph peered inside to discover a solid gold brick. Will ran to the mill to fetch someone, with Ralph staying behind to guard the sack. The man who accompanied him back appropriated the gold bar without so much as a "thank you."

The Gallachers Get Started in Ranching

In March 1900, the *White Oaks Eagle* ran the following items of interest: "Mrs. Jane Gallacher has purchased the Sheep Spring Ranch at Mesa del Gallo [Patos Mountain] and her sons, Harry and John, will go there and stock it with sheep.... John ... is out looking after the ranch his mother recently purchased from Spence Bros. in Mesa del Gallo."[18]

Also in 1900, Charley Zeilander, a racehorse owner, stayed at Jane's boarding house in White Oaks.[19] The older two Gallacher boys noticed that one of Zeilander's mares would soon foal, and in a few days she gave birth to the prettiest filly they had ever seen. All the boys fell in love with the new foal. Zeilander approached the two older boys, asking, "You fellows want that filly? I don't have any way to take care of the little rascal, and I can't take it with me. I've got to get rid of it." John and Harry, already fond of the foal, lost no time in taking Zeilander up on his offer. Then came the challenge of how to divide ownership. John finally proposed branding the horse with a "J" for John and an "H" for Harry, with a cross (+) between the two letters. Thus the J + H brand of the Gallacher ranch was born.

In 1902, Harry began raising Angora goats on a small ranch at Castle Garden Mesa near Lone Mountain.[20] On September 24, 1903, he married Marian Evalena Grumbles.[21] This marriage produced one son, Vaden Grumbles Gallacher, before ending in divorce.

In 1906, the Gallachers bought Indian Tank farther out toward the malpais (lava beds) from Billy McNew,[22] former associate of Oliver Lee and a man indicted, though never tried, for the murder of Col. Albert Fountain. It was decided around this time that young Will should go off to college at Las Cruces at what was then called New Mexico Agricultural and Mechanical College to better prepare himself for life as a rancher. He majored in mechanical engineering and graduated in 1908. Afterward he said, "The only engineering I ever did was in pumping water out of the ground."[23]

In 1909, shortly after Will graduated, he, his brothers, and his mother, Jane, filed a claim for a homestead at Indian Tank twenty miles northwest of the railroad town of Carrizozo and at the north end of the vast malpais.[24] With the help of the boys, Jane proved up on the claim. As the years went by, the Gallachers gradually bought more land to add to the growing ranch. They bought the George Spence outfit[25] and became sheep ranchers.

Jane and her mother stayed on in White Oaks while Jane's three boys ran the Indian Tank place. Mother Malcolm was always at Jane's side helping where she could until she died in 1913. The *Carrizozo Outlook* reported on Malcolm's death:

THE PASSING OF A PIONEER

On last Friday night near the midnight hour the grim reaper called from active life the soul of Mrs. Margaret Malcolm at the age of 84 years.... She died at the residence of her daughter, Mrs. Jane Gallacher, now occupied by Mr. and Mrs. Will Kelt [Jane's daughter and her husband].... The funeral was held from the Congregational Church, Rev. Wheeler of the Methodist, officiating.... Mrs. Malcolm was born in the town of Kilbirnie, Ayrshire, Scotland, and came to this country over 40 years ago. She raised a family of seven children, four of whom are now living, and passed away universally respected and beloved.[26]

In 1914 after her mother died, Jane moved to the ranch at Indian Tank, where she took over managing the household. She also bought sheep, improved the land, and helped build fences and tanks. Jane Gallacher Shafer, Jane's granddaughter and namesake, recalls how hard her grandmother worked: "Her hands were never stilled. She was always baking and she was known for her porridge, her boiled meat, her soup-like stew, her raisin pie . . . lemon cake and . . . doughnuts by the ton. At Christmas she'd make clootie [Scots for "devil"] pudding, or Scotch dumpling, a suet pudding with hard sauce. And it was plenty hard!"[27] In the midst of all this work, she clung to the old Scottish custom of afternoon tea and shortbread at four o'clock each day. Her family and the cowboys, hands, and any travelers, joined in. Another old Scottish custom for Jane that was observed daily was a "little toddy" every morning after she got up.

The Gallachers Expand Their Holdings

Wise about money, John always gave his youngest brother, Will, sage advice, "Buy land, kid; every chance you get, buy land!"[28] When homestead claim holders starved out or simply gave up, the Gallacher brothers stepped up and bought the land. They also bought sheep, improved the land with fences and water tanks, and kept right on buying out the nesters.

In those early years, strangers would often pass by the ranch and stop. The guests were a motley group—maybe Indians going to and from the Mescalero reservation, maybe a sheriff's posse, always some sort of travelers. There was an "eternal" pot of *frijoles* at the ranch house, and all were welcomed and offered something to eat. Even outlaws would stop; the law might be close on their heels, but they too would "sit a spell" and eat before pressing on.

During these years, the notorious New Mexico outlaw Baldy Russell worked for the Gallachers for two or three years.[29] Baldy's hand was never out of reach of his six-shooter; his wife never went to town and always had a .30-30 within reach.

Ranch life had its moments of high drama. The *Carrizozo News* reported, "John and Willie Gallacher made a record run in from their Indian Tank ranch on Tuesday morning. Willie had come in contact with some poisoned water and it was imperative that a doctor . . . be reached soon. In a few hours after treatment by Doctor Paden, he was much improved and he and John returned home the same day."[30]

Oracio "Rasho" Corona

Sometime in the early 1900s, with as many as fifteen thousand sheep to herd, the Gallacher brothers found that they badly needed a herder. John finally told Will to try to find someone in White Oaks.[31] Not long after Will arrived at the Oaks, he saw a young dark-skinned Indian walking down the sidewalk with his little white dog. He approached the man and asked if he was looking for a job. The man's name was Oracio "Rasho" Corona; he and his dog had herded many sheep, and he was looking for any kind of work. Will told Corona to get his things together because he had a job on the J + H. Born in Tularosa in 1885 to parents from Sonora, Mexico, Rasho proved to be an excellent ranch hand. He spent his next fifty-five years at the J + H herding sheep, taking care of cattle, hauling timber, cutting posts, and doing all kinds of odd jobs. People said he could do more work in a day than any two ordinary men.

In the days when the Gallachers still ran sheep, Rasho would sleep with the ewes in a pasture close to the ranch house to help care for the newborn lambs. He saw the little Gallacher homestead of a section or so become a large working cattle ranch. Over time Rasho became looked upon as a member of the family, much loved by everyone who knew him.

One day in the spring of 1961, Rasho was critically injured in a bad accident out in the yard. Will ran him in to Carrizozo to the hospital, but he died on March 25. The place was never quite the same with Rasho gone. Will Gallacher said that he "never saw his equal anywhere for faithfulness and dependability. He was always willing for any job, big or small, day or night. He became one of the best riders in the country. On the other hand, he was gentle and kind to our children."[32]

The World War I Years

World War I brought further changes. The *Carrizozo News* reported, "A large crowd gathered last night on the court house lawn in a farewell reception to our boys who leave tonight for Camp Funston, Fort Riley, Kansas. . . . Neither speaker [Governor McDonald and District Attorney Hamilton] attempted to minimize the danger into which we were sending our boys . . . for they were sending them to fight oppression [and] . . . to make the world safe for democracy. . . . Around the board were the following: William Gallacher and his mother."[33]

Will went off to war with the military police of the

Tenth French Army. Because he spoke fluent Spanish, after a couple of weeks of intensive training the military informed him that he could now speak French. He later spent four months in the hospital after breathing poisonous gas and being wounded by shrapnel.[34] Will was mustered out as a Top Duty Sergeant at Fort Bliss in El Paso.

Edna Gray Marries Will and Comes to Live at the Ranch

Will continued his busy life on the J + H. Visitors came often, but still he was lonely at times. Then he met the woman who would change his life. People often gave parties in various houses around White Oaks and Carrizozo. One of those party-dances attracted Will, and soon after the dance got under way, he noticed a girl who had come to the party with Tom Carr.[35] Cornered, she was being offered a drink of homemade raisin wine by another partygoer when Carr walked over and took the glass out of the young lady's hand. Will thought he might like to meet that girl, and soon he asked her to dance. Her name was Edna Eva Gray. That night he saw Edna home, and a romance began. Will and Edna were married March 12, 1921.

Edna became half of a great team. As newlyweds, Will and Edna made their first home eight miles above Indian Tank. Rasho hauled load after load of rocks and laid them together to finish Will and Edna's rock house (this house still stands today). There Edna ran a trapline along with her regular hard work on the ranch. A rancher's daughter, Edna was used to the saddle, and her horseback skills on the J + H made her a legend around the White Oaks–Carrizozo country.

The Grim Reaper Comes for John

John Gallacher married Elizabeth Truman, a girl from Ohio, on June 20, 1925.[36] In late May of 1929, John was standing on a ledge in Águila Canyon at Chupadera Mesa looking for a predator that had been killing his lambs, when the ledge, soaked from recent rains, crumbled under him. He fell and landed hard across a big rock at the bottom of the bluff, crushing his kidneys.[37] Rasho found John and hurried to ranch headquarters for help.

Will saw at once that John was critically injured. He and Rasho loaded John into their pickup and raced for Carrizozo, where Doctor Paden packed John's abdominal cavity with cotton and sent him (with Will driving) down to Hotel Dieu, an El Paso hospital. It was two days before they could reach El Paso, and by then it was too late; gangrene had set in and John died on May 31 at Hotel Dieu.

Grandma Jane Moves to Carrizozo to Send Little Jane to School

In 1933, Jane Gallacher moved into Carrizozo from the ranch to send her granddaughter and namesake, Jane, to school. At the time the two came to town to live, the elder Gallacher had never used a telephone in her life.[38] The first time it rang, she was in the house alone and did not know how to stop the ringing. She never liked any of what she termed "new-fangled" inventions anyway and moved out of the house as soon as possible.

Granddaughter Jane once asked her grandmother why she had never married again after her husband was killed in the mine accident, since she was then still a young woman. Grandma Gallacher, perhaps recalling the long years of drudgery raising her children alone, said, "I'll tell you, daughterrr . . . I wouldn't of had a man again if his arrrse was of gold!" Jane spent most of her time in Carrizozo while her granddaughter was in school but visited the ranch often. In later years, she returned permanently to the J + H to live with Will and Edna. In the 1940s, she made it over to Ruidoso to watch the races and commented, "They are such pretty horses with the wee, wee riders."[39]

Tragedy Claims Harry Gallacher

On April 18, 1918, Ida Collier, who had grown up on a ranch near White Oaks, married Harry Gallacher.[40] They moved first to Roswell and later to Oklahoma City. When this marriage ended in divorce, Harry returned to Indian Tank.

With the approach of Christmas in 1938, Harry asked Will (who saw to the finances of the ranch) for some money to celebrate over the holiday, and Will obliged him by arranging for a couple of hundred dollars.[41] This was the last time Will would see his brother alive.

Again, it was Rasho who on December 23 came hurrying into the ranch yard, saying, "I think Harry's dead, Mr. Gallacher. Somethin' terrible has happened to him. He's out there in his bed and he can't get up! I think he's dead."[42] No one has been able to determine exactly how Harry met his death. Rasho believed it could have been murder. He had seen two young women with Harry earlier in the evening; later Harry came back to his house, followed by the two. Believing that they had killed Harry, Oracio thought they might have given him some poisoned food. There was indeed cyanide poison on the ranch that was put out to kill coyotes.

Whatever had happened, Harry was dead, all the money gone, and the two young women nowhere to be found. A note was found that read, "Don't blame anybody for what I have done as I am tired of living and my troubles will be over here on this earth as life has only been a dream. H"[43]

The J + H Becomes a Cattle Ranch

Back in the early 1900s when the Gallacher brothers filed on their homestead, they ran sheep. Over the years the land area grew as they bought out homesteaders, and the sheep herds expanded. In 1939, Will discussed with Edna the idea of getting rid of the sheep and stocking the entire ranch with cattle.[44]

By this time the couple had three children whose future they needed to consider—Jane, Bill Junior, and Anne. The couple decided to sell off the sheep as soon as they could, and Will did, selling them to Holm Bursum over at Socorro. Holm did not hesitate and bought all fifteen thousand head of sheep for a good price of about $3.50 apiece. Will and Edna were finally out of the sheep business. After looking around a while for cattle, Will headed west to Arizona and found what he considered to be the wildest cattle in Arizona in the mountains. Eventually, he transferred all of them to the ranch.

It takes time to build up a good working ranch. The Gallachers put in thirteen wells and fifty-five dirt tanks on the ranch by the 1970s. They also strung miles and miles of fencing. When a terrible drought hit the range later on, Will had to sell most of the cows, but he restocked later.

Time Moves On

Jane Malcolm Gallacher lived with Will and Edna for years and drank her toddy every morning as soon as she rose. One morning in June 1949 she got up, and the family, as usual, was ready to wait on her. Will asked if she wanted her toddy, but she refused, saying, "Son, I feel sick this morning. I don't believe I'll have one today."[45] Knowing Jane must be seriously ill indeed to turn down her toddy, the family decided to rush her to the doctor in Carrizozo. Will went into the bedroom to tell his mother, but she had died. She was ninety-three years old.

Honors Come to Will Gallacher

Besides running the ranch, Will served as chairman of the Lincoln County Commission for eight years and also headed the ration board during World War II.[46] He played a major role in bringing hospitals to Carrizozo and Ruidoso and had a hand in getting many other community landmarks built: the racetrack in Ruidoso, the annex to the Lincoln County Courthouse, the Carrizozo Country Club, and even a golf course on club grounds (where in early years the players had to kick away rattlesnakes on the dusty course). As chairman of the Carrizozo Board of Education, he also helped build the present Carrizozo high school.

He was recipient of the Annual Banker's Award of the Carrizozo Soil and Water Conservation District. Several times he was honored by his university, now called New Mexico State. He was a member of Carrizozo Masonic Lodge No. 41. And, last but by no means least, he won golf tournaments.

The Best Cattle Outfit in New Mexico

By the 1970s, the Gallachers were running around 1,200 head of cows on the range, each year selling off most of the calf crop and all the young steers. They kept from two hundred to three hundred head of heifers to replace the old cows.

Will Gallacher lived almost a century. He died December 12, 1985, at ninety-eight years of age and was the oldest alumnus of New Mexico State University. Wife Edna died three short weeks later, on January 4, 1986, at the age of ninety. Cedarvale Cemetery in White Oaks is home to the first Will Gallacher; his wife, Jane Malcolm; all three sons, Harry, John, and Will; and Will's wife, Edna Gray. Also rightfully buried in the Gallacher plot is Oracio Corona *Gallacher*, who was born in 1885 and who died in 1961 at the age of seventy-six.

As of this publication, the sprawling J + H Ranch on the north end of the lava beds near Carrizozo covers some 150 sections, or nearly 100,000 acres and is run by Jane's granddaughter Jane with husband, R. M. "Bob" Shafer; grandson Bill Junior and wife, Joy; and granddaughter Anne with husband, Wally Ferguson, along with some of their children. It remains a successful family business and has been called "The Best Cattle Working Outfit in New Mexico."

NOTES

1. Adams and Adams, *In the Shadow of the Malpais*, ix. This forty-eight-page book contains several stories of the Gallacher family that do not appear in this present chapter; readers who wish to learn more about the Gallachers may visit the Lincoln County Historical Society for access to a copy of the book.

2. Jane Gallacher Shafer, interview by Haldane, October 16, 2001, J + H Ranch.

3. Birdsong, "Jane Malcolm Gallacher."

4. Flanagan, *Old West*, 231.

5. Birdsong, "Mrs. Jane Malcolm Gallacher, Lincoln County Pioneer."

6. John J. Kelt, Sr., Genealogy Records, Family Group Sheet, copy in author's collection, furnished by Jane Gallacher Shafer. (Kelt was a former long-time resident of White Oaks who moved to Tucumcari.) These records list the birth, marriage, and death dates and places for William Gallacher of Scotland, his eight brothers and sisters, and the four children of William Gallacher and his wife, Jane Malcolm.

7. Kelt, Genealogy Records, Family Group Sheet.

8. Glen Ellison, "If You Can Walk With Kings," *Lincoln County News*, April 6, 1978.

9. Ad in the *Lincoln County Leader*, February 11, 1888.

10. Ad in the *New Mexico Interpreter*, June 6, 1890.

11. "Major step for White Oaks," *Lincoln County News*, June 12, 1969. This article quotes the speech of John Kelt of Tucumcari, son of Maggie Gallacher Kelt, on the occasion of Cedarvale Cemetery's dedication as a state memorial on Memorial Day 1969.

12. Robert Leslie to Haldane, December 27, 1999.

13. Shafer, interview.

14. Adams and Adams, *In the Shadow of the Malpais*, x.

15. Shafer, interview.

16. Birdsong, *1991 Lincoln County Pony Express Calendar*.

17. Story adapted from Adams and Adams, *In the Shadow of the Malpais*, xii.

18. *White Oaks Eagle*, March 22, 1900.

19. Story adapted from Adams and Adams, *In the Shadow of the Malpais*, 3–4.

20. Birdsong, *1998 Lincoln County Calendar*.

21. Greer Society, *Cousins of Thomas Alfred Bragg*, 39.

22. Shafer, interview.

23. Ellison, *Lincoln County News*, April 6, 1978.

24. Birdsong, *1998 Lincoln County Calendar*.

25. Ellison, *Lincoln County News*, April 6, 1978.

26. *Carrizozo Outlook*, November 24, 1914.

27. Birdsong, *1998 Lincoln County Calendar*.

28. Adams and Adams, *In the Shadow of the Malpais*, 17.

29. Ellison, *Lincoln County News*, April 6, 1978.

30. *Carrizozo News*, January 29, 1915.

31. Adams and Adams, *In the Shadow of the Malpais*, 27.

32. *Lincoln County News*, March 31, 1961.

33. *Carrizozo News*, September 7, 1917.

34. Ellison, *Lincoln County News*, April 6, 1978.

35. Adams and Adams, *In the Shadow of the Malpais*, 21–26.

36. Greer Society, *Cousins of Thomas Alfred Bragg*, 45.

37. Adams and Adams, *In the Shadow of the Malpais*, 31.

38. Birdsong, *1991 Lincoln County Pony Express Calendar*.

39. Shafer, interview.

40. Greer Society, *Cousins of Thomas Alfred Bragg*, 35.

41. Adams and Adams, *In the Shadow of the Malpais*, 32–33.

42. Ibid., 31.

43. Greer Society, *Cousins of Thomas Alfred Bragg*, 45.

44. Adams and Adams, *In the Shadow of the Malpais*, 31–35.

45. Ibid., xi.

46. Obituary for William Gallacher, Sr., *Lincoln County News*, December 19, 1985.

BENJAMIN GUMM AND SONS
Builders of the Gumm House, the Hoyle House, and the White Oaks Schoolhouse

In 1894, Benjamin Franklin Gumm, patriarch of the large Gumm family of five boys and one girl, wrote a biographical sketch of his early life for the benefit of his children.[1] The diary covers from 1827 through November 1859—the first thirty-two years of his life. According to his diary, Gumm was born in Rockingham County, Virginia, on August 30, 1827, of Welsh and Irish stock on his father, John's, side and of German blood on his mother, Annie Shoup's, side.

Gumm's parents moved from Virginia to LaSalle County, Illinois, just after the close of the Black Hawk War in 1833. They settled on the Vermillion River near the farm of his father's brother, Norton. John Gumm rented a farm on the south side of the Vermillion River. That first year the crops failed, and, worse, the entire family (including the seven children) suffered from a "fever" of some sort so crippling that they were unable to take care of each other. After the fever came whooping cough. The Reeders, good neighbors of the Gumms, gave Ben's brother Joseph work and sold the family enough items to make it through the winter. The second summer saw good crops.

In the fall of 1837, the Gumms moved onto the farm of another neighbor, Martin Reynolds. There they lived for four years. Then the hotelkeeper in the nearby small town of Utica persuaded John Gumm to go into the hotel business, after which everything went further downhill. John became so distraught over his finances that he fell sick and died in 1846, leaving his wife and the two younger children (Ben and Elizabeth) to fend for themselves (the older children, Joseph, Melvina, Eli, Will, and Catherine, were now on their own).

Ben and older brother Will ran a little farm successfully for about three years. This would only last until October 1851 when Ben came home at sundown one day to find his new house and all its contents burned to the ground. Standing in the yard and surveying the smol-

Benjamin Franklin Gumm, age seventy, ca. 1897. Photographer Max H. Koch. Courtesy White Oaks Historical Association.

dering ruins of his hard work and sweat, Ben resolved to go to California and try his luck in "Golden Country." He left his mother and sister Elizabeth in the care of relatives and joined the thousands already streaming to the California goldfields in the wake of the forty-niners.

California, Here I Come!

TO IOWA ON THE
FIRST LEG OF THE TRIP
On March 29, 1852, Ben set out from Ottawa, Illinois, on a dangerous two-thousand-mile journey across the American Great Plains. He undertook the trip in a com-

pany of sixty-three travelers captained by Joseph Green. For a fee of one hundred dollars, Green promised to deliver his party to Placerville, California, furnishing everything in the way of provision and transportation. Each man had to agree to walk all the way unless sick.

The weather stayed cold and disagreeable throughout the trek across Iowa to Kanesville (present-day Council Bluffs, Iowa), a Mormon town on the east bank of the Missouri across from the mouth of the Platte River and present-day Omaha, Nebraska. The travelers arrived at Kanesville on May 8 to find that they could not cross the Missouri because hundreds of other travelers were registered on the ferryman's books before them. Green began to grow impatient, and on May 19, he moved his train a short distance down the river where a steamboat ferried parties across for five dollars a wagon. The Green train was the first to cross after the steamboat arrived. The Green party divided into groups of six, each caring for three yokes of cows and three yokes of oxen. The men took turns driving, save for the one detailed to cook. Guard duty came around once a month except when the party was among presumed hostile Indians; then the guard was doubled. The company had no trouble with Indians on the entire trip. Another factor in the train's success was the large number of experienced pioneer men in the outfit.

A MEMORABLE FOURTH OF JULY 1852

On Independence Day 1852, the party organized a feast near present-day Willow Creek, Colorado, several miles east of South Pass in the Rocky Mountains, to celebrate its successful crossing of the Great Plains. All the men had gone hunting and brought in seventeen antelope, one buffalo, and many sage hens. A Mrs. Keysor and her daughter organized a magnificent dinner produced by the three women in the party. The travelers brought out many things from their hiding places in the big wagons, among them a small keg of rare old wine. The women made pies and cakes and set out preserves and other dainties so that the table looked like a proper Fourth of July in "God's Country."

AT SALT LAKE CITY

After Willow Creek, the party made Green River in Wyoming, where the ferry would take across only the wagons. The travelers had to swim all their stock across the river, which was deep and swift, but they lost only one cow out of eighty. Indians shadowed the wagons from daylight until dark until the party neared Fort Bridger.

Reaching Salt Lake City on July 15, the captain decided to lay over a week at a place ten miles south of the city and rest the stock on the excellent feed in the surrounding valleys. In this lull, many of the party hired out to the Mormons to help with their harvests. On July 25, the group continued on and, after many days along the Humboldt River and much suffering from lack of water and wood, made "The Meadows," no doubt a reference to Lassen's Meadows, Nevada, below present-day Winnemucca and some distance above the Humboldt Sink. There, the Green party broke up into squads of three, four, and five and traveled from then on at their own speeds.

CROSSING THE DESERT

Ben Gumm's squad included Jacob Goff, Lafe Mackey, Mack Simmons, and a "little Swede" called Chris. Each had blankets, and they shared a frying pan, coffee pot, and food. Leaving Lassen's Meadows on August 29, 1852, and traveling all day, they camped for the night at the Humboldt Sink. Beyond the Sink, overland travelers had to cross the deadliest and most dreaded part of the California trail, the Forty Mile Desert, referred to in Gumm's journal as the "Great Sandy Desert." They rested up the next day for their start across this waterless stretch. On Ben's twenty-fifth birthday, the travelers filled their canteens with water, strapped packs onto their backs, and headed into the desert with the goal of reaching the Carson River.

Traveling by night in a hard slog over desert sand, the travelers were exhausted when they eventually reached the Carson River at Ragtown, where they rested briefly. The next day they pushed on to what is now Carson City, Nevada. There they left the main emigrant road and took a cutoff that they had been told would be fifty miles shorter than the wagon road, but that turned out to be nothing more than a pack trail. By now the friends had become used to hard walking, and they made good time as they crossed the Sierra Nevada, though often short on food.

CALIFORNIA, AT LAST

On the evening of September 8, 1852, the party arrived in Hangtown (Placerville),[2] California, ragged, footsore, and dead broke. As the travelers huddled together to discuss their situation, the one they referred to as the little Swede, Chris, ripped open his watch pocket with a sharp knife and pulled out a twenty-dollar gold piece. He put it in the "pot," saying his mother had sewn the gold piece in his pocket to be used in an extremity, and this was "vat I calls dis." That night the young men feasted, but Chris had gone overboard and bought a good bottle of whiskey upon which he got "gloriously drunk" and had to be put to bed.

A FIFTY-FIVE MILE WALK TO SACRAMENTO, CALIFORNIA, FOR THE MAIL

The next morning Ben started walking from Hangtown to the post office at Sacramento, California, fifty-five miles away to see if some of the loved ones at home had written or sent money. On the way, he stopped at a wayside hotel.

While wandering around Sacramento trying to figure out what to do next, Ben ran into a Mr. Norris, who had plastered Ben's house that later burned down back in Illinois. Mr. Norris offered Ben one hundred dollars a month to work for him. Visions of all the sackfuls of gold Ben and his companions were to take out of the California ground led him to refuse this offer.

PANNING FOR GOLD

Back in Hangtown, the young men (minus the little Swede, who had gone to another mining camp) found a boarding house. Then they bought two shovels, three picks, one gold pan, and a cradle on credit. In a little gulch close to Hangtown, they panned diligently but found no gold. A man by the name of Alva Inman told them of rich mines on the Middle Fork of the American River north of Hangtown. They should all, Inman informed them, immediately go up there with him and check out the Middle Fork.

It had to have been the most foolhardy trip ever undertaken by any party of miners at that season of the year. It was now the fall of 1852, and they would have to cross several very high ranges of mountains to reach the place described by Inman.

A WILD-GOOSE CHASE

On September 28, the party of prospectors and a few packhorses started for "Inman's El Dorado on the American River." After five days of hard climbing over high mountains and through nearly inaccessible gorges, they arrived at the goldfield.

There was only one trail by which they could leave mountain country, and no food to be had nearer than the town of Gray Eagle thirty miles from camp. Two weeks after they arrived, the food ran out and someone had to walk to Gray Eagle for a new supply of bacon, flour, and beans.

While the men were used to walking fifty miles a day, the horses could not make good time because the mountain trail was so steep that in many places the horses were forced to catch hold of the bushes with their teeth to pull themselves along. When they arrived, the group worked hard and earnestly with renewed vigor to make their fortunes, but their hoped-for wealth accumulated very slowly.

A week later, the miners had to retreat due to the fast-approaching winter. By now their clothing had been reduced to rags and the remaining horses were mere shadows of their former selves. Five days of hard travel brought the ragtag miners back to Hangtown to board (on credit). Their landlord soon put them onto a claim that some forty-niners wanted to sell. Ben and his friends paid $265 for the claim and built a miner cabin nearby.

A NEW CLAIM NEAR HANGTOWN

On December 23, 1852, the rainy season set in; it rained eleven days and nights in torrents, filling the miners' gulch with water. Soon the men were at work placer mining, and by Christmas Day their mining amounted to a worth of $52. Off to Weber a mile away they went, to return with a gunnysack filled with bacon, beans, coffee, and another little necessity—a bottle of Old Rye. Constant rains made the road from Sacramento to Hangtown impassable and so provisions soon became very dear: flour now cost $275 a barrel. Nonetheless, the small band worked diligently until early March, making enough gold dust to pay for the claim and their store debt in Hangtown. The partners were left with $200 each.

ON TO MACOSMA RIVER

At the end of March 1853, the miners packed up and moved twelve miles away to the Macosma River, where they installed 700 feet of flume to carry the water of the river when it became low enough (about May 15). A sawmill at Wisconsin Bar sold the men 17,000 feet of lumber to complete their flume, at a cost of $100 per thousand feet. By July 1, 1853, the flume was completed and the men began to build a dam to turn the water into the flume. After all this work, the claim proved worthless. A new claim brought as much as $16 each a day and never less than $5 a day. The men worked all that fall and winter and until late spring 1854. Ben sold his interest in the Macosma claim for $175; by this time he had cleared $1,200 over and above expenses.

KNOCKING ABOUT

Tiring of mining, Ben found odd jobs in Hangtown driving a delivery wagon for a bottling business and working on an irrigation ditch. He turned out to be such a good workman that he was promoted to boss of his fifteen-man gang. This ditch cost a Sacramento company $800,000 to complete. Then Ben returned to the Macosma River and mined again until November 1854.

A letter from old friend Jacob Goff, now at Galena Hill in the "Northern Mines," summoned Ben from Sacramento to Comptonville. Jacob owned a rich mine with his brother, and they offered Ben $4 a day to work for them. He worked for the Goffs until February 1856 when he bought a tenth interest in a company mine twenty miles from Galena Hill, paying $3,250 sight unseen for his interest.

Trekking toward the new camp, Morrison's Hill, Ben had to wade through twelve to sixteen inches of snow. There was no need for snowshoes since the snow was crusted over heavily. On the second day, he passed over places where the snow was twenty-five to thirty feet deep in canyons on the north side of the mountains and crusted so hard that often he could barely stand upright. On finally arriving at the camp, he found snow eight feet deep all over the country.

By the time the snow melted, on March 27, 1856, there was enough water to run one set of hydraulic hoses nearly all day. By May 10, the miners had three full heads of water and were washing off ten thousand tons of top dirt daily with the hydraulic hoses. The ore lay eighty feet down in its deepest place, so all the dirt and rocks had to be sluiced off before they could reach bedrock and gold. The work was urgent; by June 15 the snow would be melted and the water supply gone.

Over the summer's run the miners made $45,000, which after expenses left Ben $1,500. He paid off his first note and had about $250 left. The company hired him to stay on the rest of the year at $150 a month. By March 1, 1857, everything was ready for another run. Cleanup[3] this year came to $55,000, and after all expenses were paid Ben had about $1,100.

A FORAY AS SALOON OWNER

Ben and a new partner then bought a saloon in Morrison and stocked it with liquor and a billiard table. The partners soon sold their saloon, and with his $500 share, Ben took off for Jacob Goff's ranch. There he spent the winter of 1857 hunting bear, elk, and black-tail deer.

On March 1, 1858, Ben and Jacob set out for Petaluma near the Russian River, seventy-five miles away, riding in an improvised cart made of the forewheels of a wagon. On the way, the men stopped in Healdsburg to take in a dance. A "little misunderstanding" resulted in the death of two men on the dance floor. Another man had all the tendons of his left wrist cut off by a knife-wielding belle of the ball. This small disruption did not break up the dance; the dead men were carried out and the dance went on.

YET ANOTHER CLAIM AT MORRISTOWN

Ben left Petaluma, boarded a small steamboat to San Francisco twenty-five miles away, and took passage on the steamboat *New York* for Sacramento 125 miles up the Sacramento River. There he boarded another boat for Marysville at the mouth of the Feather River, where he caught the stage to Comptonville and then reached Morristown traveling by pack train and foot. The next trip was to nearby Onion Valley. A chance encounter brought him to William Price, a man who had worked in the coal mines in years past for Ben's brother Joe. They talked all night and after a good breakfast set out on horseback through very mountainous terrain to try to reach the home of a friend of Ben's whom he had known back in Illinois. Ten miles out it began snowing so hard they could no longer follow the trail and had to turn back. When the weather cleared, they tried again and made it. In July 1859, old chum Jacob Goff showed up and talked Ben into accompanying him back to Illinois.

Back to New York by Way of Panama

In San Francisco, Ben and Jacob awaited a steamer to carry them to New York. They bought cabin passage and a railroad ticket across the Isthmus of Panama. He and Jacob boarded the *Panama Steamer* along with 250 other cabin passengers and 300 steerage passengers. Only one stop was made on the entire trip from San Francisco to the isthmus; that was at Acapulco, a coaling station in Mexico.

Ben was wretchedly seasick most of the time, except when ashore at Acapulco, and hated the trip. However, the travelers had splendid weather, and there was a fine brass band aboard that played almost every evening. On August 18, 1859, they arrived safely at the old town of

Panama on the Pacific side of the isthmus, having lost only one passenger, who committed suicide by jumping overboard.

The train took five hours to reach the east side of the isthmus because of a heavy upgrade in several places where the train had to go back five miles to get a running start at a high grade. The train arrived at the new American town of Aspinwall near midnight. Early the next morning everyone transferred to the steamer awaiting them on the Atlantic side bound for New York.

From New York to Illinois to Kansas

Ben arrived safely in New York twenty-two days and eight hours after leaving San Francisco; Jacob and Ben then hopped the Michigan Southern Railroad to Toledo, Ohio. Beyond Toledo, their train was blockaded by a mob of one thousand striking railroad workers armed with everything from pitchforks to rifles. The mob held them hostage for seven hours, then commandeered the train and ran everyone back to Toledo.

Finally arriving at Chicago, Ben took the Alton & St. Louis Railroad for Ottawa, Illinois, arriving September 6, 1859, after an absence of more than seven years. At Reading, twenty-five miles south of Ottawa, his family eagerly welcomed him back.

Ben soon became disgusted with Illinois. A man could not earn fifty cents a day in wages, and many of his schoolmates were either dead, married, or gone to some other state or territory. Jacob Goff and Ben decided to push west together to Kansas. The two arrived at Jacob's brother's place in Kansas on November 30, 1859.

Only bare-bones details are known of the two decades Ben spent in Kansas. In 1860, soon after he arrived in Fort Scott, Kansas, Ben married Martha Goff and began a family of five boys and one girl: Joseph "Joe," Wallace "Wall," Elmer "Pete," John, Roy, and Edith LaVenia "Vena."[4] Ben became a prominent figure and pioneer in Fort Scott near the Kansas-Missouri border. During the Civil War, he commanded the local militia as its captain. After the war, he served two terms as sheriff of Bourbon County.

The Gumms Build a New Life in White Oaks

His never-ending quest for gold led Ben Gumm to White Oaks at a time of life when most men his age, now fifty-three, were looking for comfort and a slower pace. In 1880, in company with twenty-two-year-old William C. McDonald (future first governor of the state of New Mexico), Mayor Peck, and Charley Bull, Gumm arrived at the gold-mining camp of White Oaks.[5] There he located the Miguel Otero mine and received the very first mineral patent issued in Lincoln County. Gumm was also a charter member of the Masonic Lodge at

Gumm family, ca. 1897. Top row, *left to right*: Sons Joseph, Wallace, Elmer "Pete," daughter LaVenia "Vena." Bottom row, *left to right*: Martha Goff Gumm, sons John and Roy, Benjamin Franklin Gumm. Photographer unknown. Courtesy White Oaks Historical Association.

Uniontown, Kansas, and he continued Masonic affiliation when he became a charter member of Lodge No. 20 in White Oaks.[6] During the decade 1880 to 1890, Ben gathered his fine crop of five sons around him and began building thriving businesses in White Oaks. The boys were sports-crazy; they played baseball, hunted, and generally brought energy to the community life of the town.

Early on the Gumms constructed an ice plant.[7] By 1890, Gumm brothers Pete and Wall had taken over the two local sawmills. Later they added a planing and shingle mill and a woodworking factory.[8] In March 1892, the Gumms held open house for their new planing mill; during a newspaper reporter's visit, they demonstrated a planer at work, turning out flooring at the rate of eighteen thousand feet per day.[9] They also demonstrated a shingle and a lathe machine, each with a capacity of nineteen thousand feet per day.

Finally, the family built their own residence north of town with their own hands. It was the largest and most conspicuous in White Oaks.

John and Pete Gumm near Gran Quivera after successful hunting trip, October 1897. Photographer unknown. Courtesy White Oaks Historical Association.

The Gumm House, Hub of Social Life in White Oaks

A houseful of attractive young people naturally became the focus for community life. The boys participated in the drama club, playing parts in the many plays staged by the White Oaks Dramatic Club in Bonnell's Hall. Wall's magnificent mustache always lent panache to his roles.[10] Vena, the popular daughter, had a beautiful soprano voice and was one of the most ardent workers in the Congregational Church.[11]

If there was fun to be had, the Gumms were part of it. They organized camping outings on the South Fork of the Río Ruidoso in summertime. Often they camped for as long as three weeks by the river, fishing, hiking, and having a good time.[12] On one such outing in 1894, Robert Leslie drove his team and wagon loaded with camp gear, and the vacationers all piled into a second wagon. To make it over Nogal Hill, the passengers had to get out of the wagon and walk up. Farther on, both teams had to be hitched to each wagon to make it over the even steeper Alto Hill.[13] At camp, Roy Gumm—great hunter that he was—shot gray squirrels. Those and trout from the Ruidoso made up most of the camp fare.[14]

Later Years in White Oaks

THE "WHITE OAKS 8"

In February 1898, New Mexico territorial governor Miguel Otero—on learning that Spanish troops had sunk the battleship USS *Maine* in the harbor at Havana, Cuba—readied a cavalry regiment for volunteer service. The Spanish-American War was officially declared on April 25, 1898.[15]

Gumm family and friends at the Gumm home, 1901. Standing, *left to right*: LaVenia Gumm McCourt, unknown, John McCourt, Morris Parker, Benjamin F. Gumm. Sitting, *left to right*: unknown man with dog, Jennie McCourt Parker and baby, Wallace Gumm, Ben McCourt, Joseph A. Gumm with Dorothy McCourt in lap. Photographer unknown. Courtesy White Oaks Historical Association.

Another group of U.S. volunteers made up Company E of the First Territorial Infantry. Included in this infantry group were eight volunteers from White Oaks. Pete Gumm was one of the "White Oaks 8."[16] This contingent, trained in Georgia, fought in the Philippine Insurrection phase of the Spanish-American War. Lasting from 1899 to 1902, this insurrection is the least known of U.S. wars and also one of the most vicious, with many atrocities on both sides.

Peace finally settled in, and on July 4, 1902, President Roosevelt officially declared the war ended.[17] The United States lost 4,234 lives, and $600 million had been spent. Sixteen thousand Filipino soldiers were killed, and 200,000 civilians died.

THE GUMM SAWMILL BLOWS UP

In August 1899, the Gumm sawmill in the Capitán Mountains blew up, seriously wounding the engineer and two or three others working around the mill. The boiler was leaking and exploded just as an attempt was being made to draw the fire and run down the steam.[18] As a result of the blast, the mill had to be shut down.

TIME TO LEAVE

Pete and Wall stayed through worsening times in the 1890s, but left White Oaks around the turn of the century. Vena and her husband, John McCourt, moved to Montana. Eldest brother Joe moved to Benson, Arizona, to farm and in time became the "Onion King of Benson."[19]

Wall married Elizabeth Austin, a schoolmarm from Missouri, on June 29, 1899. Elizabeth had come to Lincoln County in 1894 to her father's home on the Bonito.[20] She was a graduate of New Mexico Normal College and taught school in White Oaks for three years before her marriage.

Ben and Martha stayed on in their Victorian mansion. As money became scarce, the couple lost some of their property in White Oaks to a public sale after defaulting on a mortgage held by Paul Mayer.[21] Next year Martha fell victim to pneumonia on May 2, 1904, at sixty-one years of age;[22] she is buried in Cedarvale Cemetery. Ben Gumm went to live with his unmarried son, John, in Silver City, and died at the age of ninety-one on May 24, 1918.[23]

The "White Oaks 8" in civilian clothes, January 1899. U.S. Volunteers, Company E, First Territorial Infantry, Albany, Georgia, bound for the Spanish-American War. Photographer unknown. Courtesy John and Wesley Mottinger.

The "White Oaks 8" in uniform, January 1899. *Left to right*: Sgt. Art H. Norton, J. H. Reel [Reed?], Cpl. Mel S. Graves, Cpl. E. J. "Pete" Gumm, E. A. Kelley, J. Cavan, Harry S. Comrey, Sgt. P. S. Pate [Tate?]. Photographer unknown. Courtesy Lincoln County Historical Society.

A Lasting Legacy: Three Magnificent White Oaks Buildings

Were he alive today, Ben Gumm would take great pride in knowing that all three of the remarkable buildings that the Gumm family and their 1890s workers built in White Oaks have survived for more than a century and are still in use. His buildings, stronger than any of the rest, stand unbowed with time as proud relics of the grandest era of the gold-mining town.

THE GUMM HOUSE

This beautiful Victorian two-story frame mansion, built around 1890 on the north side of town, has fourteen rooms and typical Victorian trim on the outside.[24] The Gumms filled their house with period mahogany furniture and installed stained glass, the colors of which matched the decor of their rooms.

The entry hall has a stairway on the right leading upstairs to four large bedrooms and a smaller maid's room next to the stairway. To the left on the first floor is a front parlor, and behind that parlor a sitting room. Straight ahead is the dining room and to the right a cheerful, light-filled sunporch. A butler's pantry stands between the dining room and the kitchen, with the kitchen directly behind the pantry. Behind the wood-burning stove in the kitchen is a back stairway by which the maid could descend to the kitchen. A bathroom was added later behind the stairway.

After the Gumms left White Oaks, the house passed through several owners' hands.[25] Anne New owns the mansion at the time of publication, and she and Michael New have restored the house to much of its former glory.

The Gumm House as it is today, in winter. Photographer unknown. Courtesy Anne New.

The Gumm House, ca. 1890. Ben Gumm and his wife, Martha, stand on porch. Note hammock hung to the porch posts of second floor. Photographer unknown. Courtesy John Kelt and Don Ward.

Every self-respecting old mansion must have its own ghost story. Fran Mack, a onetime occupant of the house, said that in the mid-1990s, soon after she and her husband moved in, they began hearing the sound of a ringing bell late in the evenings.[26] This continued for three nights, becoming irritating. Mack also told of another odd occurrence, happening in the parlor. When Fran or Jim placed an object on a table, it would disappear within seconds. Once, as Fran was preparing to hang a picture and had the nails, hammer, wire, and picture hook on a table next to where she was seated, she picked up the hammer in one hand and reached to the table for the picture hook, but it had vanished. She looked everywhere, but it was nowhere to be seen. Frustrated, she walked into the kitchen, then back into the parlor to sit down again on the chair. To her amazement, the picture hook had reappeared on the table.

THE HOYLE HOUSE

Watson Hoyle arrived in White Oaks in the fall of 1890 and had this dream home built for the bride he hoped would come out West to marry him. She rejected him for unknown reasons, and ever after the locals have called the mansion "Hoyle's Folly."

Construction of this Victorian masterpiece duplicated the Gumm house plan, only inverted. It was built in the Imperial-Federal style with plaster walls. Begun

The Hoyle House from the front. Note the widow's walk atop the house, the double-high windows, and the lighter-colored sandstone belts edging the redbrick construction. Date and photographer unknown. Courtesy Larue Lane Wetzel.

in 1892, the house was completed in 1893 at a cost of $42,000.

Watson Hoyle left White Oaks and a mortgage in default soon after the Old Abe could no longer produce gold ore. Lawyer Andrew Hudspeth bought the house eventually. Later, following years of vacancy and decay, the Kirkpatricks acquired the house and began restoration. In 1968, Mr. and Mrs. Charles L. Wetzel bought the property. After her husband died in a truck accident in 1973, Larue Wetzel stayed on in her beloved mansion. (Larue is the granddaughter of White Oaks pioneer Dr. Alexander Lane.)

THE WHITE OAKS SCHOOLHOUSE

Before a real schoolhouse was built, White Oaks children got their schooling in various ways. Many students attended school in a tent located between Doctor Paden's drugstore and the Paden house.[27] Some families, like that of Morris Parker, tutored the children at home with the mother as teacher. Mrs. Parker sent to Saint Louis for textbooks, and the three Parker boys spent regular hours and days in study and recitation.[28] Boys might be sent off for advanced schooling to relatives in their parents' hometowns, or by the turn of the century to the New Mexico Military Institute in Roswell. The *White Oaks Eagle* of September 1900 mentions that Morgan "Brent" Paden and Richard Taliaferro would be entering their second year at the institute that month.[29]

Dr. Alexander Lane, a man intensely interested in education, for a time taught a regular elementary day-school in a room in his home that he outfitted with desks and benches. He also served as Lincoln County school superintendent during the mid-1880s.[30] When he and other residents decided they must have a schoolhouse, Doctor Lane served as treasurer for the group of men selected for this job. They gathered about $10,000 to fund construction.[31]

On July 1, 1893, the White Oaks School Board awarded the White Oaks Building and Lumber Company the contract for the new schoolhouse.[32] The school was to occupy a plat of ground in the Kempton addition. Designers drew the plans for the forty-eight-by-sixty-foot, two-story building to include a basement under the east half, four schoolrooms—each twenty-eight by forty-five feet—on the first and second floors, and nine-by-forty-five-foot hallways. A wide stairway led from the first to the upper floor, and double stairs led from the rear hall door on the second floor to the ground. An eight-by-seventeen-foot front roofed porch with an attached open deck was built and a handsome eight-by-eight-foot bell tower was affixed atop the building. The company planned to use brick manufactured in White Oaks, with cut-stone trimmings consisting of water tables, caps, sills, and two belt courses, also cut-stone corners. The plan called for the white sandstone for the foundation and the cutwork to be obtained from local quarries. Designers planned to cover the roof with Courtwright Metal Roofing patent lock shingles. All the interior finish was white pine painted in particolors, and the blackboards in each room were made of a very hard adamant wall finish.

In 1895, the White Oaks Schoolhouse opened with an enrollment of thirty-five students. John Allen Haley was the first to teach in the new school. (Haley later served as superintendent of Lincoln County Schools from 1899 to 1900.) Haley also organized the Teachers' Normal School in this same building from 1901 to 1904, where the normal school program graduated a goodly number of teachers.[33] This program was a four-week course taught by Haley in 1902 and 1903, and by a Professor Stevenson in 1904.[34]

Over the years, the schoolhouse has served many purposes. In the summer of 1907, men even laid down a nice floor in one room of the schoolhouse for use as a skating rink.[35] The school finally closed in the 1940s,

Schoolchildren and teachers, White Oaks Schoolhouse, ca. 1896. Photographer unknown. Courtesy White Oaks Historical Association.

but it still serves the community in a variety of ways. It is currently a museum, and in summer visitors troop through the rooms looking for traces of ancestors or just witnessing a piece of one of the last frontiers of the Old West. Historical societies meet there, and weddings and art exhibits have been held. On occasion the old kitchen stove is fired up for potlucks and dinners.

NOTES

1. Evelyn Lynch of Tucson, Arizona, donated a copy of Benjamin Gumm's diary to the White Oaks Historical Association and a copy to the Lincoln County Historical Society. (Evelyn's husband was "Boots" Lynch, a grandson of Pete Gumm.)

One copy is a thirty-page typed transcript of Gumm's handwritten diary; the other is a longer copy of the original handwritten diary. Unless noted otherwise, most of the material and quotations from this chapter are taken from the diary, written "for the use and benefit of his boys" by Ben Gumm in October 1894. The diary is a fascinating account of the California gold rush days of the mid-nineteenth century, and anyone interested in learning more of what it was like to dig, pan, sluice, and otherwise search out gold in rugged places in those long-ago times will be well rewarded for reading it. Ben Gumm walked from his Illinois home west across great distances and through formidable obstacles, including the American Great Plains, the Rocky Mountains, the Great Salt Lake Desert, and several other mountain ranges, to Placerville, California. Once there, he continued to walk and work his way across much of northern California. His was an epic journey. Ben spent seven years in California until the day of the lone miner working his gold claim ended and mine employees took his place.

2. American Automobile Association, *California, Nevada Tourbook*. Placerville, originally known as "Old Dry Diggin's," became so prosperous and lawless that lawbreakers were hanged first one at a time, then in pairs. As a result, the gold camp was nicknamed "Hangtown."

3. A mining term denoting the ore collected in a periodic cleanup at a mine.

4. An introductory page to Ben Gumm's diary by an unknown author briefly summarizes the twenty years Ben spent in Fort Scott, Kansas, including his involvement in politics, his marriage, and the birth of his six children.

5. The same unknown author is the source for the names of Gumm's companions who accompanied him to White Oaks in 1880.

6. Ronald A. Brinkman, Grand Secretary, Grand Lodge of Ancient, Free and Accepted Masons of New Mexico, to R. E. Lemon of Carrizozo, August 3, 1971, copy in author's collection. Brinkman lists the 1891 membership of White Oaks Lodge No. 20, which includes the name of Benjamin Franklin Gumm.

7. *Lincoln County News and Carrizozo Outlook*, July 14, 1958, and July 20, 1958, in a look back at old White Oaks days.

8. Parker, *White Oaks*, 70.

9. *Lincoln County Leader*, March 12, 1892.

10. Parker, *White Oaks*, 86.

11. Ibid., 71.

12. *White Oaks Eagle*, July 18, 1901.

13. Robert Leslie to Haldane, August 22, 2001.

14. *White Oaks Eagle*, July 18, 1901.

15. Crutchfield, *It Happened in New Mexico*, 74.

16. Marrin, *Spanish-American War*, 145–46.

17. Ibid., 166–67.

18. *White Oaks Eagle*, August 10, 1899.

19. Evelyn Lynch has in her possession a photo of Joseph Gumm holding several of the very large onions he grew in Benson.

20. Haley, *1913 Year Book*.

21. *White Oaks Eagle*, February 6, 1902.

22. Record of Deaths at White Oaks and Vicinity, 1880–1905, by J. B. Slack and (later) M. H. Koch.

23. *Lincoln County News*, October 30, 1931.

24. Birdsong, "Gumm House."

25. Robert Leslie, undated listing of occupants of the Gumm House.

26. Garcez, *Adobe Angels*, 207–209.

27. Birdsong, "White Oaks School House."

28. Parker, *White Oaks*, 57.

29. *White Oaks Eagle*, September 6, 1900.

30. *White Oaks Golden Era*, August 20, 1885.

31. Larue Wetzel, undated communication with Haldane.

32. *Old Abe Eagle*, July 13, 1893.

33. Birdsong, "Schoolhouse Tales."

34. Ibid.

35. John Kelt, undated communication with Haldane.

JOHN A. HALEY
Printer's Ink Flowed in His Veins

John A. Haley was born to Thomas and Margaret Maxwell Haley on a farm called Grange Hall near Cleburne in Johnson County, Texas, on October 24, 1868.[1] Both of his parents were devout Baptists. He never departed from his early religious training, always held women in high regard and the ministry in respect, and practiced an unwavering honesty. After an education in the public schools of Johnson and Comanche Counties in Texas, he graduated from Howard Payne College in Brownwood. For a while, he taught in Comanche County and acquired some newspaper training on the side.

To White Oaks and a Career as an Educator

In July 1893, Haley, accompanied by Texas friend S. M. Wharton, came to White Oaks in the "glory era" of that small town.[2] Production of gold at the Old Abe Mine was at its peak. Haley's primary purpose in moving to New Mexico was to teach school, which he did from 1893 to 1895 and again from 1901 to 1903. That first winter of 1893 found Haley teaching school upstairs in the Brown building. White Oaks' new two-story redbrick schoolhouse had not yet been built.

Haley served as assistant principal in the White Oaks Graded Schools with his friend Wharton as principal. Once, the two men donated solid gold medals to deserving students, earning a commendation in the *Old Abe Eagle*. One medal was awarded for both attendance and deportment and the other for deportment only. Haley served as county superintendent of schools from 1899 to 1901. In 1908, he was appointed to fill the unexpired term of county superintendent and then was elected for his own two-year term in 1909. One of his more important educational accomplishments was the creation of a Teachers' Normal School housed in the White Oaks Schoolhouse.[3] Haley also earned money setting type on the *Old Abe Eagle* and its successor, the *White Oaks Eagle*, foreshadowing his future life's work.

John A. Haley, 1895. Photographer unknown. From *Illustrated History of New Mexico* (Lewis Publishing: Chicago, 1895). Courtesy Chris and Jack Harkey.

On May 5, 1897, in Lincoln, Haley married Jennie Mae Lesnet, daughter of a pioneering Lincoln County family who owned the Dowlin Mill in present-day Ruidoso until 1892.[4] The couple became parents of one boy (who died young) and two girls, Lorena and Aileen. Jennie Haley died in 1914, and Haley married Meda C. West in 1922.

The Eagle *Moves to Capitán and Then to Carrizozo*

Haley forsook teaching as a career and instead partnered with S. M. Wharton to buy the *White Oaks Eagle*.[5] In 1903, the associates moved their newspaper to Capitán, renaming it the *Capitán News*. Before long Haley bought

John Haley's print shop, ca. 1902. *Left to right*: Ollie Grumbles, Mildred Taylor (?), Bessie Lesnet, Zella Grumbles (?), Georgia Lesnet. Photographer unknown. Courtesy John and Wesley Mottinger.

out Wharton to become sole owner and operator of the paper. Then Haley abruptly left for unknown reasons and experimented with a six-month trial as a newspaperman in Alamogordo for the *Journal*, but before long he returned to his *Capitán* paper.

In 1908, Haley moved his newspaper to the thriving new railroad center of Carrizozo and renamed it the *Carrizozo News*. Since Haley was both editor and owner until 1925, the paper was suspended for a few months that year while he served on the state tax commission in Santa Fe. In 1926, he reopened the newspaper office in Carrizozo and began publishing again under the name *Lincoln County News*. Haley published this paper until his death, and as of this publication it survives as a traditional home paper reporting on the happenings of Lincoln County and surrounding counties.

Politics Beckons

In 1896, three years after leaving Texas for White Oaks, Haley was elected as deputy treasurer and collector of Lincoln County and continued in this office until 1899. From 1907 to 1911 he was both member and treasurer of the New Mexico Bureau of Immigration. Ralph Emerson Twitchell credits Haley as the recognized leader of Lincoln County's Democratic Party.[6] Twitchell recognized Haley's ability to exert widespread influence in southern New Mexico through his editorial and newspaper work.

The apex of Haley's political career came in 1911, when, as delegate from Lincoln County to the first Democratic state convention, he had the honor of placing the name of his friend William C. McDonald before the convention as Democratic candidate for state governor.[7] McDonald won the election and became New Mexico's first elected governor.

In 1913, Haley was appointed Carrizozo postmaster and was reappointed in 1918. Bigger political honors came his way in 1924 with an appointment as delegate to the Democratic National Convention in New York City. Haley, sometimes called "Judge Haley," also belonged to Carrizo Lodge No. 11 of the Knights of Pythias of Carrizozo. During his membership in the organization, he served at various times as Grand

Chancellor of New Mexico and Supreme Representative to the supreme lodge.[8]

1913 Year Book of Lincoln County, New Mexico

Haley researched, wrote, edited, and published the *1913 Year Book of Lincoln County*—an encyclopedia of Lincoln County that details its history, statistics, physical characteristics, climate, population, schools, and mining, agricultural, and industrial resources. It also contains photos and biographies of then-prominent Lincoln County men and women. It may be that Haley would not have thought of this book as one of his major legacies at the time he published it in 1913, but that is exactly what it turned out to be—an invaluable tool for Lincoln County historians.

End of the Line

Sometime in the early 1930s, Haley developed cancer of the esophagus. Despite operations, so-called serums, and radium treatments, he died on March 18, 1932, at the relatively young age of sixty-four. His own newspaper ran his obituary under the masthead "Jno. A. Haley Answers Call": "His family is heartbroken; his friends are crushed; we of the NEWS are heavy-hearted and don't know how to carry on without him."[9]

At the time of his death, Haley was chairman and secretary of the Lincoln County Democratic Committee. The editor was further eulogized by a member of his staff as: "a gifted writer, a talented speaker, an experienced newspaper man and exerted a strong influence in behalf of his party through editorial work. He was an optimist and could always see the bright side of everything; and paradoxical as it may sound, even his enemies were his friends. He was a gallant warrior, too, and we have seen him after fighting a hard campaign which ended in defeat for his party, come up smiling and hoping for better things next time."[10]

NOTES

1. Twitchell, *Leading Facts of New Mexican History*, 302–303; Haley, *1913 Year Book*.

2. Paul Baker, "John A. Haley," Ramblin' Around Lincoln County, *Lincoln County News*, November 29, 1957.

3. Dorothy Leslie, "Tribute to John Haley," taken from the 1993 White Oaks Historical Association financial report. This tribute provided background information concerning the donation of $5,000 given by Aileen Haley Lindamood in memory of her parents, Mr. and Mrs. John Haley.

4. Haley, *1913 Year Book*.

5. Obituary, *Lincoln County News*, March 25, 1932.

6. Twitchell, *Leading Facts of New Mexican History*.

7. Ibid.

8. *Lincoln County News*, November 29, 1957.

9. Obituary, *Lincoln County News*, March 25, 1932

10. Ibid.

JOHN Y. HEWITT
First Citizen of White Oaks

John Y. Hewitt, one of the most important men in White Oaks as well as the wealthiest, played many roles during his long life there: lawyer, owner of the famous Old Abe Mine, newspaper editor, and politician.

Hewitt was born in West Farmington, Ohio, on October 11, 1836.[1] He grew up on a farm and attended public schools before entering Western Reserve Seminary in West Farmington. In 1857, he left Ohio for the timber camps of Oshkosh, Wisconsin. Hating timbering and hoping the West would offer greater opportunity, he settled in Ohio City, Franklin County, Kansas Territory, in 1859. There, he taught school for three terms until the outbreak of the Civil War. While teaching school, he studied law and clerked for lawyers.

Civil War Service

Hewitt enlisted as a private in the Second Kansas Cavalry of the Union army on October 14, 1861, and eventually rose to the rank of regimental commissary sergeant.[2] He fought in the battles of Fort Wayne (Indiana), Canehill, and Prairie Grove, among others, during his four-year tour. On entering the frontier service, he was posted west to Fort Union in the territory of New Mexico, where he took part in many skirmishes and running fights. On January 4, 1864, Hewitt reenlisted and served until discharged at Lawrence, Kansas, in September 1865. His whereabouts for the next two years are unknown.

Ruined by the Panic of 1873

Hewitt returned to Kansas and the town of Ottawa in 1868, and he dealt in real estate until the Panic of 1873.[3] The Panic ruined him and thousands of other men in Kansas who had invested heavily in property. He hung on until 1877 before finally giving up on real estate. Then he caught gold fever.

Gold had been discovered in 1874 in the Black Hills, and Hewitt wanted a share. In 1877, he joined the gold rush to the Black Hills but was unsuccessful. Hewitt then recalled the territory of New Mexico from his military days on the frontier; in January 1879, he rode the train over Ratón Pass to Las Vegas, a town not far from Fort Union, where he had been posted in Civil War years. For the next fourteen months he did hard manual labor cutting railroad ties for the newly arrived Atchison, Topeka, and Santa Fe Railroad. Yet the gold virus ran deep in his blood.

White Oaks Beckons

Hewitt heard the news about the White Oaks strike in Las Vegas. He bought a humble burro and some gear

John Y. Hewitt as a private in the Union army. Hewitt served in the Second Kansas Cavalry. Date and photographer unknown. Courtesy Joe Spence.

John Y. Hewitt, 1895. Photographer unknown. From *Illustrated History of New Mexico* (Lewis Publishing: Chicago, 1895). Courtesy Chris and Jack Harkey.

and headed out as fast as the burro could be coaxed. He arrived in the mining camp on March 20, 1880, already forty-three years old.[4]

By now Hewitt's cash had dwindled to less than ten dollars. The first year in camp he was forced to do any work he could find to keep body and soul together while prospecting on the side. In August 1881 in Lincoln, then the county seat of Lincoln County, he finally passed his bar exams.[5] He began practicing law at White Oaks and continued this occupation throughout his life. George Zimmerman and Arthur Rolland were his first clients.[6]

Early Years in White Oaks

In his first years at the Oaks, Hewitt formed a close friendship with another lawyer, an Iowan some twenty years younger named Emerson Hough.[7] Three years in a gold camp overrun by other lawyers convinced Hough that the law was not his profession, and so he returned to Iowa and his first love: writing. Over the years Hough had autographed seventeen volumes of his books for Hewitt. Hewitt took great pride in his library collections and in art, and later donated a noteworthy historical library to the schools of Lincoln County. His law library alone consisted of one thousand volumes.

In 1882, Hewitt married Mrs. Amelia Rawlins, a native of Ohio.[8] Rawlins, a widow, must have been wealthy, because until his marriage Hewitt had been impoverished. His fortunes improved dramatically; the couple built a residence at the corner of Lincoln and Pine Streets and Hewitt began investing in mining and real estate. As he grew wealthier and more respected, he became known as "Judge" Hewitt.

Over the years the Hewitts hosted many social events. In July 1892, they gave an elegant party for two hundred guests at which attendees from as far away as El Paso listened to light opera and arias from the old masters.[9] Mrs. Hewitt also once held a reception for forty lady friends. Her veranda, covered with a massive Virginia creeper, other ivy, and potted plants, was shaded by awnings and spread with rugs. In late afternoon, the ladies were invited into the dining room to partake of ices, cakes, and fruits.[10]

Amelia C. Rawlins Hewitt. Date and photographer unknown. Courtesy John and Wesley Mottinger.

Law office of John Y. Hewitt. Hewitt ran his law office from 1888 to 1918. *Left to right*: Hewitt, Tim Brown (?), William Watson, William C. McDonald. Note the astounding array of law books on the wall. Date and photographer unknown. Courtesy John Kelt and Don Ward.

Part Owner of the Old Abe and Other Mines

Failure by three Baxter Gulch placer miners to continue assessment work on the Old Abe Mine, first located in 1879, resulted in loss of their title after three years. This left the claim open to others. In January 1884, Hewitt and his partners H. B. Fergusson and William Watson located two new claims called the White Oaks and the Robert E. Lee, covering in part the former Abraham Lincoln. The three partners each owned one-third of the Old Abe, with Fergusson and Hewitt having put up two-thirds of the one-hundred-dollar assessment fee in cash. For his part, Watson promised to do the actual assessment work.[11]

Around the first of November 1890, Watson and his nephew, Watson "Watt" Hoyle, did the required assessment work that finally opened up the Old Abe to development. The mine began production at the start of 1891. In time it would prove very profitable for this coalition of local men. Hewitt assumed management of the Old Abe as president; he also owned one-third of the Little Mack and other mining interests.

Prominent Businessman of White Oaks

Hewitt continued to dominate White Oaks largely through force of personality. Hewitt put most of the wealth that the Old Abe produced back into the town; he invested in real estate and eventually owned two business blocks built of stone and brick, besides considerable other income property. He built what is now known as the Brown building in the early 1890s on the corner of White Oaks Avenue and Pine Street. At one time he also owned a half-interest in the *White Oaks Eagle*, then

Hewitt block: the Exchange Bank and the Little Casino, ca. 1917. The Exchange Bank is on the left with "Hewitt" chiseled in stone above the entrance; the Little Casino saloon is to the right of the bank. Photographer unknown. Courtesy Sara Jackson, My House of Old Things, Ancho, New Mexico.

Interior of Exchange Bank. Raymond Lemon is standing behind the teller's cage. Mr. Sager is seated; Mrs. Crutcher is to the right. Date and photographer unknown. Courtesy Marilyn Lemon Ingley.

the only paper in town, and acted as its editor.[12] In January 1892, he bought a lot on White Oaks Avenue and let a contract to Ben Gumm to erect a stone storeroom twenty-five by sixty-four feet with an iron and plate glass front, to be occupied as a dry goods and clothing store. He also built the Exchange Bank in 1893, the crown jewel of the White Oaks main street.

The bank was a two-story building at the corner of White Oaks Avenue and Placer Street. It was originally established as a branch of the San Miguel County Bank of Las Vegas. Joshua Reynolds of Las Vegas was the first president of the White Oaks bank with Hewitt named vice president.[13]

Hewitt let a contract in June 1891 to Best & Treverton of Las Vegas[14] to erect the two-story bank building on the northeast corner of White Oaks Avenue and Placer Street. Stone for the bank's foundation, quarried from the South Fork of the Bonito River thirty miles away, was supposed to be marble but turned out to be alunite. A more suitable kind of stone—sandstone—was located in the nearby hills. Huge blocks of these stones were quarried and hauled into White Oaks by wagon to line the front, back, and east sides of the building. Bricks laid three thick made up the wall of the fourth side and also formed the interior partition.

Hewitt hired an expert mason from Las Vegas named James Young to oversee the masonry of the new building.[15] The stones were all hand-chiseled even though some weighed as much as five hundred pounds. Several had elaborate designs. The rocks below the outside windows were cut in a U-shape to fit the windows, and the glass itself was a quarter-inch thick. Judge Hewitt planned for his new building to house a general mercantile business and post office as well as the bank, and on the second story a series of offices for lawyers. Future governor of New Mexico William McDonald occupied one of these offices.

By January 1892, Hewitt assumed the presidency of the Exchange Bank, with William Watson as his vice president (later George Ulrick), and I. W. Zollars as cashier. Hewitt saw to it that his half-brother, Frank Sager, who had joined him in White Oaks, was appointed

Exchange Bank vault door blown off during 1892 robbery. Photographer unknown. Johnson Stearns Papers, Archives and Special Collections, Ms98.3, New Mexico State University Library.

secretary-treasurer of the new bank.¹⁶ The bank was completed in 1893.

The Exchange Bank moved from White Oaks to Carrizozo in May 1907.¹⁷ The White Oaks Exchange Bank was locked down permanently the night of April 30, 1907, and the Exchange Bank in Carrizozo opened the morning of May 1.

Politician and Civic Leader in Lincoln County and New Mexico

Hewitt was a very active Democrat in the politics of Lincoln County. The state Democratic convention prior to the presidential election of 1896 opened in Las Vegas on May 1. Judge Hewitt, by now considered the elder statesman in Lincoln County and Territorial politics, was elected a delegate to the Chicago convention where William Jennings Bryan emerged as the dominant figure in the Democratic Party for many years to come.¹⁸

At the annual meeting of the territory of New Mexico's legislature of 1909, Judge Hewitt was floor leader for the Democratic minority.¹⁹ In 1923, at the age of eighty-seven, he won the Democratic contest for county assessor of Lincoln County.²⁰ Politics remained a major interest throughout his life.

Hewitt was also connected with Kearny Post No. 10 of the Grand Army of the Republic and served as departmental commander in New Mexico Territory from 1887 to 1888.²¹

A Ninetieth Birthday Party to Remember

In 1926, the biggest party ever thrown for a citizen of Lincoln County was held in White Oaks to celebrate Judge Hewitt's ninetieth birthday.²² Old friends from all over as well as White Oaks residents turned out en masse. The hall overflowed with people, and automobiles lined all the roads approaching town.

Every woman in White Oaks and for miles around baked a birthday cake. The judge himself led the grand opening march with vivacious Nellie Reily, an old friend from Carrizozo, at his side. Dancing followed to the strains of an excellent orchestra, with many old-timers who had not danced for years joining in. The Judge danced until the wee hours to waltzes, schottisches, varsoviennes, and fox-trots.

After the dance had gone on for some time, a halt was called while the judge was presented with a large bouquet of ninety flowers. Judge Hewitt addressed the guests, saying he would cherish the magnificent cluster

Judge John Y. Hewitt in later years. Date and photographer unknown. Herman B. Weisner Papers, Archives and Special Collections, Ms249, PB49, New Mexico State University Library.

of flowers as the dearest memory of his life. Every one of the dozens of birthday cakes brought by Lincoln County ladies was then lit. Each cake bore ninety candles; the hundreds of candles shed so much light in the hall that the electric lights paled in comparison. Everyone partook of coffee, sandwiches, and cake until the orchestra stuck up more tunes. The merrymakers celebrated until early the next morning.

The Railroad

White Oaks had always cherished hopes of a railroad. For years the newspapers fanned talk of this or that railroad driving north from El Paso to establish a terminus at White Oaks, whereupon the town could not fail to blossom as the leading commercial center of southern New Mexico.

In April 1897, two first-class entrepreneurs, Charles B. Eddy and his brother John, of the southeast New

Mexico town of Eddy (now Carlsbad), began speaking to investors from the East whom they hoped to interest in backing a railroad to be built from El Paso to Clayton, New Mexico (a huge cattle shipping point).[23] The brothers plotted an extended pleasure excursion for their prospective investors, starting from El Paso and making the rounds of much of southern New Mexico. The easterners were shepherded by Albert Bacon Fall, George Curry, and William A. Hawkins, three prominent men in the territory who joined the Eddy entourage at San Antonio. Well-known Oliver Lee and a young (only twenty-two years old) Andrew Hudspeth also came along to orchestrate a Wild West roundup especially for the occasion.

The easterners traveled in splendid hacks and coaches drawn by four horses and pretended to be real westerners for days, sleeping at night in the open on bedrolls and eating from the back of a chuck wagon. They toured White Oaks, Nogal, Salado, the Mescalero Indian reservation, Tularosa, and many other places along the proposed route of the railroad. All the while the Eddy brothers and their associates bombarded their guests with glowing accounts of mining properties, the cattle industry, and timbering possibilities. When the entourage stopped at White Oaks, hopes of the locals reached fever pitch. Returning to El Paso to their luggage and special Pullman car called the Newport, the investors dined at an elegant hotel and astonished El Pasoans by dressing in formal evening attire for dinner.

The Eddy brothers' plan worked. By October 1897, the eastern capitalists notified Charles Eddy that they would back his railroad from El Paso to White Oaks. Eddy named this railroad the El Paso and Northeastern. By June of 1899, the line had reached Alamogordo and was poised to extend farther north. In negotiations with White Oaks business leaders, at first the railroad asked for vacant acreage near the west edge of town and offered a cash incentive of $50,000. Shortly afterward, the railroad abruptly canceled the offer. In 1900, John Eddy also left the team of developers, selling his interest in the New Mexico holdings to brother Charles and moving to Colorado.

Did Judge Hewitt Kill White Oaks?

There are suspicions that the judge headed a coalition of White Oaks leaders who refused to compromise with the railroad and would not spend even a dollar to lure investors. Some say these men stubbornly held fast to the idea that the railroad *had* to pass through White Oaks as the shortest, easiest, and cheapest route with the lowest grade over the summit and therefore in their greed grossly inflated the price of land needed for the right-of-way.

Perhaps, instead, the Eddy interests thought the grades would be too steep, resulting in unacceptably high operational costs. The investors may have been aware of the full extent of the gold crash at White Oaks. Maybe the people controlling the coal mines near White Oaks demanded excessive royalties for their coal.

There could have been many reasons for the failed venture. The Eddys counted on coal output from the new Salado coalfields at Coalora near Capitán (known as Gray until 1900), but by May 1901 their hopes were dashed as much larger coalfields of northern New Mexico at Dawson began coming on line. Additionally, a good supply of the water so necessary for steam locomotives had always been a problem for White Oaks.

Whatever the reason for the failure, it soon became apparent that the railroad would bypass White Oaks in favor of a new railroad terminal to be built on the Carrizo flats twelve miles south of White Oaks. This decision signaled the death of White Oaks. The end actually arrived in August 1899 as the rails reached the site of present-day Carrizozo.

Judge Hewitt's Last Years

Judge Hewitt refused to abandon the dying town he had come to think of as his own fiefdom. Long after the mines closed and the railroad bypassed the Oaks, he stayed on, spending most of his money in futile attempts to revive mining interests. His wife, Amelia, died in 1906 at the age of sixty-nine and is buried in Cedarvale Cemetery.

In 1930, Hewitt fell at his home and was not found until the following day.[24] Different neighbors tried caring for the aged man at home, but he had become cranky and ill-tempered and no one would stay long. Mary Lou Townsend Welch's parents were living in the Hoyle House at the time. Mary Lou, her husband, and their two babies also lived in the house with her parents, the Townsends (Mary Lou's little girl was the only baby ever born in the Hoyle House).[25] Mrs. Townsend took care of Hewitt at his home until the house caught fire and burned down. She and her husband then drove the old man to the Carrizozo Eating House to lodge.

On March 31, 1930, one cold and windy Sunday afternoon Paul Mayer of Carrizozo drove up to the

Hoyle House and deposited Judge Hewitt inside with his suitcase. Hewitt announced that he was moving in if he had to sleep on the floor. There was nothing to do but relocate the young Welches and their babies upstairs, leaving the judge ensconced in the front parlor of the house. He lived on for two years in the parlor, with Mrs. Townsend, Mary Lou, and Dave Jackson all taking turns caring for him.

The judge had a habit of cursing constantly, vastly enriching the vocabulary of Mary Lou's two-year-old toddler. The boy would repeat the new words he had learned to Mary Lou and his grandmother. Once he came into the kitchen to observe, "That old son-of-a-bitch is cussin' again."[26] And Hewitt possessed another trait especially trying to the Welches and Townsends: he didn't believe in God. He also fussed about how slowly Mary Lou's mother was feeding him, forcing the three caregivers to change off to keep peace. Finally, he died in January 1932 in the front parlor of the Hoyle House, the Welches, Jacksons, and Townsends all at his bedside. He is buried in Cedarvale Cemetery.

Lawyer Andy Hudspeth Becomes Judge Hewitt's Partner

In 1902, after a young lawyer named Andrew H. "Andy" Hudspeth moved to White Oaks from the Cree family's VV Ranch at Angus near present-day Ruidoso, Hewitt and Hudspeth formed a law partnership. For many years, they shared a law office in the Exchange Bank, and Hudspeth followed Judge Hewitt as the leading citizen of White Oaks.

John Hall, a lawyer who practiced in Carrizozo before moving to Albuquerque, named Judge Hewitt as his favorite old-timer.[27] When Hall visited the judge a short time before his death, Hewitt told Hall that after the Old Abe explosion it was impossible to jumpstart the mine due to the higher costs of operating and the fixed price of gold, but the old man insisted that there was still a lot of gold left in the deep mine. Hall probated Hewitt's estate and bought his old rolltop desk.

Hewitt headstone in Cedarvale Cemetery. Photograph by Bob Workhoven. Author's collection.

NOTES

1. "John Y. Hewitt," 513–15.

2. Ibid.

3. Ibid.

4. Robert Leslie, interview by Haldane, August 17, 2002, Carrizozo.

5. Keleher, *Violence in Lincoln County*, 369–70.

6. Robert Leslie, interview by Haldane, August 17, 2002.

7. Mrs. Tom Charles to Associated Press, telegram, n.d.

8. Robert Leslie, interview by Haldane, August 17, 2002.

9. Parker, *White Oaks*, xiii.

10. *Lincoln County Leader*, August 29, 1891.

11. Parker, *White Oaks*, 88, 90. Regardless of names given by owners, the Abraham Lincoln mine has always been known popularly as the "Old Abe."

12. *Lincoln County Leader*, January 9, 1892.

13. Birdsong, *1992 Lincoln County Pony Express Calendar*.

14. *Lincoln County Leader*, June 13, 1891.

15. *Lincoln County Leader*, August 15, 1891.

16. Birdsong, *1992 Lincoln County Pony Express Calendar*.

17. John Kelt to Amanda Treat, May 14, 1969, copy in author's collection.

18. Curry, *George Curry*, 93.

19. Ibid., 231–32.

20. Charles, telegram.

21. Biographical note, John Y. Hewitt, Lincoln County, New Mexico, Collection, 1885–1976, copy in author's collection.

22. *Lincoln County News*, October 15, 1926.

23. Birdsong, *Tracks North*, 74–75.

24. Charles, telegram.

25. Mary Lou Townsend Welch, interview by Haldane, August 18, 2002, Lovington, New Mexico.

26. Ibid.

27. Hall to Haldane, August 22, 2001.

"JOLLY JERRY" HOCKRADLE
Age Cannot Weary the Prospector

Jerry Hockradle (also Hockeriddle, Hokeriddle, Hockraddle) was one of Lincoln County's earliest prospectors. He was born in Michigan of German ancestry; his father had immigrated to this country from Württemburg in southwest Germany.[1] Jerry came to Lincoln County in 1868, settling first along the Río Hondo in Lincoln County. The 1880 census lists him as married to nineteen-year-old Louisa, who was born in Texas of parents from Saxony in central Germany.

In Hockradle's early New Mexico years, his location and the times combined to ensure that he would be caught up in the Horrell and Lincoln County Wars.

Jerry and the Horrell War

On March 14, 1873, the Horrell brothers—Ben, Martin, Merritt, Sam, and Thomas—gunned down five Texas state policemen in a shootout at a Lampasas saloon.[2] To escape the pursuing Texas lawmen, the Horrells packed up their families and livestock (except for Ben, who left his family behind) and headed for the wild southeast New Mexico gateway town of Seven Rivers in September. They bought a homestead in the Ruidoso valley of Lincoln County just in time to run into a race war already primed to explode.

Resident Hispanics on the Ruidoso resented newcomer Texans, claiming they were no better than bandits. In their turn, the Texans regarded native Hispanics as less-than-human "greasers" on a par with their opinion of recently freed blacks of Texas. Clashes between the two groups stoked the Lincoln-Ruidoso population into open warfare. The Horrells claimed that Hispanics had robbed the gold from the sale of their Texas cattle. The Hispanics countered that one of the Horrell brothers had shot and killed a neighbor as he was digging a ditch.

In December 1873, Lincoln constable Juan Martin was shot dead after attempting to disarm and arrest Tom Horrell and his companions. Close on the heels of that killing came the shootings of Ben Horrell and Jack Gylam in cold blood by "hired Mexicans."[3] To restore order and stave off a bloody race war, Sheriff "Ham" Mills led a posse of local men (mostly Hispanics) to Horrell headquarters on Eagle Creek, where they demanded the surrender of the Horrells.[4]

A bloodless daylong fight ensued, after which the posse abandoned its siege. Within days, the Horrells and their supporters thundered into Lincoln and shot up the town, killing four Hispanics. Leading citizens of Lincoln begged Governor Giddings for help. In response, the governor issued a proclamation offering a reward of one hundred dollars each for several men, including three Horrell brothers.[5] Sheriff Mills and a posse of sixty men set off to Eagle Creek to arrest the brothers. The posse shot or drove off all the Horrell horses, but during the night the besieged men escaped to Casey's Mill on the Río Hondo. They stayed about ten days at the Casey place.

Young Billy Casey, not yet fourteen, reported events from a different perspective: "Major Hunter from Fort Mason, Texas . . . suggested they go up after the [Horrell] women and children. Old man [Heiskell] Jones who was at my father's place, Jerry Hokeriddle [Hockradle], Frank McCulloch [McCullum] and my father sent up [to Lincoln], saw Ham Milles [Mills] [and told him what they were going to do], and brought all of the women and children down to our place fourteen miles below where the Horrells were surrounded."[6]

The Horrells then hired early settlers Frank McCullum, Heiskell Jones, Billy Casey, and Jerry Hockradle to freight their cattle, food, and household goods to nearby Missouri Bottom below the little village of Picacho. This group of four was attacked by a mob of Hispanics while on their way from Casey's Mill to Missouri Bottom with a six-yoke ox team and wagon and a trail wagon. The mob made off with all the Horrell possessions but left the quartet unharmed. Billy Casey recounted, "Jerry Hokeriddle and McCulloch drove mules to some wagons. Mrs Jones drove horses and a small wagon with a cooking outfit. We passed through San Patricio and got permission from Ham Mills at Lincoln, to get the Horrells' stuff. We loaded up their household goods, gathered their hogs and food, and started back.... On the way back the Mexicans stopped us and took everything we had ... Jones, McCulloch, and Hokeriddle went on and reported to the Horrells, and they said 'they'd get lots of the sonsofbitches.'"[7]

These setbacks forced the Horrells to head back to Texas. But in January 1874, before they could make the move, a mob led by Jimmy Dolan rode to the Ruidoso and burned down the Horrell ranch house. On hearing of this latest outrage, the Horrells headed back up the river, bent on killing every Hispanic in the area. At Picacho, for some unknown reason, they murdered a young Anglo settler named Haskins. At Casey's Mill, the Texans split into two groups. Enraged, one group slaughtered five harmless Hispanic freighters hauling corn; the second group stole stock from three ranches. Then they headed for the Beckwith ranch at Seven Rivers and finally reached Hueco Tanks east of El Paso, Texas.[8] They never returned to New Mexico.

The Hockradle Mine in Nogal's Dry Gulch

Luckily, Jerry escaped all of the bullets and continued life as usual, buying necessities in Lincoln from Murphy and Company, whose accounts receivable ledger rated his credit "good." On Gov. Lew Wallace's list of Lincoln County names of interest, Jerry was called a "Murphy man but honest."[9] The always-restless Jerry also began working his mine in the Dry Gulch of Nogal, some fifteen miles south of White Oaks. The *Las Vegas Daily Optic* of June 1880 reported, "The Nogal district lies just above the reservation line. The rock has good mineral showing, and ... will run strong in silver although several claims show traces of gold.... In Dry Cañon is a mine, generally termed the Jerry Hocradle [*sic*] mine, on which there has been about $1,500 worth of work done. It was located in 1877, and an assay of over $1,600 per ton claimed."[10]

Jerry appears to have been well thought of by his neighbors on the Hondo. He managed to survive the Lincoln County troubles even after being called for grand jury duty in 1878 to investigate the murders of John Tunstall, William Brady, and George Hindman.

Jerry and the Lincoln County War

On April 8, 1878, district court began its regular session in Lincoln in spite of the building storm of Lincoln County troubles. According to historian Maurice Fulton, "By the thirteenth, Judge Bristol had the grand jury organized and functioning. On it were: Joseph H. Blazer, foreman; Juan B. Patrón; Crescencio Sánchez; Vicento Romero; Camilo Nuñez; Wesley Fields; Robert M. Gilbert; Francisco Romero y Valencia; Desiderío Zamora; Jerry Hockradle ... and Francisco Pacheco. With the grand jury organized, Judge Bristol delivered a comprehensive charge, reviewing the recent lawlessness."[11]

The grand jury finished its investigations and reported on April 18 to the Honorable Warren Bristol, associate justice of the supreme court of New Mexico Territory and presiding judge of the Third Judicial District, "The murder of John H. Tunstall, for brutality and malice, is without parallel and without a shadow of justification.... We equally condemn the most brutal murder of our late sheriff, William Brady, and George Hindman. In each of the cases, where the evidence would warrant it, we have made presentments."[12]

After his duty was done amid real danger, Jerry settled down again to his ranch on the Hondo and prospected on the side.

Jerry's Ranch in the Jicarillas, Last Stop on the Las Vegas–to–White Oaks Highway

The 1879 White Oaks gold strike lured Jerry from Hondo to the Jicarilla Mountains, a strategic location north of White Oaks where he could profitably "mine" the pockets of travelers along the Las Vegas–to–White Oaks road. Water from a spring in the mountains had to be hauled to his ranch. The *Las Vegas Daily Optic* observed, "Jerry Hocradle, eighteen miles out from White Oaks, certainly has a fatter take than Uncle Dick Wooten ever had at his toll gate in the Raton pass. Jolly Jerry is reaping a harvest of wealth by charging five dollars a barrel for water at his place.... The bottom will soon be knocked out of Jerry's water monopoly by the action of the Las Vegas people, who have arranged for the immediate sinking of wells."[13]

But the "Las Vegas people" did not sink any wells. Meanwhile, Jerry got involved in Lincoln County troubles again, if only from the sidelines, after the Greathouse-Kuch ranch shootout at the end of November

1880. The shootout, between a White Oaks posse and Billy the Kid and some of his gang, ended with the killing of White Oaks blacksmith Jimmy Carlyle: "Discouraged by Carlyle's death and suffering severely from the cold and from lack of food and water, the posse rode off to Jerry Hockradle's ranch, about twenty-five miles distant. The besieged men [Billy the Kid and his companions] ... walked off."[14]

Many tired and dusty travelers stopped at Jerry's place before their final push into the Oaks. In March 1881, young Paul Mayer abandoned his former trade of cigar making in Covington, Kentucky, and headed for new opportunities as a livery stable owner in White Oaks. He stopped at Jerry's place in the Jicarillas to water his team.

In late summer 1882, the Erastus Wells Parker family left Saint Louis for the management of a newly completed mill at the South Homestake Mine at White Oaks. A light buggy carried the Parkers (father, mother, and three boys) from Las Vegas as fast as their horses would allow because of fear of Apache attack. They arrived at Hockradle's place on the afternoon of the fifth day. According to son Morris Parker, Jerry and his "extra-fat, good-natured wife" treated the weary travelers well but "did not refuse the customary fee for water."[15] After the railroad forged its way south from Las Vegas to Socorro and San Antonio, travelers to White Oaks began bypassing Jerry's ranch despite his drastic cut of water prices to twenty-five cents a bucketful. They preferred the faster route by rail to San Antonio, then overland by stage. In the late 1890s, the Hockradles moved to Nogal nearer his Dry Gulch mine.

Chasing That Pot of Gold in California

Life continued quietly until May 1899 when the "extra-fat" Mrs. Hockradle was seized with a heart attack and toppled dead out of her buggy.[16] Jerry stayed on in Nogal for three more years until he heard of a gold strike in the Lower California peninsula in 1902. Jerry left Nogal and headed west by way of El Paso. A reporter caught up with him long enough in that border town for the following interview, published in the *White Oaks Eagle* as a news item from an earlier *El Paso Daily News* edition:

PROSPECTOR 40 YEARS

Jerry Hockradle Cleans Up a Little Fortune and Returns to the Pacific Coast.

IS GOING TO CALIFORNIA

Jerry Hockradle ... passed through this city today en route from his home at White Oaks, in Lincoln county, to San Diego. Though about 70 odd years of age [actually 60], he is hale and hearty and a typical prospector, in appearance. He said to a News reporter today:

"Yes, guess I am the oldest resident of [southeast] New Mexico, for I went there in '68 when there was nothing in that section but mountains, wild animals and Indians, and there were plenty of the latter.

"That was about the time that Pat Garrett first became known in that country. We had just gone through with one war between Mexicans because I had sheltered their White enemies [referring to the Horrell War] when the Kid came into prominence as a killer. He seemed to have the sheriff pretty well bluffed and the people were getting pretty tired of him anyway, when they came to me to run for sheriff to clean him out. I said that I would not take the job under any circumstances, but told them there was a likely young fellow in a cow camp who was a deputy sheriff, named Pat Garrett, who could do the work. Then we had the convention and nominated Pat. And I will say that he made the best sheriff Lincoln county ever had. . . .

"Now I am quitting New Mexico for good and am going over into the Muddy River country region on the railroad that Senator Clark is building from Salt Lake City to Los Angeles. I passed over the country with a Mormon wagon train away back in '66 when I was coming east to New Mexico. The place is off the Mohave desert and when I passed over there, there was alkili so thick on the ground that you could have skated on the crust. Then the water holes were 65, 72 and 85 miles apart.

"It was on this spot that we had a bloody battle with 150 Indians, Piutes [Paiutes]. The Indians had stolen some of the horses which belonged to the Mormons I was traveling with; when the men of the party tried to recover them they took position on a rocky knoll and fired several volleys into the camp.

"Our party then made a detour and got behind the rock. After the smoke cleared away we found 19 redskins whom we buried the next morning. It was a narrow escape for us for there were only about six good guns among us.

"While we were passing over that section I saw some mineral stains and now that the railroad is being built through there it will be a fine place for prospecting.

"I expect to stop there on my way to San Diego and see what the stains amount to. . . . I made a good stake out of the White Oaks claims and I hope will have good luck in that country. I know it is there, for a cousin of mine told me that he got enough gold out of a flour sack full of dirt to buy several sacks of flour."[17]

NOTES

1. U.S. Bureau of the Census, *United States Census of 1900*, White Oaks.

2. Nolan, *Bad Blood*, 16–26.

3. Ibid., 51–52, 54–55.

4. Nolan, *Lincoln County War*, 51.

5. Ibid., 52.

6. William Casey, interview by J. Evetts Haley, March 30, 1932, J. Evetts Haley Collection, Haley Memorial Library and History Center, Midland, Texas.

7. Nolan, *Bad Blood*, 80–81.

8. Ibid., 86, 88–89.

9. Notes from Mullin Collection.

10. *Las Vegas Daily Optic*, June 1, 1880.

11. Fulton, *History of the Lincoln County War*, 196.

12. Ibid.

13. *Las Vegas Daily Optic*, undated copy in author's collection.

14. Rasch, "Billy Wilson," 60–61.

15. Parker, *White Oaks*, 15.

16. Lincoln County Historical Society, "Jerry Hockradle," 21.

17. *White Oaks Eagle*, October 23, 1902.

EMERSON HOUGH
Lawyer, Conservationist, Author of Heart's Desire

In the late 1800s and early 1900s, some of the Southwest's finest literary talents emerged from White Oaks. Unlike the town itself, they would never "go ghost." Among the writers were Eugene Manlove Rhodes ("Bard of the Tularosa" revered by New Mexicans for both his poetry and his books about the Tularosa Basin, especially *Paso por Aquí*), Jack Thorp (collector and publisher of *Songs of the Cowboys*), and Emerson Hough. It was Hough who immortalized the gold-mining boomtown of White Oaks in his 1905 novel *Heart's Desire*. As a writer, he is most remembered as an interpreter of the frontier experience in the American West. Though he stayed in White Oaks a little less than a year (June 1883 to February 1884), it was a period that profoundly affected the rest of his life.

Emerson Hough was born June 28, 1857, in Newton, Iowa, to Joseph and Elizabeth Hough in a time of great change.[1] The Civil War was but four years away. In Hough's lifetime, seventeen more states would join the Union, countless mills and factories would emerge out of the industrial revolution, railroads would span the continent, and the First World War would be fought. This era also saw the United States changing from an agrarian to an industrial society.

Growing up in Newton, the boy's greatest influence was his father, a man with a strong sense of family tradition. Joseph Hough had spent his youth in Virginia, where he developed the moral code of the Virginia gentleman derived from a highly romantic vision of the South. The son soaked in his father's influences and internalized them.

Hough graduated from Newton High School in 1875 and taught a year in a country school before attending the State University of Iowa. He majored in modern languages, played football, and edited the college newspaper. After graduating in 1880, Hough took a job as a civil engineer, surveying parts of central Iowa

Emerson Hough, 1880. Photographer unknown. Courtesy Emerson Hough files, Special Collections Department, University of Iowa Libraries, Iowa City, Iowa.

for a new road. Then he began reading law under the Honorable H. Winslow of Newton and was admitted to the bar in 1882, the same year that saw publication of his first article, "Far from the Madding Crowd," in *Forest and Stream*.

Looking West

After a quarrel with the sweetheart Hough had met at State University, she broke off their engagement. At precisely that time a friend in New Mexico, Eli H. Chandler, invited him to join his law firm. A free railroad pass from the magazine *American Field*, given to him in return for writing some sketches on the Southwest, provided the only other necessary excuse: he would go West.

Restless, young, and ambitious, Hough alighted in White Oaks on June 1, 1883. It was his first real venture away from his family. He recalled, "I went to White Oaks, N.M. God knows why, for it was the last place in the world anyone else would of [*sic*] gone to practice law. In a year or so, we lawyers got the town all tied up, and most of us walked back to the states."²

Lawyer Hough

Newly minted lawyer Hough arrived in White Oaks in the company of several other young lawyers, all broke, all hopeful, and all carefree bachelors. He bunked with George Ulrick, who would later practice law in an office above the Exchange Bank, and William McDonald, New Mexico's first governor after statehood in 1912. The three bachelors shared one good white shirt that traveled from one to the other for important occasions, the only rule being that it must be washed by the user before passing it on.

Tongue in cheek, Hough describes an imaginary ad for hungry Southwest lawyers in his book *The Story of the Cowboy*: "John Jones, Attorney-at-Law, Real Estate and Insurance. Collections promptly attended to at all hours of the day and night. Good Ohio cider for sale at 5 cents a glass."³

The Law Firm of Beall, Chandler, Hough, and Thornton

Hough participated in one of only two White Oaks cases of any significance at that early time. *Patterson vs. Baxter Gold Mining Company* stemmed from a complaint brought by Henry Patterson of White Oaks against the Baxter Gold Mining Company.⁴ It was argued in 1884 before the territorial supreme court. Harvey Fergusson, another of Hough's friends, took the Patterson side, and Hough's firm—Beall, Chandler, Hough, and Thornton—took the Baxter side.

While this case was important, many other minor fights over mining claims were constantly being litigated by legions of newly arrived lawyers. The miners grew so tired of all the litigation that they threatened to form lynch mobs to hang any lawyers who did not get out of town. One firm that did not leave White Oaks was Beall, Chandler, Hough, and Thornton.

Patterson vs. Baxter paid little, and Hough, like many of his companions, resorted to shooting a great number of cottontails to keep body and soul together. The Baxter case caused Hough to lose heart with all the legal claptrap. He also became disillusioned with the legal and political shenanigans throughout New Mexico Territory at that time, saying that "There is no use for these things [sportsmanship, honor, fair play] in the practice . . . of the law. It is the most unsportsmanlike of all callings, and the most unfair of all businesses. As seen with us it was also the most corrupt."⁵

Besides, the sun was shining on the mountain slopes outside, the sky and trees beckoned, and his passion was growing for hunting, fishing, and the wilderness. As writer John Sinclair put it: "The case of *Patterson vs. Baxter*? The hell with it!"⁶ Hough never again practiced law.

Fledgling Writer

With almost no legal work at hand, Hough began spending most of his time as a reporter and typesetter in the offices of the White Oaks weekly newspaper, *Golden Era*. He noted local happenings and sought out interesting people in Lincoln County for interviews. Hough's apprenticeship with this paper served as "kindergarten for a young writer who eventually achieved a national reputation as a novelist."⁷ Hough also began contributing articles to *American Field* and *Forest and Stream*.

On February 7, 1884, he wrote a short notice for the *White Oaks Golden Era* about the death of one James "Stumpy" Wicks. Stumpy had once confided to him, "I used to be a preacher back east, and a dam'd good preacher, too." Hough reported, "Died. At 1 a.m. of the morning of the 6th inst., of pneumonia, James Wicks, of White Oaks. Mr. Wicks' face was a familiar one to our people. He has gone over the range, to the miner's better country."⁸

Back Home in the Midwest

The serious illness of Hough's mother called him home in February 1884. He would not return to White Oaks for almost two decades. However, he was unable to get Stumpy out of his mind and wrote an article about the funeral for the April 1, 1884, *Peck's Sun* published by George W. Peck at Milwaukee, Wisconsin. "Over the Range" was quickly reprinted in the April 10 issue of the *White Oaks Golden Era*.⁹ The article is so descriptive of its place and time that it is reproduced in full in a sidebar to this chapter.

OVER THE RANGE
by Emerson Hough

Stumpy Wicks was dead. The mountain fever had killed him a few days after he had started off into the hills, telling the boys that he would find them something rich, or never go out again. He did not find anything rich; and he never went out again. The fever laid its grip upon him, and in three days he was dead. He had "gone over the range," the boys said.

It became necessary to bury Stumpy Wicks. And how was he to be buried? By his relatives? He had no relatives. By the town? There was no town. By his pard? He had no pard. Forty years ago Stumpy Wicks had left his home—no one knew where—and his people—no one knew whom—to wander alone in the west. His wife, his mother, his sister, if he had one, will never know where he died, or what hands laid him in his grave.

It was the boys. They got together and made a coffin out of a box or two and covered it with black cloth. They put Stumpy into it, with a clean floursack over his poor, dead face. They chipped in and hired an ex-parson, who for some years had abandoned his profession, to "give Stumpy a sendoff." They dug a grave to a good and honest depth in the tough, red earth. They went out and found a flat rock for a head-stone, and on it, with an engineer's graver, they scratched the brief epitaph, "Stumpy Wicks." Then they followed the coffin-wagon to the grave, walking through the mud and rain.

There were forty men in that funeral procession, and not one woman. Almost no one was drunk, and nearly all had taken off their six-shooters. There were forty men who stood around the open grave, and not one woman to drop a tear, as the ex-parson read a brief portion of the Episcopal burial service and offered a short prayer for the safe journey of Stumpy's soul over the range. There was no history of Stumpy's life. No one knew that history. It was doubtless a sad enough one, full of slips and stumbles; full of hope, perhaps, before he finally "lost his grip." They found a woman's picture, very old, and quite worn out indeed in Stumpy's pocket, and this was buried with him. This was probably his history.

There was not a tear shed at Stumpy's funeral. Not a sob was heard. But neither was there any oaths or any laughter.

When the time came to fill up the grave, ready hearts assisted ready hands, and the experienced miners quickly did the work. They rounded up the mound, and fitted in the headstone. When the ex-parson stepped back from the grave he stumbled over the head-stone of Billy Robbins, the gambler, whom Antonio Sanchez knifed. There were a good many of the boys resting there. The bullet, the knife and the mountain fever had finished them, except those whom the committee assisted. It was the committee who put Antonio Sanchez at the foot of Billy Robbins' grave.

There was no green thing in this graveyard, no living plants, no little flowers. It lay, red and bare, upon a red and bare hillside. There were no white stones to mark the homes of the sleepers; those used were of the rough, red granite.

The boys were quiet. They were thinking, perhaps. They looked up at the sky, which, strangely enough for a sky in New Mexico, had in it no tinge of blue, and the sky, in pity that no tear was shed, wept some upon them.

As the funeral procession broke up and moved back to the saloons, one was heard to say that it was the "d---dest mournfulest plantin' he ever had a hand in." In fact the camp did not get back to its normal condition until the next day. There was something too sad even for these rough souls in the lonely, broken life, the lonely, unwept death of Stumpy Wicks. It made them think, and I wonder if some of them did not reach out their arms from their blankets that night, and hold them up and call out softly, "Oh, Stumpy, Stumpy! What is it that you see over the range? After a wretched, broken life, what is there for a man, over the range?"[1]

1. Emerson Hough, "Over the Range," *Peck's Sun*, April 1, 1884. Reprinted in *White Oaks Golden Era*, April 10, 1884.

The success of articles like "Over the Range" encouraged the fledgling writer. He began selling brief pieces on hunting and fishing to magazines and thought of becoming a writer full time. In Iowa, he got a job as business manager of the *Des Moines Times*. White Oaks kept up with him for a while. The *Golden Era* predicted, "The paper will prosper under his management. Hough is a royal good fellow, and his New Mexico friends will regret that he has concluded not to return to New Mexico."[10]

Six months later he became associate editor of *Sandusky Register* in Sandusky, Ohio, and the *Golden Era* again notified its subscribers: "Mr. E. Hough . . . is now associate editor of *The Register*, Sandusky, Ohio. This is a streak of good luck, well deserved, and we from the land of Montezuma send kindly greetings to our old friend and associate whose ken now looks out over the bright blue waters of lake Erie."[11]

Hard Years as a Writer

Hough formed a partnership with an artist he had known in Des Moines, and, soon after, they were hired by a Chicago publisher to write and illustrate county and city histories in the West.[12] When the publisher fired them both, Hough found a new partner and went into the newspaper business in Wichita, Kansas. He reconciled with his college sweetheart and planned to marry her on June 27, 1888—but in a few months he was sitting in a Chicago room contemplating suicide. His new partner had fled to Mexico with all the funds of their Wichita newspaper. Hough went to Iowa to break the bad news to his fiancée, then returned to Wichita to try and save his paper. He succeeded, but his fiancée refused to be married to him for unknown reasons.

Back in Chicago and now thirty-two years old, Hough returned to writing. He got a job as western editor of *Forest and Stream*, where he managed the office, edited the "Chicago and the West" column, and solicited ads, earning fifteen dollars a week. He did that kind of job for fifteen years. It worked out well—his boss, George Grinnell, was a thousand miles away in the East. Hough made extra money by writing for two or three newspapers, and in 1897 he actually published *The Story of the Cowboy* as part of the History of the West Series published by D. Appleton and Company. This marked a shift from writing articles and short pieces into writing books.

Hough married Charlotte Cheesebro ("Miss Lottie") in 1897, the ideal helpmate for life. Lottie managed his business affairs and left him free to travel, hunt, fish, and write. In 1903, Hough left *Forest and Stream* and took a job for a year and a half with *Field and Stream*, ending his tenure there in February 1905 (he had to sue both companies for back pay). He continued to write short stories and novels and finally produced his first financial success. It was published by Bobbs-Merrill and was called *54-40 or Fight*.

Success, at Last

His reputation growing, Hough found the first decade of the twentieth century to be fairly happy and calm. The turn of the century marked a brief return to his old days of the frontier. He and Miss Lottie toured Europe and England in 1909 and 1910. The coming of World War I barely interrupted the flow of his writing, though in 1919 Hough received a commission in the U.S. Army Reserve.

Hough wrote a best seller, *The Covered Wagon*, published first in serial form in the *Saturday Evening Post* beginning April 8, 1922, and as a book published by Appleton later the same year. The book dealt with the movement west of the wagon trains over the Oregon Trail and the role of the pioneer woman. It was made into a silent film in the fall of 1922, opening at the Criterion Theater in New York in March 1923. Hough received $8,500 for the movie rights.

Hough sold the movie rights to *The Man Next Door* and *The Broken Gate* to Vitagraph and Tiffany, respectively, and *North of 36* to Paramount. *North of 36*, written a year after *The Covered Wagon*, was also a popular and lucrative film. Its subject was the first successful trail drive.

Finally, Hough had reached national recognition and financial security. He and his wife were able to move out of rented apartments and buy a beautiful home in Highland Park, a fashionable Chicago suburb on the shores of Lake Michigan. They paid $35,000 cash for their house.

Conservationist

Hough was also an early advocate of conservation and worked to preserve wildlife and to establish national parks.[13] In the winter of 1893, as one of his projects for *Forest and Stream*, he went as a guide with two Fort Yellowstone soldiers to tour Yellowstone Park on snowshoes in subzero weather in order to photograph game animals and count bison. When the men found that the park's great bison herd had been reduced by poachers from five hundred to fewer than one hundred, Hough fired off an angry report to the government about the situation. His document moved Congress to enact legislation on May 7, 1894, extending federal protection to the park's herds. Credit for stopping the wholesale slaughter of the Yellowstone bison herd belongs to Hough.

Later Hough worked on behalf of a congressional bill to protect game birds and to establish national wildfowl preserves. Two ideas motivated him. As a sportsman and journalist writing for other amateur sportsmen, he advocated game-preservation laws to safeguard the "wealth of the people at large." He believed in the concept of *harvesting* game. To him, hunting was like farming: seeds must be sown before crops can be raised. Hough wanted to raise awareness of the country's natural resources and preserve certain areas with their original animals.[14] Like many other conservationists, he also saw a need to set aside recreational areas, asking, "What are our American Jewels? They will be our splendid mountains, our wildernesses, our sporting out-of-doors, where a man can wear a blue shirt and swear by the nine gods."[15]

Hough also became the western representative of George Grinnell's *Forest and Stream* policy.[16]

A Dark Side

The decade between 1900 and 1910 was hard for Hough, who did not believe in unrestricted immigration. That decade saw over eight million immigrants from southern Europe and Russia arriving in the United States. Hough was appalled. In a letter to Samuel Merwin, he wrote, "Three things no sane editor will touch . . . immigration, Jews, and labor. These, with the Negro Problem, cover pretty much all the big in's and out's of Americanism today. If we could wipe out all the Jews, Negroes, Immigrants, and Labor Unions in the world, we would have a fairly decent country to live in."[17]

In 1919 he wrote "The One Hundred Per Cent American," attacking the idea of hyphenated Americanism. He believed that people should abandon their diverse heritages and declare undivided loyalty to America. The same year, his novel *Sagebrusher* attacked Bolsheviks and Jews.

White Oaks Reunion

In 1904, Hough revisited New Mexico in a trip back to the locale of his early first western adventure to collect material for his forthcoming book *Heart's Desire*, which was to be modeled on his early 1880s life in White Oaks.[18] He hooked up with an old friend, former Sheriff Pat Garrett, for a nostalgic tour of what used to be old Lincoln County. Revisiting the days of their youth, they stopped in White Oaks and passed through Roswell on their way to Fort Sumner. Garrett showed Hough where he had killed Billy the Kid, the Kid's burial place, and other places linked to the outlaw's life.

Hough published an account of that visit in the *Field and Stream* article "The West After Twenty Years." In that article, he admits that again seeing the beautiful little valley in the mountains rekindled his undying love for the Rocky Mountain West, and he recalls some opinions formed from that time: "There were able men in abundance. As to physical courage, it was at times more than a drug on the market. It was a peppery, fighting little village . . . where everybody wore a gun, and where nobody pulled one until he meant business." He goes on to talk about White Oaks as a town: "As we . . . entered the lower edge of town, I admit that I had a catch in the breast. My heart nearly choked my throat, as I looked over the little town which I had not seen again during the best part of my lifetime. . . . The same glorious sunshine was there; the same unspeakable clearness of the air; the same brooding, peaceful tranquility." "The boys" still living in White Oaks planned a little reunion. Reminiscing was the order of the evening, and many tall tales were told. John Hewitt, the lawyer who stayed in White Oaks, had given Hough his first drubbing in a lawsuit. He was no older, simply a little more intense, a "little more *Hewitt*" than two decades before. Jones Taliaferro was there, and L. Rudiselle, George Ulrick, Frank Sager, Eugene Stewart, Paul Mayer, Charley Bull, G. R. "Dick" Young, Colonel Prichard, and Doctor Paden. They talked cows, mines, railroads, and the social system of the land. They talked about absent friends: "Major Caffrey, editor of the *Leader* and archrival of the *Golden Era*, dead since 1893, the same year Uncle John Brothers died. Dye, the litigious, who jumped every claim on Baxter Peak, now in Joplin, Missouri. W. C. McDonald, court surveyor and mining engineer, now foreman of the Carrizozo ranch . . . old law partner George Beall, dead in 1885, drank himself to death. Colonel Fountain also dead; he and son murdered by unknown parties years ago. Old friend Louis Monjeau, with whom Hough used to hunt cottontails for breakfast, dead for 20 years. Johnny Hudgens killed him."[19]

Many others of "the boys" were absent, some dead, others doing time in the penitentiary: "The best of men will sometimes get in prison. Owing to death—and the penitentiary—it was hardly as large a circle as used to meet sometimes in the past in the emporium of Johnny Hudgens.... Twenty years! Great God, what I have missed!"[20]

Heart's Desire, the Novel Based on Hough's White Oaks Experience

Heart's Desire grew from a look back at Hough's White Oaks days in a series of short stories for the *Saturday Evening Post*. The book blends local humor, farce, and romanticism and explores how the frontier West is destroyed by the introduction of civilization. The reader can find authentic White Oaks details in the book. Many of the characters are based on "the boys" Hough knew personally; in some cases even their names are unchanged. The main character is a cowboy named Curly.

In one episode in the novel, the men of *Heart's Desire* try out the new game of croquet. Tough cowboys line up for their turn at a "new style of shootin'": "There were certain features of the contest . . . which were perhaps not usual. . . . I do not recall ever to have seen any other game of croquet in which two of the high contracting parties wore 'chaps' and spurs and the other two overalls and blue shirts."[21]

The players in this cowboy game of croquet stand around and kibitz from the sidelines as each of "the boys" swings his mallet:

"Can't you hit that stake?"

"I could if you'd let me take a six-shooter or a rope," said Curly. "I ain't fixed for this here tenderfoot game. . . . If it wasn't for that there spur, I'd have sent Doc's ball plumb over Carrizy Mountain that last carrom. You watch me when onct I get the hang of this thing."

"You can't get the hang of nothing," said McKinney. "A cowpuncher ain't got no sense except to ride mean horses and eat canned tomatoes."[22]

Another time, Tom Osby tells Madame Donatelli the name of their town. She replies, "Heart's Desire? What an exquisite name! Where is it? What is it? That sounds like heaven." Tom quickly brings the conversation down to earth, saying, "It might be, ma'am, . . . but it ain't. The water supply ain't reg'lar enough."[23]

Heart's Desire, though a financial disaster, was Hough's most artistically successful novel. He returned to Bobbs-Merrill after Macmillan did not even recoup the advances paid to him on publication.

The Covered Wagon

The Covered Wagon, the only of Hough's books made into a film during his lifetime, made film history in 1923. It was such a success that it ran for fifty-nine weeks, breaking the previous record run of forty-four weeks established by D. W. Griffith's *Birth of a Nation*.[24] James Cruze, raised in southern Utah, directed the film. He selected a site in Snake Valley, Nevada, to shoot eight weeks on location. He hired three thousand people, including one thousand Arapaho, Shoshone, Crow, and Navajo Indians, and handled the logistics of managing six hundred oxen, five hundred mules, one thousand horses, hundreds of covered wagons, and a tent city of five hundred tents.

Over the Range

Hough was assured of future royalties from the sale of movie rights for *The Covered Wagon* and other novels, but in the summer of 1922, he began to run out of time. He went to the mountains of Colorado to work on a novel about old cattle trails. Falling ill, he moved to Saint Luke's Hospital in Denver in August, where he stayed into September, dictating a novel as he improved. Hough left the hospital September 25 but stayed in Denver. He was now at the top of his game, marked by the opening of the movie made from *The Covered Wagon* at the Criterion Theater in New York in March 1923. The movie, a financial and critical success, was chosen as one of 1923's best pictures. Hough saw the movie not long before he died.

In March of 1923, he entered Saint Joseph's hospital in Chicago and seemed to recover quickly, managing to address the Society of Midland Authors on April 23, 1923. On April 26, he reentered the hospital because of an intestinal obstruction. Following an operation, Hough died of heart and respiratory failure on April 30, 1923. He is buried at Hope Cemetery in Galesburg, Illinois. His estate was probated at $113,857,[25] more than the financial goal he had set for himself. He and his wife had lived together in their magnificent new home for only a month before his death.

NOTES

1. Wylder, *Emerson Hough* (1981); Wylder, *Emerson Hough* (1969), 2. Many of the details of Hough's life in this chapter are taken from Wylder.

2. Hough to Laurance Chambers, February 8, 1909, Emerson Hough Collection.

3. Wylder, *Emerson Hough*, 3 (quoting from Hough's *The Story of the Cowboy*).

4. Sinclair, "Great Bypass," 18–19.

5. Wylder, *Emerson Hough* (1981), 21.

6. Sinclair, "Great Bypass," 19.

7. Keleher, *Fabulous Frontier*, 157.

8. *White Oaks Golden Era*, February 7, 1884.

9. *White Oaks Golden Era*, April 10, 1884.

10. *White Oaks Golden Era*, June 12, 1884.

11. *White Oaks Golden Era*, November 20, 1884.

12. Johnson, "Emerson Hough's American West," 3. Also found at http://www.lib.uiowa.edu/spec-coll/Bai/johnson.htm.

13. Phillips and Axelrod, *Encyclopedia of the American West*.

14. Wylder, *Emerson Hough* (1981), 90–92.

15. Hough, "In the Jewel Box," 84–85.

16. George Bird Grinnell, owner of the magazine *Forest and Stream*, was a famous American naturalist and conservationist, who founded the Audubon Society in 1886. (At different times, Hough later was offered the position of superintendent of Yellowstone Park and Grand Canyon Park. He turned down the offers.)

17. Hough to Samuel Merwin, June 19, 1908, Emerson Hough Collection.

18. Bonney, *Looking over My Shoulder*, 67.

19. Hough, "West after Twenty Years." Emphasis added.

20. Ibid.

21. Hough, *Heart's Desire*, 124.

22. Ibid., 125.

23. Ibid., 156.

24. Johnson, "Emerson Hough's American West," 6.

25. Ibid.

MATTHEW WATSON "WATT" HOYLE
Partner in the Old Abe Mine and Builder of Hoyle House

The lofty Hoyle House continues to dominate the south side of White Oaks, despite being well over one hundred years old. The mansion was built with profits from the Old Abe Mine.

The Old Abe Mine Partnership

The Old Abe mining location northeast of Baxter Gulch was originally discovered in the winter of 1879–80 by J. M. Allen, O. D. Kelley, and A. P. Livingston.[1] Failure to continue assessment work for three years caused them to lose title to the claim. (U.S. law requires annual assessment work on an unpatented mining claim in the public domain in order to maintain title to a claim.)

In January 1884, a trio of lawyers became the new claimants to the Old Abe Mine: Harvey Fergusson, lately of White Oaks and then engaged in a rising political career in Albuquerque; John Hewitt, foremost lawyer of White Oaks; and William Watson.[2] Early in 1890, the three men formed the Old Abe Company to further develop the mine, with Hewitt as president, Watson as vice president, and Fergusson as member.[3]

The company's first order of business was the annual assessment work. Fergusson and Hewitt agreed to put up one hundred dollars each to finance the project, with Watson doing the actual work as his one-third interest. Watson then recruited his recently arrived nephew Watt Hoyle to help. (Matthew Watson "Watt" Hoyle, born in 1860 in Ohio, had been named after his uncle.) In the fall of 1890, Watson gave Hoyle half the available cash and one-fourth of his own one-third interest in the mine, or one-twelfth of the entire interest, in return for his labor. Uncle and nephew must have been struggling

William Watson, 1896. Watson was vice president of the Old Abe Company. Photographer unknown. From *Southwest Illustrated Magazine* 2, no. 4 (April 1896).

William Watson house built in the early 1890s. Watson house was built on Lincoln Avenue and Carrizo Street and still stands today. Photograph by Bob Workhoven. Author's collection.

for money since both worked with pick and shovel on county roads for three days that same fall rather than pay the three-dollar poll tax required for voting.[4]

The Old Abe Proves a Winner

In November, Watson and Hoyle turned their attention to the agreed-upon assessment work. The easiest and cheapest place to do the work was just below the dump from the previous work, in line with the tunnel and about seventy-five feet from the road. The two men sank a ten-foot hole. Continuing until the hole was thirty feet deep, the men piled up a dump on the surface that assayer Morris Parker tested to show an average of twenty-seven dollars per ton.

Using a hand windlass, Watson and Hoyle continued excavation to seventy-five feet, finding the vein increasing in both value and width. They struck pay dirt at the end of 1890, with the recovery of gold ore surpassing

Partners, Old Abe Mine, ca. 1882–1884. Back row, *left to right*: William Watson, John Y. Hewitt, Watson Hoyle. Front row, *left to right*: B. H. Sigafus, Harvey B. Fergusson, B. B. Dye. Photographer unknown. Courtesy Robert and Dorothy Leslie.

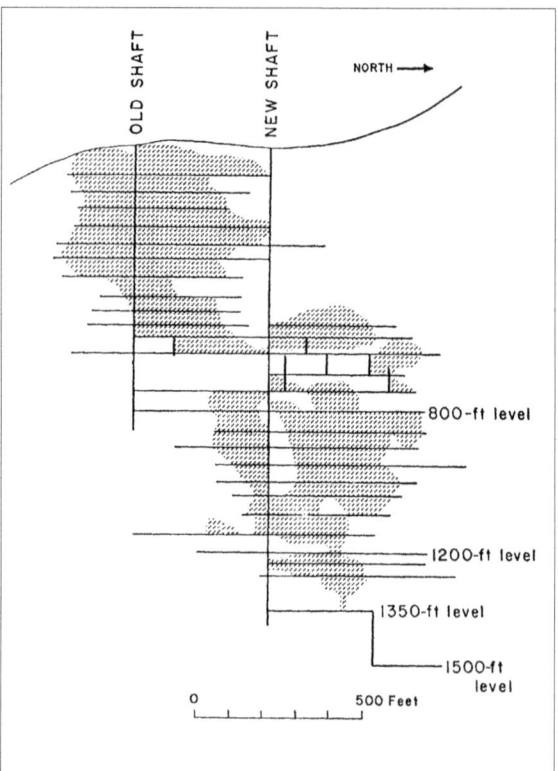

Vertical longitudinal section of the Old Abe Mine. From George B. Griswold's *Mineral Deposits of Lincoln County, New Mexico*, Bulletin 67 (1959), 32. Courtesy New Mexico Bureau of Mines and Mineral Resources, a division of New Mexico Institute of Mining and Technology, Socorro, New Mexico.

View of the Old Abe Mine, ca. 1900. On the left is the stamp mill, which had sixteen-inch-square timbers extending from the ground to the top of the building to hold a bank of stamps. In the middle is the shaft and on the right the cyanide plant. The mill, which produced some $3.5 million in gold, closed around 1910. Photographer unknown. Courtesy Donald M. Queen.

anything known in White Oaks up to that time. From the discovery in December 1890 of the new vein until May 1, 1892, the Old Abe produced 11,064 tons of ore, yielding the sum of $296,817.[5] Of that figure, $113,584 was spent for machinery, mill rental, mining, and other expenses.

In August 1891, the newly affluent Watson wed the daughter of a preacher and arranged to take the stage to Carthage to start the couple's honeymoon. Packed in equal shares among the bride's trousseau in four trunks were sixty-five pounds of gold worth over $16,000 sealed in four packages—one for each Old Abe Mine partner. At Carthage the station agent retrieved the four sealed packages, weighed them on an ordinary platform

Billie Donan at the Old Abe Mill, 1904. Photographer unknown. Courtesy Larue Lane Wetzel.

Mining at the one-thousand-foot level, Old Abe Mine. Standing, *left to right*: Al Curtis, Forrest Smith, Oliver Peaker, George Wilkerson. Sitting, *left to right*: A. P. Green and John Y. Hewitt. Date and photographer unknown. Courtesy Wayne Van Schoyck and Leslye Cooper Van Schoyck.

Head frame of the Old Abe Mine, ca. 1949. *Left to right*: Edward L. Queen, Forrest Queen, Donald John Queen. Photographer unknown. Courtesy Donald M. Queen.

Miners at the Old Abe Mine, ca. 1900. Photographer unknown. Courtesy White Oaks Historical Association.

Glass stamp mill. Date and photographer unknown. Courtesy John and Wesley Mottinger.

scale, and placed them in a Wells Fargo safety container for shipment. It was the largest single shipment of gold out of White Oaks until that time.

With profits rolling in, Watson and Hoyle installed a hoist, enlarged and timbered the shaft, and ran more drifts; now a mill was badly needed. As a stopgap measure, the North Homestake Mine leased its mill for use at the Old Abe; the partners also leased the old Glass ten-stamp mill.[6]

Eventually, both mills treated about fifty tons of ore daily, with net profits approaching $1,000 a day. Then both the Glass and North Homestake mills were closed, and the South Homestake Mill was leased to the Old Abe until its owners could build their own. Watt Hoyle took charge of the new twenty-stamp amalgamation mill erected in April 1893.

During this period, White Oaks quickly reached a hitherto unknown level of prosperity. The *New Mexico Interpreter* reported that "the Old Abe mine is proving a bonanza.... The grade of ore has so far improved that the milling has been brought up to an average of over $60 per ton from the start.... If the mine contained no more than now in sight the four owners [Watt Hoyle was now also a part owner] would have respectable fortunes assured."[7]

In 1898, the mill was expanded to include use of the cyanide process recently introduced to the United States. The most productive era for the Old Abe was from 1890 until 1904. By 1904, workings on the mine had become the deepest in New Mexico, extending to 1,450 feet.[8] Typical of the district's mining geology, the Old Abe never encountered significant water even at its lowest level, making it also the deepest dry mine in the world.[9]

On July 1, 1891, six months after the Old Abe started producing, a mine fire erupted at the South Homestake in which two miners perished and the hoist machinery and shaft timbers were totally destroyed. This fire caused Saint Louis stockholders to abandon their mining investments in White Oaks. Thereafter, local financiers/owners, instead of outsiders, continued mine development and operations from earlier profits.

The Worst Mine Fire in White Oaks History

The worst disaster in the history of the White Oaks mines occurred when the Old Abe hoist house and shaft caught fire and burned on March 9, 1895.[10] On Saturday morning at three o'clock, the entire town was awakened by the blowing of whistles and the firing of guns. Before rescuers could even reach the mine, the shaft house, sawmill, and wood and blacksmith shops were already a mass of cinders, with flames rocketing skyward from the shaft.

Twenty men were working underground at the time; eight managed with superhuman effort to run up the ladders and stagger out of the opening, leaving twelve men trapped below. Four of the remaining men fought their way through an air shaft to the surface.

The first survivor had just emerged from the dressing room and was being lowered in an ore bucket. The bucket had reached the ninth level when the fire was discovered. The hoist man abandoned the bucket and ran, dropping the man two hundred feet to the bottom, where the bucket turned over and ejected its passenger into the

Memorial to the eight men who died in the Old Abe in 1895. Photograph by Bob Workhoven. Author's collection.

Marble steps in front of the Hoyle House, 1959. Note the magnificent stained glass archway over the wide window in the front parlor. This archway was imported from France in the early 1890s. Photographer unknown. Charles Family Papers, Archives and Special Collections, Ms18.366, New Mexico State University Library.

drift. Miraculously, the miner was unhurt. Looking up the shaft, he spied the fire and at once bounded up the ladders, sounding the alarm as he ascended an amazing distance of eight hundred feet to the surface.

The main shaft quickly collapsed. Wives, children, and friends of the remaining eight trapped men gathered around the collar of the shaft to keep a constant vigil as they handed out nonstop rounds of food and hot coffee to rescuers. By evening, the fire had been brought under control enough to allow rescuers to descend into the mine for the first time and look for any survivors. The first body was recovered the following morning from the third level; five others were soon found. The final two bodies were not recovered until late Monday at the thirteenth level, 1,200 feet below the surface. White Oaks hung under a pall of gloom for months. However, work resumed after repairs were made and a mill was finally completed near the mine.

Watt Hoyle Builds a Showplace

Newly flush from his one-twelfth share in the Old Abe in 1892, Watt Hoyle decided to build the finest house in town. Fittingly, he would site his mansion on Grand Avenue. According to the *Lincoln County Leader* of November 1891, Watt's brother William (born in 1856 in Ohio) arrived that same month from Ohio with wife, Louisa, and small daughter Ida. Will hoped that the drier air of the Southwest would cure his wife of consumption. The Hoyles all lived together in John Wilson's brick home while the Hoyle House was being built.[11] Watt Hoyle contracted with the five Gumm brothers of White Oaks to build his mansion. Plans called for a nineteen-room, four-bedroom Victorian masterpiece of epic proportions.[12]

The red-brick Hoyle House. The Hoyle House is on the east side of White Oaks. The small white house in front of the mansion is the old Dave Jackson residence. Photograph by Bob Workhoven. Author's collection.

Still standing today, the mansion features a red-brick exterior with belts of sandstone quarried from nearby hills. The bricks, sun-dried and never fired in a kiln, came from San Antonio, New Mexico.[13] Pillars of sandstone form the major supports and extend into solid bedrock; all twenty-five hand-hewn cornerstones are identical. An entire year was required to cut the stones before building could begin—the Gumms imported master stonemasons from Saint Louis to cut the slabs. Inside, there are both front and back parlors, a dining room, a kitchen, a pantry, an entrance, and back halls. Stairways lead to the upstairs bedrooms and bathroom and to the basement downstairs. Two swing-doors open from the kitchen to the dining room, designed so that servants would not collide.

The front parlor in those days served as a formal place for entertaining guests, while the family used the more intimate back parlor much as modern families use their living rooms or dens. An especially interesting feature of the dining room and the back and front parlors is the "pocket" doors that slide into openings in the wall. A pair of these pocket doors divides the dining room and back parlor; another pair divides the back parlor from front parlor. Between the front parlor and the front hall is a single pocket door. With all doors open, a generous space is created that can accommodate large gatherings.

Double-high windows admit sunshine into every room and afford gorgeous mountain views from all sides. The large window in the front parlor frames a glimpse of Patos Mountain to the southeast. Over this window is an exquisite archway of stained glass. The largest stained glass window was imported from France, with smaller stained glass windows coming from Canning, New York. The front stair landing boasts colored-glass panes, while the plain glass on the back stair landing looks onto views of the town site. Dave "Jack" Jackson, longtime White Oaks resident, said that on the day the large parlor window arrived by freight wagon, Hoyle rounded up fourteen men who happened to be drinking in a bar at the time and promised more drinks in exchange for helping to install the windows. Amazingly, none of the windows were broken during installation.[14]

Fireplace mantel, front parlor of the Hoyle House. This mantel is hand-carved from cherrywood. Photograph by Bob Workhoven. Author's collection.

Hand-carved balustrade on stairs of the Hoyle House. Photograph by Bob Workhoven. Author's collection.

Except for the hand-carved fireplace mantels of cherry wood, the woodwork throughout the house—including the sliding pocket doors—is California redwood. All of the floors are also of redwood with pine subflooring. The edifice also boasted the very first indoor bathroom in the territory of New Mexico, outfitted with a tin bathtub, flush toilet, and marble-topped wash basin.[15] Hoyle also had a large trunk room built and hooks installed in a spacious closet for his mansion's hoped-for future mistress to hang her long floor-length skirts from the ceiling.[16]

As a final touch, atop the house is a graceful observation platform with a metal railing—the southwestern version of a widow's walk, except that the platform is merely decorative and not big enough to sit on. There are no stairs to the platform, only a ladder.

Watt Hoyle never completely finished the interior of the house, nor was it ever completely furnished, prompting locals to dub it "Hoyle's Folly." Work on Hoyle's Folly ended in December 1893 at a total construction cost of $42,000.[17]

Legends Surrounding "Hoyle's Folly"

Many legends surround the mansion. The one most often repeated is that Hoyle built it for a bride-to-be "back East" from Ohio who never arrived; the lady refused to leave her home and journey so far west. It is known that Watt Hoyle returned to Ohio in April of 1891, perhaps to ask the girl of his dreams to marry him. Yet the *Lincoln County Leader* offered a different story: "M. W. Hoyle, one of the Old Abe syndicate, left on a pleasure trip to Ohio this week. He says there is no truth in the report that he is going to import a woman."[18]

A variant of the legend has it that the lady did make the long stage trip to White Oaks, but before she and Hoyle could be married she fell in love with a local cowboy, married her lover, and moved with him to southwest New Mexico.[19] Another tale has a jealous Hoyle killing both the girl and her cowboy and burying them in the basement, with their ghosts haunting the house to this day. Still others maintain that, when his bride-to-be wrote Watt that she was not coming, Watt walked up to the North Homestake and leaped to his death one thousand feet down the main shaft. If any of these rumors are true, the fault possibly could have been with Watt instead of his eastern bride-to-be. Mrs. Kirkpatrick, a resident of the Hoyle House in the 1960s, described Watt as "not a tall, swash-buckling, virile fellow . . . No, not at all. He was short, slight and homely. Not only physically a disappointment; his personality was lacking in most appealing traits."[20] White Oaks citizen Dave "Jack" Jackson said that he was "tremendously overbearing, given to illusions of grandeur. . . . The fact he held this ideal of a fine mansion to befit the girl of his dreams is not a bad thing. Only he didn't have her acceptance to marry him when he began it. It seems it was all in his mind, this enchanted palace, a love nest, an idyll."[21] A latter-day ghost story comes from Clara Snow, former resident of White Oaks. As a child walking past the Hoyle House on her way to school the day Watt Hoyle's sister-in-law died, she swore she saw an angel atop the widow's walk high on the roof.[22]

Watt Hoyle Leaves White Oaks

According to mining man Morris Parker, Watt Hoyle did not immediately abandon White Oaks but continued to live in his house for years, leaving only when it became clear that he needed to seek a living elsewhere.[23] Watt eventually left for Denver when the Old Abe faded and ceased to operate after the turn of the century. He was also heavily in debt; he further mortgaged his mansion to the Exchange Bank in White Oaks for $1,500.[24] Before departing, Hoyle gave the Exchange Bank his prized gold-headed cane as additional collateral. Four gold-headed canes had been made in the heyday of the Old Abe Mine as Christmas presents for the company's four partners.[25] These fancy canes were not made with hooked hand grips for practical walking support but as showpieces. Instead of a separate knob made entirely of hard metal, gold from the Old Abe was formed around the top of the wood cane.

In Denver, Hoyle lost his remaining money to real estate sharks. The last known trace of Watt Hoyle comes from a *Carrizozo News* piece of June 1915 in a reprint of an article from a Denver newspaper. Watt's bad luck with women dogged his steps even in Colorado: "Watt Hoyle, formerly White Oaks mining man, and his bride, Mrs. Sarah Bell Dingey Hoyle, were traveling near Boulder in a car to be married when they were thrown from the car. Both injured, they sat in their wheelchairs and were married."[26]

Hoyle had driven his automobile off a twelve-foot bank. The couple survived, nursing cracked ribs for Hoyle and a broken collarbone for Mrs. Dingey. Undaunted, and attired in hospital "negligees" and

Boston Boy Mine. Date and photographer unknown. Courtesy Larue Lane Wetzel.

sporting black-and-blue bruises in lieu of the traditional bride's blue, the couple wed in the hospital.

Will Hoyle's wife died of consumption on December 31, 1901.[27] Will himself was killed some years later in an accident at the Boston Boy Mine in White Oaks. While he was climbing to the top rung of a ladder, he slipped and plummeted down the mine shaft to his death.[28]

The Hoyle House Today

As of this publication, the Hoyle House is home to Larue Lane Wetzel, granddaughter of the first physician to arrive in White Oaks, Dr. Alexander Lane. Larue spent her early childhood in the Oaks. As a small child, she relocated with her widowed mother to California, grew up, married, and moved with her husband, Charles, to El Paso. When El Paso became an increasingly unsafe place to raise five small boys, Larue and her husband determined to move back to White Oaks.

The Wetzels arrived with an entourage of five boys, a dog, a cat, and horses in 1967. The family camped out until October of that year, when they had the unexpected chance to buy the Hoyle House from the L. E. Kirkpatricks.[29] After 1973, when Charles was killed in a truck accident, Larue kept working as a teacher in the Carrizozo school district. She retired after twenty-one years. Her five boys have become adults while she continues to live in her fabulous mansion. Dave Jackson prophesized on the day Larue had to leave with her mother and sister for California, "Don't cry, 'Rudy,'" he consoled, "you'll get to come home some day."[30]

NOTES

1. Parker, *White Oaks*, 88.

2. Ibid., 88, 89.

3. *Denver Rocky Mountain News*, May 29, 1892.

4. Parker, *White Oaks*, 90.

5. *Denver Rocky Mountain News*, May 29, 1892.

6. Parker, *White Oaks*, 91–93.

7. *New Mexico Interpreter*, April 24, 1891.

8. Griswold, *Mineral Deposits of Lincoln County*, 30; Jones, *New Mexico Mines and Minerals*, 172–73.

9. Parker, *White Oaks*, 127.

10. Charles L. Wetzel, "Mining in White Oaks, New Mexico," undated twelve-page paper in the collection of Larue Lane Wetzel; undated three-page account of the Old Abe fire by Frank J. Sager, bank official of the Exchange Bank in White Oaks, copy in author's collection; Parker, *White Oaks*, 96–99.

11. *Lincoln County Leader*, November 21, 1891. Nettie Lee Lemon, a White Oaks old-timer, said that the Hoyle household eventually came to include a Hoyle sister and Mrs. Hoyle's mother, Mrs. Kominsky.

12. *Mesilla Southwestern Old Times*, April 1971, Mullin Collection.

13. Katherine Blackson, "Mysterious 'Hoyle's Folly' Has Not Died: Fabled Mansion Built for Lost Love," *El Paso Times*, October 11, 1964.

14. Ibid.

15. *Mesilla Southwestern Old Times*, April 1971.

16. Joanna Dodder, "Hoyle's Folly," *Ruidoso News*, October 30, 1995, copy in author's collection, courtesy of Larue Lane Wetzel of White Oaks.

17. Looney, *Haunted Highways*, 64.

18. *Lincoln County Leader*, April 25, 1896.

19. John Kelt to Amanda Treat, January 1, 1970.

20. *El Paso Times*, October 11, 1964, quoting Mrs. L. E. Kirkpatrick, then occupant of the Hoyle mansion in White Oaks. (The newspaper article incorrectly identifies Kirkpatrick as Fitzpatrick.)

21. Ibid. In the article, Mrs. Kirkpatrick relates this story told to her by Dave Jackson of White Oaks. The view that the fiancée moved to southwest New Mexico is supported by a letter dated January 1, 1970, to Amanda Treat from John Kelt: "I am searching high and low trying to find out the name of the woman that Hoyle was supposed to marry when he built the big house, all the stories that have been written about this no where have you ever seen her name. Uncle Herman [Kelt] said that one time many years ago J. C. Reasoner and his wife come [*sic*] to White Oaks from Silver City to make pictures of the old mansion for this lady and that she was teaching school in the Silver city area at that time."

22. According to Larue Lane Wetzel, present occupant of Hoyle's Mansion.

23. Parker, *White Oaks*, 116.

24. Larue Lane Wetzel, undated communication with Robert Leslie of Carrizozo.

25. Larue Lane Wetzel, interview by Haldane, September 1, 2006.

26. *Carrizozo News*, June 9, 1915.

27. Record of Deaths at White Oaks.

28. *El Paso Times*, October 11, 1964.

29. Wetzel, "The House Built for Love."

30. Larue Lane Wetzel, interview by Haldane, May 29, 2001.

WILLIAM AND JOHN HUDGENS
Double Trouble from Louisiana

William "Will" Harrison Hudgens, Jr. (born October 14, 1851), and his brother John Newton Hudgens (born December 12, 1854) were natives of Louisiana.[1] Before the Civil War, the Hudgens family had owned a plantation near Athens in Claiborne Parish in north-central Louisiana, the largest and richest upland parish in that state.[2] The brothers were sons of William Harrison Hudgens, Sr., and Julia Ann Cargill. Siblings included Juliette, Margaret, James Henry, Laura Louiza, and Martha Ann.

In Claiborne Parish, Will must have known Pat Garrett, future sheriff of old Lincoln County in the territory of New Mexico, who would become famous for killing Billy the Kid. Garrett (born in 1850) was a contemporary, and he had come to Claiborne Parish at the age of three after his father moved there from Alabama and bought a plantation in 1853.[3] The odds of same-aged sons of Claiborne Parish plantation owners eleven miles apart getting to know each other in a time of sparse population would have been excellent. Not long after the Civil War, in 1867 or early 1868, the Hudgens clan left the ruined South for good. They headed for Eagle Pass in southwest Texas just over the border from Piedras Negras in Coahuila, Mexico. On April 20, 1868, the elder William Hudgens died in Maysfield, Texas.

In 1872, Will married fifteen-year-old Mary Taylor, a Texan. Two sons were born in Texas to Will and Mary, Willie in 1873 and John in 1876. On November 5, 1876, and still in Eagle Pass, Will's brother John married Mary's fifteen-year-old sister, Margaret Taylor. A daughter, Nora, was born to them the following year in Bandera, Texas. Before long, the Hudgenses, including younger brother James, moved again and headed north to the territory of New Mexico. They probably arrived in Lincoln County early in 1878.

Selman's Scouts Wreck the Old Brewery Saloon near Fort Stanton

Later that same year, Will and John joined forces to open a saloon called the "Old Brewery" just off post limits at Fort Stanton about halfway between the fort and the town of Lincoln. The saloon building had formerly housed a brewery belonging to L. G. Murphy, post trader at Fort Stanton from 1866 to 1873.[4] Operating a saloon made it easy for the Hudgens brothers to become acquainted with many of the men living in or passing through Lincoln County, including the soldiers stationed at Fort Stanton.

In September 1878, with outlaws and hired guns thick on the Hondo, John Selman's gang, Selman's Scouts, rode to the Old Brewery and tried to force Will to buy ammunition for them from the post trader at Fort Stanton.[5] When Will refused, the gang sent four men to the post trader and succeeded in buying bullets there. Colonel Dudley, the post's commanding officer, had the outlaws arrested and marched back to the post trader, where they were made to return the ammunition. Then the colonel kicked the four off his post with a stern warning not to come back.

A man named Dan Lemons led and reconvened the gang nearby in the Old Brewery to lick their wounds and—blaming Will Hudgens for their troubles and fueled by yet more whiskey—went on a rampage. They wrecked the saloon, raided the store's supplies of groceries, and assaulted Will's wife, Mary, her sister, and a man trying to defend the women. Then the outlaws thundered on down to Lincoln to raid and menace the citizens there before riding on. Taking his family to Fort Stanton for protection, Will told Colonel Dudley that he had recognized two Texans in the gang: Reese Gobles, an escaped convict from a Texas penitentiary, and John Selman, alias "John Gunter."[6]

In October 1878, after the dust settled from the attack by Selman's Scouts, John Hudgens, Gus Gildea, and others joined a long cattle drive to help move some of the Beckwith brothers' cattle to Fort Stockton in Texas.[7]

Troubling Times

More troubles arrived later on in 1878. On December 13, former Lincoln County sheriff John Copeland shot and wounded nineteen-year-old Johnny Mace (Juan Mes/Maés) but was acquitted after appearing before justice of the peace Juan Batista Wilson.[8] Will then swore in an affidavit on December 16 that a mob was planning to take Mace and John Hurley, who was nursing the boy to health, and lynch them. Acting on this tip, Colonel Dudley brought both Hurley and Mace to his fort for protection. Mace recovered from his wounds and turned to stealing horses. A little over a year later, Mace repaid Will's kindness by robbing his store. Deputy Sheriff Tom Longworth led a posse that killed Mace before he could flee the country.[9]

At some point during the Lincoln County War, Will Hudgens made the acquaintance of Billy the Kid and gained the Kid's trust. During negotiations for a mock arrest between the Kid and Gov. Lew Wallace in March 1879, after which the Kid would provide information to the governor in secret, the Kid sent a messenger with a note to Juan Batista Wilson in Lincoln, saying, "it may be he [Governor Wallace] has made different arrangements if not and he still wants it the same to Send 'William Hudgins' [Hudgens] as Deputy, to the Junction tomorrow . . . with some men you know to be all right."[10] The governor replied in a note on which he had crossed out the following line: "If you still insist upon Hudgins, let me know."[11] The Kid appeared at the prearranged meeting place for a surrender and mock arrest. He and Governor Wallace conferred for some time. Possibly Will had extricated himself as go-between before the Kid was arrested on March 23; his name appears no more in these negotiations.

During this time, John was facing his own difficulties. A grand jury under foreman Isaac Ellis convened April 17, 1879, and, among other business, charged three men as horse thieves: John Hudgens, Tobe Hudgens, and Robert Henry.[12] All three eluded the sheriff, and Tobe Hudgens was not heard from again in Lincoln County. (Tobe, a cousin of Will and John's, was about sixteen years old.) James Hudgens, Will and John's younger brother, died May 6, 1879, under

William Hudgens, ca. 1880. Photographer unknown. Herman B. Weisner Papers, Archives and Special Collections, Ms249, PB37E, New Mexico State University Library.

unknown circumstances while a prisoner in the Fort Stanton prison and is buried in the private Fort Stanton cemetery.

John found himself in trouble with the law again in July 1879. The territory of New Mexico brought charges against him on July 25 for grand larceny (what he is accused of stealing is not listed). John, Will, and J. S. Redman posted John's bond, and there is no record of this case ever coming to trial.

From Fort Stanton to White Oaks and the Pioneer Saloon

By now Will Hudgens had more pressing concerns. In the company of another twenty-nine-year-old—the same Jim Redman who helped bond out brother John—Will went over the mountains to White Oaks to try his hand at gold mining at the start of its gold boom. Redman, in a letter some time later to a local White Oaks newspaper, said, "I came to what is known as White Oak's Spring in the fall of 1879, in company with William H. Hudgens and a bottle of whiskey. . . . We visited the now famous Baxter Gulch and found imbedded [engaged in mining] in the side of a hill...Jack Winters, better known as 'old blue skin.'"[13]

Partners Hudgens and Redman soon thought up a better way to get their hands on White Oaks gold—mining the miners. Redman explained, "In the spring of '80, Mr. Hudgens and myself put up a saloon on the flat where White Oaks stands now, and . . . stocked it with Dowlin & DeLany's tanglefoot and pure Havana cigars . . . and I tell you we did a rushing trade. But the town commenced to grow and we concluded to build the old Pioneer Saloon."[14]

The partners moved into their brand-new, two-story building in May 1880.[15] George Curry (later a territorial governor of New Mexico) said of the Pioneer Saloon, "The principal saloon, a palace of lights, mirrors and boasting a long mahogany bar, was owned by W. H. Hidgins [*sic*]."[16]

Will, Redman, and J. A. Sweet continued to invest in mines. Early on they put money in the Little Mack and Little Nell gold mines.[17] The *Las Vegas Daily Optic* noted in April 1880, "Hudging [Hudgens], Redman & Co. are jubilant over the rich finds in the Large Hopes and No Man's Friend."[18] A reporter identified only as "S"

Hudgens brothers' ad for the Pioneer Saloon in the *Golden Era*, December 6, 1883. Author's collection.

> HUDGENS'
> PIONEER SALOON,
> AND CLUB ROOMS,
> Corner Placer Sree and White Oaks Avenue
> All Kinds of Drinks In Season,
> BEST OF WINES,
> BOURBON WHISKYS,
> CHOICEST CIGARS
> Saloon Refitted in First-Class Style.

told the *Daily Optic* a short time later that "four different parties struck free gold yesterday . . . [including] the Large Hopes, owned by J. A. Sweet, Hudgens and Redmond."[19] Life in White Oaks began to look better and better. A son, John William, was born to John Hudgens in White Oaks on June 16, 1880. However, Will would soon encounter trouble.

Will Kills I. T. McCray, a Renegade from the Cherokee Nation

On May 31, 1880, soon after the Pioneer Saloon opened, Nogal prospectors Joseph Askew and Virgil Cullom rode in to White Oaks from their camp outside town and became extremely intoxicated.[20] They began shooting off their pistols into signboards, houses, and, worst of all, into a boarding house filled with women and children. In retaliation, the townspeople formed a line of battle and returned fire on the mounted pistoleers, wounding young Askew, who fell from his horse after a bullet shattered his right arm. Askew was taken to a tent in town and Doctor Lane called to treat the wound. (Askew recovered and would later become a prominent member of John Kinney's gang of rustlers in southern New Mexico.) Joel Fowler, another White Oaks saloon owner, was quickly deputized to quell the uproar. He pursued Cullom and shot him in the right lung. Cullom died early the next morning, after confessing that he was so drunk he remembered nothing of his actions the previous night. Meanwhile, Will Hudgens ran into Fowler's saloon and asked for a gun. Two pistols were available inside. A man named Alex Colvin, claiming that one of the guns belonged to him, refused to allow Hudgens to take it, whereupon Hudgens shouted at Colvin that he was no better than the men who had shot up the town. Hudgens grabbed another pistol and ran out of the saloon.

While Will was trying to obtain a weapon, one I. T. McCray showed up inside the tent where a crowd had collected to gawk at the wounded Askew. When Hudgens also entered the tent, McCray warned him that an insult to Colvin was an insult to him and dared Will to fight.

Mistakenly thinking McCray meant to have a fistfight, Will started to unbuckle his gun belt when McCray jerked out his pistol and fired at Will five times, missing every shot. Will was much more accurate; he raised his gun and put a bullet through McCray's heart, killing him instantly. Colvin ambled over from the bar to tell Will that if he had taken McCray alive he would have been paid $1,600 in reward money from the federal government; McCray was wanted for killing a U.S. marshal in the Cherokee nation. Will was later tried for the killing and acquitted on a plea of self-defense.[21]

Chasing Billy the Kid at Coyote Springs

Will was soon appointed deputy sheriff. On November 22, 1880, at the request of new sheriff, and possible old friend, Pat Garrett, Hudgens raised a posse of eight men that included brother John, Jim Carlyle, and James Bell to capture Billy Wilson, Billy the Kid, and Dave Rudabaugh. The Kid's gang was being sought in connection with the theft of horses from merchant Alexander Grzelachowski at Puerto de Luna and from John Bell on the outskirts of White Oaks.[22]

The Kid had trusted Will Hudgens as the only man he would surrender to in negotiations with Gov. Lew Wallace only a year earlier; now Will was heading up a posse to capture the outlaw. After a fruitless search at Blake's sawmill near White Oaks, that evening the posse surprised the Kid's gang at Coyote Springs. In a desperate fight, both Billy Wilson and the Kid had their horses shot from under them. John Hudgens's horse was also shot and killed, but John's life was saved when the horse raised its head and took the bullet meant for John.[23] The Kid's gang broke up, abandoned everything at camp, and fled through the night in bitter cold—Wilson and the Kid were on foot, the others on horseback. The Kid was forced to make his way without overcoat or gloves.

The posse looted the outlaws' camp and made off with most of their camp equipment. White Oaks blacksmith Jimmy Carlyle also apparently helped himself to the Kid's brand-new gloves left behind in the confusion and drew them on in the cold.

Later that night the Kid's gang slipped into White Oaks and put up at West and Dedrick's livery stable. The next night, mounted again on horses, the outlaws emerged to hurrah the town. They shot at and missed Jim Redman in front of Hudgens's saloon as they left.[24]

Shootout at the Greathouse-Kuch Ranch

The gang rode off to Greathouse-Kuch ranch about forty miles away. On November 27, a second pursuing posse was raised from White Oaks that also included the Hudgens brothers. The posse fought a daylong battle at the Greathouse-Kuch ranch during which one of their own was killed: Jim Carlyle, a popular young blacksmith from the Oaks.[25]

Will Hudgens played a key role in the Greathouse affair; it was he who had sent Joe Steck (teamster and cook for Greathouse and Fred Kuch) into the tavern with a note demanding surrender.[26] After Carlyle's death, however, the posse lost heart and withdrew.

Will Is Elected to the White Oaks City Council; John Leaves for Tombstone

Amid all the chaos, there was a town to be organized. The *Daily Optic* reported a new structure for White Oaks with J. S. Redman elected as mayor, and William Hudgens, Jimmy Dolan of Lincoln, John Hudgens, and others named to the city council.[27] Town lots were laid out.

Sometime in 1881, John Hudgens traveled over to Tombstone, Arizona. He thought better of moving there, however, because the *Tombstone Daily Nugget* reported: "John N. Hudgens left yesterday morning for the new mining camp of White Oaks, New Mexico, where he has several claims, and also town property."[28]

Ranching outside White Oaks

In 1883, the Hudgens brothers bought a ranch around Coyote Canyon and Red Lake eight to ten miles northwest of White Oaks.[29] They first started out raising sheep in August 1883, but in February 1884 they decided to go into the cattle business.

Will traveled to Fort Worth and other places in Texas seeking good cattle for the ranch. In Colorado City in the Texas Panhandle, he bought 4,500 head of cattle at eighteen dollars per head.[30] More than 1,000 of the head were steers two to four years old. The plan was to place the steers on Will's ranch in the Sacramento Mountains and the rest of the cattle on the White Oaks ranch.

These were relatively calm years for the Hudgenses. In March 1884, the *Golden Era* reported that Will was planting three hundred cottonwood trees on his home ranch.[31] He and his wife entertained in style at their ranch, as the *Golden Era* also reported in March, "A large number of the friends of Mr. and Mrs. Will Hudgens were heartily welcomed by them at their ranch on the evening of the 22d and made the recipients of hospitalities rarely met with in this wild, half-civilized country.... Dancing and whist were the chief amusements until 10 o'clock, when supper, consisting of an endless variety of cakes, pies, cold meats, salads, tea, coffee, wine, etc., was served, and much to the credit of the hostess, which was prepared entirely by her own hands."[32]

Will and John worked to build up their herd of cattle

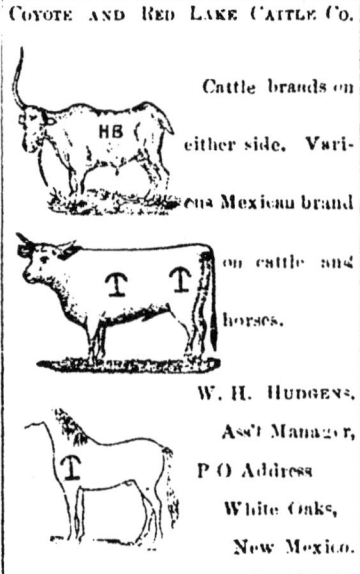

Brands for the Coyote and Red Lake Cattle Company in the *Golden Era*, July 30, 1885. Author's collection.

and to stock the Red Lake range. The *Golden Era* noted in June 1884, "Two new brands appear on the second page of this week's paper. Hudgens Bros., with their brand of HB and post office address at White Oaks, are well known in Lincoln county.... All cattle on Hudgens Bros.' range are first class and they won't have any other kind."[33]

Something went awry with the cattle business soon after, though, because in July 1884 Will placed a notice in the *Golden Era* announcing that he had sold his entire brand of cattle to the Coyote and Red Lake Cattle Company.[34] However, he said that he would assume full management for the company during the absence of a Dr. Wm. Y. Provost (apparently the buyer).

Yet the *Golden Era* in July 1884 reported that John was in Texas buying cattle for *his* ranch in Lincoln County. One month later the same newspaper observed, "J. N. Hudgens has returned from Texas where he has been making arrangements for cattle to stock his ranch in this county. He secured 500 head which are now on the trail and will reach this point about Sept. 10th."[35]

In August, there was further news of Will: "W. H. Hudgens ... arrived at the hub [White Oaks] Saturday evening. Will is one of the pioneers of the Oaks and a betterhearted gentleman never lived. He purchased Isaac Ellis' fine bay two-year-old colt.... Fine stock is his hobby and he is rapidly filling up his range with imported cattle and horses."[36] Will continued as manager of the Coyote and Red Lake Cattle Company for a time; the *Golden Era* chronicled him visiting the big cattle markets of Kansas City and Leavenworth in September 1884.[37]

John Escorts Prisoners to Kansas

Meanwhile, John was having his share of adventures. In early 1884, the *Golden Era* reported that, as appointed lawmen, John and his partner, Brent, were escorting a group of territorial prisoners by train. When the train stopped for supper at Coolidge, Kansas, a Kansas cowboy named Bill Logwood boarded the train as the two deputies were eating supper at the station and began abusing the prisoners. He said that he would "shoot the handcuffs off of 'em and then lick 'em" and that he "could shoot the lights of the car plum out the first round."[38] In the midst of this harangue, Brent and Hudgens stepped back into the car holding the prisoners, quickly grasped the situation, and just as quickly captured the cowboy. That night and the next day the boastful young man was said to be as quiet and inoffensive as any cowboy ever seen.

Lawsuits Galore

Life never permanently settled down for the Hudgens brothers. In February 1883, the future first governor of the state of New Mexico, W. C. McDonald, brought a criminal action suit against Will before W. Blanchard, White Oaks justice of the peace.[39] McDonald claimed that Will had given him cause to fear being wounded, killed, or otherwise harmed by personal violence. A warrant was issued for Will's arrest; he showed up in court later that day and posted a $500 bond. Nothing more was heard of the matter.

In late 1884, Will was again summoned to court, this time by the territory of New Mexico.[40] On December 9, George Ulrick charged that Will had challenged James A. Alcock (manager for an English syndicate of the Carrizozo Cattle Ranche Company formerly owned by L. G. Murphy) to fight a duel with deadly weapons. Dueling was now illegal in the territory. Will appeared in court the next day and, after posting a $4,000 bond, went about his business.

In August 1885, Will was subpoenaed on behalf of the territory of New Mexico as a witness against James A. Alcock of the Carrizozo Cattle Ranche Company near White Oaks.[41] Alcock had been charged with willfully and maliciously intending to "kill, maim, and disfigure sheep belonging to Dionicio Chávez." Will was cited for contempt after failing to show up as a witness for the prosecution. He finally appeared in court on September 14 and, after posting a penalty of $200, was ordered to appear again in court on September 30. At this session, he was found in contempt and fined $10, in addition to having to forfeit the original penalty of $200. Justice Collier in White Oaks dismissed all three of the separate cases on October 30, 1885. At this session, the judge also ruled that Will was no longer willful and insolent and discharged him.

Two short months later, Will was severely injured when a horse fell on him on the Carrizozo range.[42] By the spring of 1887, still recuperating from his injuries, Will opened a grocery store with a partner named Kelley. In the *Lincoln County Leader*, the partners promised "to keep in stock everything in the line of Groceries, which we will sell at prices defying competition. California Fruit A Specialty. GIVE US A TRIAL."[43]

John Hudgens Kills Louis Monjeau

John had his turn in court as well. On January 5, 1885, he was summoned before justice of the peace J. B. Collier in White Oaks[44] to answer a charge that he had shot and killed Louis N. Monjeau, whom the *Golden Era* newspaper described as the "fat, hearty and happy mayor of Manchester" (a small town a few miles west of White Oaks).[45]

A liquored-up Monjeau took the opposite side in a quarrel with his good friend, an equally drunk John Hudgens. Arming himself with a Winchester, Louis approached the Homestake Saloon, where John was, and, leaning his gun against the side of the building, opened the door and called John out. As John stepped outside, Monjeau reached for his gun to shoot at John when John quickly drew his revolver and shot Monjeau. The ball entered to the right of Monjeau's mouth, passing through and breaking his neck a little below the base of the brain. John was arrested by Deputy Sheriff Charles Bull and attended a hearing, where his bond was fixed at $500. He was bound over for trial in May. Monjeau had been known as one of the most peaceable, inoffensive citizens of White Oaks (drunk *or* sober), and the community was deeply saddened by his death. John was acquitted. Today, Monjeau Lookout in Lincoln National Forest memorializes the young man he killed.

On June 25, 1885, a second son, Constant James, was born to John and Margaret in White Oaks.

Will's Son, King, Drowns in the Rio Grande

The *Lincoln County Leader* of July 7, 1888, related another incident in Will's turbulent life from Socorro, where the Will Hudgens family was now living: "Sometime yesterday afternoon, King, the 12 year old son of Mr. Wm. H. Hudgens, of this city . . . went down to the Rio Grande to bathe in some of the pools adjoining the river . . . the treacherous stream soon drew him into a hole, and . . . he was at the mercy of the waters and finally disappeared."[46]

A crowd gathered to search for the boy, but he was not found until the next morning when the grief-stricken father entered the river at the point where his boy had disappeared and recovered the body.

Will Kills Again

It is not known how long Will Hudgens stayed in the grocery business in White Oaks. He can be placed in Socorro by the newspaper account of his son's drowning in 1888, but he had relocated to El Paso by September 1891. In 1892, he made his way to the southern New Mexico silver-mining town of Hillsboro and opened another saloon called the Long Branch. In the Long Branch on the night of February 18, 1892, Will shot a Hispanic named Alfonso Ayón.[47] Ayón died of his wound almost a week later. The cause of the shooting is by now familiar. A party of drunks began fighting in the Long Branch, and Will threw them out. When an angry mob marched on the Long Branch, Will pulled his gun and fired into the rioters; the bullet found its mark just over Ayón's eye. Will gave himself up immediately after the shooting and never stood trial.

The End of a Turbulent Life for Will

Will died in Hillsboro on October 11, 1894, at the age of forty-four and is buried in the local cemetery beneath an imposing white marble monument.

Monument to Will Hudgens in Hillsboro Cemetery. Photograph by Bob Workhoven. Author's collection.

According to the *White Oaks Eagle* of October 4, 1900, a "Mrs. Hudgins [*sic*]" still lived in Lincoln County.[48] It is not known whether this was Mary, Will's widow.

John Hudgens in Nevada and Arizona

In April 1887, John moved his family to Pioche, Nevada, by covered wagon and on horseback, crossing the Colorado River by ferry. The family lived in Nevada through 1890 but returned to Arizona for several years.

The John Hudgens family was living at McCabe near Prescott as of June 20, 1898, but by 1900 lived in Mayer, Arizona. May of 1900 found John N. and son John W., along with Joe Askew and John Kinney, as owners of the Yeager mines near Mayer. This is the same Askew who twenty years earlier had shot up the town of White Oaks with Virgil Cullom and who, on recovering from his wound, joined the John Kinney gang of rustlers in southern New Mexico. Kinney, the formerly dangerous "King of the Rustlers," operated a stock theft ring out of his ranch west of La Mesilla in the late 1870s and early 1880s. John Hudgens must have befriended Askew and Kinney (the latter's three years in prison apparently had convinced him to go straight). The partners sold the Yeager mines to a group headed by Richard Sloan, later a governor of Arizona, in 1900.

Hudgens and Askew then bought mines in the Black Mountains near Jerome, Arizona. June of 1900 found the

John Kinney's headstone in Prescott, Arizona. Date and photographer unknown. Courtesy Troy Kelley.

John N. Hudgens family living at 213 Mt. Vernon Street in Prescott; John was a mine operator. In 1903, John moved from Arizona first to Pioche and then to Ely in White Pine County, Nevada, location of one of the world's largest copper pits. The family lived in Ely for two years.

In 1897, ore had been discovered near what would later become Searchlight, Nevada, a small mining town an hour's drive south of Las Vegas and about fifteen miles from the Colorado River. John moved his family for the final time, arriving in Searchlight in 1905.

The elder John Hudgens lived the rest of his life in Searchlight. In 1909, he was severely injured by being thrown from a wagon but recovered. As he had since his White Oaks days, John continued to lease mining properties and finally found his bonanza in 1917 with the Quartette Mine, the successful mine that purportedly created Searchlight. Leasing part of this mine, he made about $40,000 that year working the surface east of the mine's air shaft.

John lived on in Searchlight until December 16, 1923. His wife died at the age of seventy-eight in 1940 and is buried in the Searchlight cemetery next to her husband.

A purported photograph of young John Kinney. Date and photographer unknown. Courtesy Chris Harkey.

Headstone for John N. Hudgens's grave in Searchlight, Nevada. Photograph by Bob Workhoven. Author's collection.

NOTES

1. U.S. Bureau of the Census, *United States Census of 1880*, White Oaks.

2. The author acknowledges a profound debt to Troy Kelley of Johnson City, New York, for making his research about the Hudgens families available both before and after their White Oaks years. His detective work unearthed the origins of the Hudgens family in Louisiana; their exodus from Louisiana to Texas; the location of Will Hudgens's grave; and the complex movements of John N.'s family back and forth between Arizona and Nevada.

Kelley is writing a book that chronicles the lives of John Newton Hudgens and his son, John William "Jack," born in White Oaks in 1880. Besides continuing the mining interests of his father throughout life, Jack became a noted and respected lawman in Jerome, Arizona, and Ely, Nevada. He led the typically turbulent life of a western lawman in his era and killed his share of badmen. In 1918, Jack also volunteered for the draft in WW I, despite being about thirty-eight years of age, and fought in France in the Argonne Forest, where he was wounded.

3. Metz, *Pat Garrett*, 5.

4. Note from Chris Harkey of Carrizozo, n.d., copy in author's collection.

5. Fulton, *History of the Lincoln County War*, 291. Fulton says that Shepard, who tried to defend the Hudgens women, got hit over the head with the butt of a revolver.

6. Ibid., 291–92.

7. From Troy Kelley's "Timeline" for the two John Hudgenses, in Hudgens family genealogy. Gildea was alleged to have been part of the John Selman gang at that time.

8. Rasch, "They Fought for 'the House,'" 148.

9. Ibid.

10. Nolan, *West of Billy the Kid*, 196.

11. Ibid.

12. Rasch, "The Governor Meets the Kid," 223.

13. "W. J. Woodland" (alias of Jim Redman), 23. Jack Winters, in partnership with John E. Wilson, was one of the original discoverers of White Oaks gold. The two men located a mining claim they named the Homestake. After the two severed partnership, Winters took the north half of the claim and renamed his property the North Homestake. Soon after the original, or undivided, Homestake was discovered, the stampede to White Oaks began.

14. "W. J. Woodland," 23.

15. *Las Vegas Daily Optic*, May 25, 1880.

16. Curry, *George Curry*, 60.

17. Rev. Stanley Crocchiola, *Santa Fe Register*, September 30, 1949.

18. *Las Vegas Daily Optic*, April 24, 1880.

19. *Las Vegas Daily Optic*, June 24, 1880.

20. *Las Vegas Daily Optic*, June 7, 1880.

21. *Las Vegas Daily Optic*, June 21, 1880.

22. Rasch, "Amende Honorable," 59–60.

23. Rasch, "Gunfire in Lincoln County," 237.

24. Ibid.

25. Ibid., 238; Rasch, "Amende Honorable," 59–60, 101; "He Rode with the Kid."

26. Rasch, "Amende Honorable," 59.

27. *Las Vegas Daily Optic*, May 17, 1880.

28. *Tombstone Daily Nugget*, September 28, 1881. From the research of Troy Kelley.

29. *White Oaks Golden Era*, August 2, 1883.

30. *White Oaks Golden Era*, February 21, 1884.

31. *White Oaks Golden Era*, March 18, 1884.

32. *White Oaks Golden Era*, March 27, 1884.

33. *White Oaks Golden Era*, June 5, 1884.

34. *White Oaks Golden Era*, July 17, 1884.

35. *White Oaks Golden Era*, July 3, 1884, and August 7, 1884.

36. *White Oaks Golden Era*, August 28, 1884.

37. *White Oaks Golden Era*, September 18, 1884.

38. *White Oaks Golden Era*, February 21, 1884.

39. Justice of the Peace Court Records for White Oaks, 1882, Criminal Action No. 2, *The Territory of New Mexico, W. C. McDonald, Plaintiff, vs. W. H. Hudgens, Defendant*.

40. Justice of the Peace Court Records for White Oaks, 1884, Criminal Action No. 42, *The Territory of New Mexico, Plaintiff, vs. W. H. Hudgens, Defendant*.

41. Justice of the Peace Court Records for White Oaks, 1885, Criminal Action No. 59, *The Territory of New Mexico, Plaintiff, vs. James A. Alcock, et al., Defendants*.

42. *White Oaks Golden Era*, December 31, 1885.

43. *Lincoln County Leader*, April 30, 1887.

44. Justice of the Peace Court Records for White Oaks, 1885, Criminal Action No. 42. *The Territory of New Mexico, Plaintiff, vs. John N. Hudgens, Defendant*.

45. *White Oaks Golden Era*, January 8, 1885.

46. *Lincoln County Leader*, July 7, 1888.

47. *Lincoln Independent*, February 26, 1892, reporting the Hillsboro shooting.

48. *White Oaks Eagle*, October 4, 1900.

ANDREW "ANDY" HUDSPETH
Foremost Legal Mind in the Southwest

Andy Hudspeth was born October 23, 1874, in Fannin County, Texas, to John and Emily Hudspeth. He was educated in the public schools of Greenville, Texas.[1] In 1890, after graduating from Greenville's business college, he began a law course at Cumberland University in Lebanon, Tennessee. Four years later he was admitted to practice law at Greenville, Texas. In December 1894, at the age of twenty, he came to Lincoln County and hired on as secretary to the Cree family of Scotland, owners of the VV Ranch at Angus. His tenure on the Angus VV Ranch enabled Hudspeth to become acquainted with many cattlemen and cowboys.

Early on, Hudspeth caught the eye of politicians and businessmen in southern New Mexico as a promising young up-and-comer. In April 1897, he was invited to join the Eddy brothers, entrepreneurs from southern New Mexico, in helping to host a Wild West roundup for a group of capitalists from the East. The Eddy brothers hoped to persuade the easterners to finance a railroad they planned to push from El Paso to White Oaks.

Hudspeth and others organized a cavalcade north from El Paso through the southern part of New Mexico Territory. In this venture, he joined prominent New Mexicans Oliver Lee, Albert Bacon Fall, George Curry, and William Hawkins. The group succeeded in securing backing for the railroad.

The Angus VV Ranch of the Cree Family of Scotland

The history of the VV Ranch goes back to the early 1870s when the United States under President Grant issued patents to Paul Dowlin, covering what later became known as the Cree ranch.[2] Dowlin died in 1877, leaving his estate to three minor daughters and appointing brother William as executor. William also died soon after, and the entire estate was sold, some of it to Frank Lesnet. Lesnet in turn sold 880 acres to James Cree of Edinburgh, Scotland. Cree added to his original acreage in 1892 when, after President Harrison issued a patent to Jesse Greenlee, the Greenlee patent was incorporated into the Cree ranch.

The Cree land, at that time one of the most important ranches in New Mexico, went from the common boundary of the Lincoln National Forest and the Mescalero Indian reservation south and west, with the forest land continuing on east into the site of present-day Ruidoso; north to Carrizo Mountain; and from the Indian reservation to south of what was in 1947 the Palmer Gateway Subdivision. The old VV Ranch headquarters was located about five miles from Ruidoso. The Crees shipped cattle all the way from Scotland to

Andrew H. Hudspeth as a young man. Date and photographer unknown. From *1913 Yearbook, Lincoln County, New Mexico* (publisher John Haley, Carrizozo, New Mexico). Courtesy Robert and Dorothy Leslie.

the railroad terminal near Socorro. Then the cattle were unloaded and brought the rest of the way to the ranch.

The elder James Cree died not long after obtaining his ranch, and, after a lengthy probate, his son James Edward took over the property. The Crees brought furniture and personal effects and established a new Scottish "kingdom" for themselves right in Lincoln County. The VV Ranch stayed in the Cree family longer than any other such land in New Mexico until 1939, when the heirs of James Edward Cree deeded the ranch to a corporation controlled by a wealthy New York investor named Arthur Kudner. Kudner and Hudspeth eventually became close friends.[3]

The Cree Meadows Golf Course at Ruidoso owes its existence to James Edward Cree, who brought his love for the game with him from Scotland. There were large corrals nearby to lock up cattle and horses overnight. Also on Cree land was Dowlin's mill for grinding corn and wheat, still to be seen in the heart of Ruidoso.

Andy Hudspeth worked as secretary and bookkeeper for the Crees until 1902. Later, with a reputation as one of the best legal minds of the Southwest, he served as attorney for the original lands bought by the Crees and continued as attorney through each new owner until his death.

Political Life in White Oaks

In 1902, Hudspeth moved to White Oaks and a law partnership with Judge John Hewitt that lasted until 1913 when Hudspeth transferred his law practice to the county seat of Carrizozo.[4] The two continued as partners unofficially until Hewitt's death in 1933.

Hudspeth was an ardent Democrat and represented his district for one term in the territorial legislature in 1907.[5] He also served as delegate to the 1908 Democratic National Convention.[6] In 1910, he was elected a delegate to the New Mexico State Constitutional Convention, New Mexico's final step on its march toward statehood. The election for constitutional convention delegates was held September 6, 1910, with Republicans capturing seventy-one slots and Democrats twenty-eight. (The one-hundredth delegate was a Socialist.)[7] Hudspeth and another ambitious Democrat, William C. McDonald, both hailed from White Oaks. McDonald began his career as a surveyor and lawyer in White Oaks and later acquired the land holdings of both the Bar W Ranch and the famous Block ranch. In 1912, as chairman of the Democratic State Central Committee of New Mexico, Hudspeth helped McDonald win election as first governor of the state of New Mexico (though Republicans continued to control both branches of the legislature).[8]

Hudspeth remained very active for years in the Democratic State Central Committee in Santa Fe as a campaign fund-raiser for both national and state elections. In 1916, he raised about $20,000—$5,000 of which was earmarked for return and use in the New Mexico state Democratic election campaign.[9]

The Navajo Revolt of 1913

In 1913, Hudspeth was appointed first U.S. Marshal of the state of New Mexico, with headquarters in Santa Fe. He served as marshal for eight years.[10] After only a few days in office as marshal, Hudspeth faced an uprising of some 150 Navajo Indian braves.[11] The revolt was said to have been sparked by the strict policies of W. T. Shelton, superintendent of the Shiprock Indian agency. Shelton had become detested because of his policy of separating Navajo children from their parents to attend the agency school at Shiprock, and he opposed Navajo polygamous marriages. Shelton's Navajo police would forcibly drag children out of their hogans and take them to Shiprock tied to a horse if need be. The outraged parents viewed these tactics as kidnapping.

The final straw for the Navajos came when one of their respected medicine men, a polygamist with three wives, was out hunting. The police arrived at his hogan and forced Hatotcli-Yazzie's three wives into agency headquarters at Shiprock. Yazzi asked his father, an aged medicine man named Bizhoshi, for advice. On September 17, 1913, Bizhoshi, his son, and nine other Navajos armed themselves and went on the warpath. They rode in on horseback and swept down onto the Indian agency at Shiprock while firing their guns into the air. As they freed the three wives, they struck an agency Indian police officer on the head with a quirt and then burst into classrooms, searching for Navajo children. Fortunately, no one was seriously injured. That night the eleven raiders were joined by many more Navajos and staged a victory dance on the banks of the San Juan River. They became further emboldened when Bizhoshi told them that the U.S. Army was "dead." Superintendent Shelton rushed back to Shiprock from a meeting in Colorado and secured indictments against the eleven raiders for rioting, carrying deadly weapons, assault, and larceny.

Marshal Hudspeth dispatched a deputy to Shiprock, who located two of the defendants. The other nine refused to surrender and led a large force of Navajo men and women, with some of their livestock, to Beautiful Mountain, thirty miles south of Shiprock. There the Navajos, many armed with weapons, defied Hudspeth and threatened to go on the warpath again.

Wild rumors about what the Navajos were up to panicked farmers and ranchers in the Four Corners. Shelton pleaded for U.S. troops. On November 19, three hundred officers and men of the Twelfth U.S. Cavalry from Fort Robinson, Nebraska, boarded a special troop train to Gallup, New Mexico. The train arrived in Gallup the night of November 23 along with horses, mules, supply wagons, Gatling guns—and a brass band. The cavalrymen camped for the night. The despised Shelton was ordered to stay behind at Shiprock, and the next day the soldiers left on a ninety-mile march north to Beautiful Mountain, their movements hindered by mud, rain, and snow.

General Hugh Scott, a seasoned Indian fighter, had arrived in Gallup in advance of his troops. Now he rode ahead of the troops in an automobile until it became mired in mud, then mounted a horse to finish the march. Marshal Hudspeth and J. R. Galusha, one of Hudspeth's deputies, accompanied General Scott ahead of the troops to meet with the Navajo leaders. The trio and their aides reached Beautiful Mountain on November 26 and camped at a nearby trading post. Father Anselm Weber, a Franciscan missionary to the Navajos at Saint Michaels in Arizona for fifteen years, also arrived along with some Navajo interpreters to assist.

Bizhoshi sent word to General Scott that he and his followers would come down off the mountain the next day to air their complaints. He asked only that the general and the others attend the meeting unarmed. November 27 happened to be Thanksgiving Day, and Scott ordered a feast prepared for the rebels at the trading post in advance of the talks. (He also handed Hudspeth and Galusha loaded pistols to conceal in case they suddenly found themselves in danger.) Chee Dodge, a highly respected Navajo-Mexican, would act as interpreter. Bizhoshi began talks defiantly, and no progress was made the first day. On the second day, General Scott adopted a mild attitude toward the rebels but reminded them that they had no choice but to obey the laws of the United States. It worked; Bizhoshi and seven of his followers turned themselves over to face charges.

The prisoners were taken to Santa Fe, where the charges were reduced to one charge of unlawful assembly. The polygamist Hatotcli-Yazzi received a thirty-day jail sentence, to be served in the Gallup jail; Bizhoshi and four others received ten days. This bloodless uprising became known as the Navajo Revolt of 1913. It was the last time in U.S. history that military forces were dispatched in large numbers against Native Americans.

In 1916, J. R. Galusha, now serving as chief of police at Albuquerque, wrote to Hudspeth, enclosing a campaign contribution to the Democratic campaign. Because of an ongoing conflict with Albuquerque's mayor and the likelihood that he would not long remain chief of police, Galusha seized the opportunity to ask Hudspeth for a job again on his staff as U.S. Marshal.[12]

Pancho Villa Raids Columbus, New Mexico

A scant three years later another crisis loomed. On March 9, 1916, the notorious Mexican revolutionary Pancho Villa raided Columbus, New Mexico.[13] Hudspeth immediately went to Columbus and took charge of the situation, overseeing patrols on the U.S.-Mexican boundary while Gen. "Black Jack" Pershing led a punitive expedition into Mexico. During this tense time, many Mexican nationals were arrested for violating neutrality laws. Gen. José Yñez Salazar, one of the Mexican generals who crossed the boundary line, was arrested and jailed in Albuquerque; he later escaped.

A Seat on the New Mexico State Supreme Court

In November 1930, Hudspeth was elected to the New Mexico State Supreme Court where he served for nine years, the last two as Chief Justice.[14] He served again when Chief Justice Thomas Mabry resigned from the state supreme court to run for governor and Gov. John Dempsey appointed Hudspeth to fill out Mabry's term in 1946. While he sat on the supreme court bench in the 1930s, Hudspeth lived in Santa Fe and once a month or so would drive back to Carrizozo to attend to business there. He always drove the entire round-trip in his Buick coupe in a very deliberate second gear.[15] For about five years, he also drove on to White Oaks and visited Lillian Lane, widow of Alexander Lane's son, Allen, and her two little daughters.

According to Larue Lane Wetzel—one of the daughters—every time Hudspeth returned to White Oaks he would fill a grocery sack with candy for the two girls, matching the kinds of candy exactly so that each girl

Group photograph of New Mexico Supreme Court justices. *Left to right*: Andrew H. Hudspeth, John C. Watson, Howard L. Bickley, Frank W. Parker, Daniel K. Sadler. Date and photographer unknown. F. W. Parker Papers, image no. 8167. Courtesy of New Mexico State Records Center and Archives.

would have an equal amount.[16] Once a month he would mail one dollar each for Larue and her sister in the form of a two-dollar check. He alternated the check's payee every month in order to teach the girls how to endorse and cash checks correctly.

Back Home to Lincoln County

For years after his stint on the New Mexico Supreme Court, Hudspeth continued to conduct legal business from his Carrizozo office, driving slowly around town on errands, always in that famous second gear. He practiced in the Lincoln County area throughout the 1920s and was renowned for his good manners and humor.

In 1923, Robert Hurt of Capitán was murdered on his way home north of Capitán. Three men were duly tried for his murder and found not guilty, but Hurt's teenaged daughter refused to accept the verdict.[17] One day she confronted one of the accused on the streets of Capitán and emptied her six-shooter at him. She wounded him only once, and that slightly. Brought to court in Carrizozo, with Hudspeth on the bench, the jury found her not guilty. Then the judge spoke his piece. He admonished the girl that the jury had deliberated and found her not guilty. But then he softened his words by saying that, had the verdict been guilty, he would have been forced to sentence her to six months . . . six months of target practice.

Ina Dow remembers Judge Hudspeth in his later years in Carrizozo as a distinguished gentleman with white hair and mustache, courtly of manner.[18] The judge often visited his old friends, Ina's grandparents (Charles and Ina Mayer), who had moved from White Oaks to Carrizozo. The judge never married. He had family far away—a sister in Texas (Mrs. J. O. Simpson) and two brothers, Oscar of Dallas, Texas, and George of Los Angeles.

The End of His Days

Hudspeth remained faithful to the dreams of the mentor of his youth, Judge John Y. Hewitt. After Hewitt's death, he continued to do assessment work on many White Oaks mining claims, paid taxes on all the Hewitt property, and carried on—always hoping that White Oaks would stage a comeback.[19]

For the last decade of his life, Hudspeth battled cancer of the esophagus.[20] He endured treatment and operations at the Mayo Clinic, at a hospital in Virginia, and at several other well-known hospitals. He would enjoy brief respites from the cancer for a time. But finally, ill health forced him to leave his beloved Lincoln County and travel to California to enter Huntington Memorial Hospital at Pasadena. Until the very day before he left for California, Hudspeth continued working in his law office at Carrizozo. When it became known that his disease was terminal and he could not return to Carrizozo, a group of Carrizozo friends left for California and a last visit.[21] Hudspeth died March 9, 1948, at the age of seventy-three, and his body was cremated. His ashes were scattered over the White Oaks mines, most of which he owned at the time of his death.[22]

NOTES

1. Haley, *1913 Year Book*.

2. "Cree Ranch Still In the News Today," *Ruidoso News*, May 16, 1947.

3. Kudner to Hudspeth, February 27, 1931, copy in author's collection.

4. Biographical note, Andrew H. Hudspeth, Lincoln County, New Mexico, Collection, 1885–1976, copy in author's collection.

5. "Judge Andrew H. Hudspeth Succumbs Tuesday In California," *Albuquerque Journal*, March 9, 1948.

6. Biographical note, Andrew H. Hudspeth.

7. Curry, *George Curry*, 255.

8. McDonald to Hudspeth, February 18, 1913, copy in author's collection.

9. Warren Gard to Hudspeth, September 26, 1916, copy in author's collection.

10. Keleher, *Fabulous Frontier*, 291–92.

11. Bryan, *True Tales*, 229–35.

12. Galusha to Hudspeth, November 10, 1916, copy in author's collection.

13. Keleher, *Fabulous Frontier*, 292.

14. Hall to Haldane, July 22, 2001.

15. Stearns to Haldane, May 1, 2004.

16. Wetzel, interview by Haldane, May 8, 2001, and May 29, 2001.

17. Sinclair, "Mother Julian," 49.

18. Dow to Haldane, October 23, 2000.

19. Keleher, *Violence in Lincoln County*, 370.

20. Ina Wauchope Mayer to Ina Dow, March 14, 1948, copy in author's collection.

21. *Albuquerque Journal*, March 9, 1948.

22. Keleher, *Violence in Lincoln County*, 370.

DAVID L. "HAPPY JACK" JACKSON
A Lesson in Do-It-Yourself Integration

On July 7, 1897, a small, wiry black man stepped off the stagecoach in White Oaks. A casual observer would have noticed nothing remarkable about this young man as he looked around, stretched, and brushed dust off his pant legs after the ninety-mile, bone-jarring trip from San Antonio near Socorro. He and other passengers had chartered a private stage rather than wait for the regular mail stage. They left the Río Grande at sunup and arrived at the Ozanne Hotel in White Oaks early the next morning after changing horses four times on the way.

Dave Jackson ("Happy Jack" or just "Jack" to his friends) was the only African American in White Oaks. He might have been more hesitant that July morning had he realized that in 1897 White Oaks citizenry included many Southern sympathizers, and the Civil War was only three decades removed. He probably knew nothing of the fate of the last black man to enter White Oaks, also named Jackson (first name Richard), who ventured into the town in 1882 and soon after was charged with being a horse thief. Richard Jackson was then taken by vigilantes from a cellar that served as a jail and hanged from a tree within two hundred yards of the outskirts of town.[1]

Born in Collin County, Texas, in 1870, Jack was the son of poor farmers who put him to work as soon as he could hold onto a plow handle. His schooling was limited, perhaps the equivalent of fourth grade. By the time he reached White Oaks at the age of twenty-seven, Jack had been on his own for thirteen years. He was fresh from a year in El Paso, where he worked for Mr. Mendenthal at his livery stable and wagon yard.[2] Wages were $1.50 a day to start, but after Mr. Mendenthal discovered his new young employee could unload and move a rail gondola load of crushed rock in a single day, the pay went up to $2. Then Jack heard of a gold-mining town in a beautiful valley up north in New Mexico.

Getting Started

Standing on the street and looking White Oaks over on that July day, Jack got down to his most urgent business: finding a job. The work of timbering a new shaft at the Old Abe Mine had just been completed, so day-labor work in the mines was not available. The Helen Rae Mine at Nogal east of Carrizo Mountain needed men, and Morris Parker of White Oaks was superintendent. Jack boarded a stage for the short trip to the Helen Rae and found Parker, who gave him a job as shovel hand.[3] That work did not last long, and Morris returned to White Oaks, taking Jack with him to do housework for the Parkers.

While waiting for a freighter to arrive and take him to try for better work in Arizona, Jack strolled down to

David L. Jackson. Date unknown. Photograph by Woodfin Camp. From *Amarillo News-Globe*. Permission of Dennis Spies, Editor, *Amarillo News-Globe*.

the cyanide plant operated by another Parker son, James. James hired the young man on the spot. After a week in the cyanide plant, Parker raised Jack's pay fifty cents a day, and Jack thereupon decided to make White Oaks his home.

Life in White Oaks

Meanwhile, Jack rented a little log cabin and stocked it with supplies. After working at the cyanide plant for a year, he sent for his sweetheart, Mary, in Galveston, Texas. She made the stage trip to White Oaks, and they were married in the fall of 1898.[4]

In time the townspeople grew to respect Jack so much that they asked him to serve on the Lincoln County Election Board. Appointed in 1898, he continued this service to the community until 1954.[5]

Jack spent the fall and winter of 1898 at the Homestake Mine, and then got a job at the Old Abe Mill sorting ore. At that time the Old Abe shaft was the deepest dry shaft mine in the world at 1,450 feet.[6] A year later Jack was promoted to firing the boiler and running the engine, and finally to amalgamating. The huge stamps that pulverize the ores in the amalgamation process make a deafening noise, and after years of working at this job Jack lost much of his hearing and had to wear a hearing aid. He also lost two fingers at the Old Abe Mill when a steam pipe broke on one of the big boilers.[7]

Small and slight though he was, Jack possessed great physical strength. He sent off for Bernarr MacFadden's (a physical fitness guru in the first half of the twentieth century) correspondence courses on bodybuilding.[8] Jack worked ten hours in the mines by day, then came home and lifted barbells far into the night. He claimed that he had never had a fight. He came to be considered the town's strongest citizen, seizing opportunities to perform feats of prowess such as lifting up the biggest men in the community and holding them over his head. He could also move heavy mining equipment that other men could scarcely budge.

The Wild Cat Leasing Company

In 1904, the Old Abe closed down and most of the men went to Arizona looking for work. It was either lease or leave. About this time, four White Oaks men got together to obtain a lease on the Old Abe and formed a partnership called the Wild Cat Leasing Company: Ed Queen and his brother George, Alfred James, and Allen Lane. James soon moved on; in a few months George

Dave Jackson and wife Mary, on the occasion of their fiftieth wedding anniversary, November 21, 1948. Photographer unknown. C. L. Sonnichsen Papers, Special Collections Department, MS 141, University of Texas at El Paso Library.

Queen also pulled out for California. This left Ed Queen and Lane shorthanded and in need of another partner. Jack was a logical choice since he knew Lane well, and these three men remained partners for many years.

The White Oaks mines were at that time controlled by the Parkers. Jack, who had known the Parker family for some years and had befriended them, thought he could get the leases. He went by train to El Paso where the Parkers were now doing business as Parker and Parker, Mining Engineers. Negotiations went on for about two months in their office in the First National Bank Building. A trip to Saint Louis also had to be made, but finally the leases came through.[9]

Allen was the master mechanic for the Wild Cats (he later built a working automobile by hand entirely from spare parts), Ed was in charge of mining operations, and Jack, besides being in charge of milling operations, assumed the job of secretary-treasurer.

How could a man who was scarcely literate, could not type, and knew nothing about bookkeeping, manage business affairs on this scale? More correspondence courses, of course. This time, Jack enrolled in reading, writing, and bookkeeping. Jack bought a typewriter, learned to type, and for more than twenty-five years handled the business of the partnership remarkably well.[10]

In their first year, the partners leased the South Homestake Mine and grossed $18,000. Next, they leased the North Homestake Mine and Mill so they would have an operating mill and hired crews to work it—as many as eighteen men at one time. Finally, the Wild Cats bought both the North and South Homestakes. Jack estimated that the partners grossed more than $300,000 from the gold and tungsten output of these mines.[11] In order to turn gold into cash, the Wild Cats poured the gold into ingots, shipped them to Denver, and received a check from the U.S. Treasurer for the amount of the gold shipment.

The first mill run was made in March 1905, and by 1913 the Wild Cats had paid for their two mines and mill.[12]

The Alto Light & Power Company

Next, Jack and his partners decided to expand their operations, which led eventually to their providing electricity to White Oaks and surrounding towns. The 250-kVA generator driven by a Corliss valve engine (part of the 750-kVA power plant used to operate the partners' mines and mill) came from the defunct Vera Cruz mine southeast of White Oaks.[13] In order to negotiate to buy the Vera Cruz generator and mining equipment, Jack traveled with a $10,000 certified check in hand to Greenville, Michigan, and closed the deal after two weeks of haggling with the major stockholders.[14]

The partners disassembled the equipment at the Vera Cruz mine located on the railroad spur to Capitán and moved it to White Oaks in 1912. They situated the generator on the slopes of Carrizo Mountain two miles east of White Oaks. To fuel the plant, the Wild Cats bought a coal mine on the same slope.

At the time the power plant was installed, the Wild Cats ran the mines with only one shift a day. Since they had excess power at night, the partners decided to sell the surplus power, and ran lines first to White Oaks,[15] then to Carrizozo and the Parson mine in Bonita Canyon near Nogal. By late 1912, the Wild Cats had installed an electric plant at the coal mine, built several hundred miles of transmission lines, installed transformer stations, and obtained franchises to furnish electrical power to White Oaks and nearby Carrizozo.

Thus the first electric lights in Lincoln County came to White Oaks in 1912 and went on to Carrizozo and Parsons in 1914.[16] Jack was fond of saying, "I found Carrizozo in darkness and left it in light."[17] In 1917, Ed Queen sold his one-third share of the mines and power plant and moved to Arizona. Jack and Lane operated the mines off and on until 1930, when they sold out to longtime White Oaks lawyer Andy Hudspeth. The Wild Cats' partnership fully disintegrated in October 1931 with the death of Allen Lane. Of the partners, only Jack was left in White Oaks. It was then that he came into his own as a beloved legend in his community. Honesty, hard work, and helpfulness had won him the lasting respect of all the White Oaks townspeople.

Good Samaritans

Jack and Mary never had children, but Jack—who adored children—began White Oaks' first Boy Scout troop, taught Sunday School for years, and cut the hair of all the local youngsters. Mary Lou Welch at one time lived in the Hoyle House and remembers that Jack and Mary often babysat her two children when she was young so that she and her husband could attend the dances around Lincoln County.[18]

Over the years, Jack and Mary took to helping anyone who needed money, food, clothing, or nursing care. They prepared the dead for burial, and Jack made it his personal mission to take care of his friends now residing in Cedarvale Cemetery outside White Oaks. He repaired fence, pulled weeds, and decorated before national holidays. Some of the graves he tended were those of William McDonald, New Mexico's first governor after it became a state; wealthy and cranky old John Hewitt; and Susan McSween Barber, former Cattle Queen of New Mexico.

After the mines had long since closed and White Oaks had become a ghost town with only a few people left, Jack still went into the mines to oil the machinery and keep it in working condition against the days when, he was sure, the mines would reopen.

Mary Jackson

If someone took sick, Jack could soon be seen carrying a pot of Mary's delicious soup in an earthen jar with a lid and handle to take to the sufferer. Mary was an excellent cook. Before she married Jack, she had been employed as a live-in nurse to two small children of a wealthy family in Galveston. During a terrible hurricane, the lower floor

Dave Jackson at eighty-six. Dave looks back on fifty-nine years in White Oaks from the White Oaks Schoolhouse grounds. Across the draw can be seen the Taylor blacksmith building and the large building that once housed both Sol Wiener's "gents" furnishings (men's clothing and accessories) and the post office and secondhand store run by John A. Brown. Date and photographer unknown. Courtesy Sara Jackson, My House of Old Things, Ancho, New Mexico.

of the home flooded and Mary had to pass the children through a window to the second floor balcony, where she tied the children to a porch pillar. Then she signaled boats that were picking up flood survivors.[19]

A timid and reserved woman of much dignity, Mary would only visit people in White Oaks where she was sure of a welcome. Sometimes the couple would entertain guests. Mary had lovely silver, beautiful china, and linens. Mary Lou Welch recalls that Mary would seat her dinner guests and serve them a wonderful meal, standing all the while and waiting on everyone.

There were a few ups and downs in the marriage. Jack's wife told the historian Eve Ball, "He was always a good man except for one bad habit: he liked to play poker.... He'd slip out at night and stay till almost morning.... One night I found him gone, got the molasses jug, and poured sorghum on the window sill. Just before daylight I heard him washing his hands, but I never mentioned it to him."[20]

Sometimes Jack liked a little nip or two, as well. Once when Jack had a bit too much to drink, Mary was waiting for him at home. She had in her bedroom on a brass bed two fancy pillows, big squares filled with goose down in snow-white shams. Upset with her husband, she was leaning back on these pillows at the head of the bed holding Jack's shotgun and awaiting his return. When he finally arrived home, somehow the gun went off, blowing up the pillows and raining down feathers all over the house. Jack, unhurt, quietly gathered up all the down in pillowcases while Mary sat crying.[21]

In later years Mary became crippled with arthritis, but her faith and optimism never wavered. A devout Christian, she read widely and loved poetry. She died in 1954.

His Friends Were Legion

Jack lived on nine more years after Mary died and was strong and healthy most of the time. He often walked the twelve miles from White Oaks to Carrizozo and back to visit friends, throwing out of the road any rocks that could cause flat tires. As winters began to feel colder, Jack traveled to El Paso to spend part of the cold

Dave Jackson and Charlie Wayne Wetzel, 1962. In the background is Jack's Peak in the Jicarillas. Photographer unknown. Courtesy Larue Lane Wetzel.

months with Mrs. Rufus Cadenhead, grown daughter of his former partner Allen Lane and his little "Baby Doll" of earlier times.[22]

Jack's many friends in Lincoln County and elsewhere began to hold annual "Dave Jackson Day" celebrations in his honor in August at White Oaks. People came from hundreds of miles around, spread picnic lunches under the oaks, and made a day of it. Later on, reunions in Jack's honor would take place in the schoolhouse.

A stroke in 1963 ended the life of this beloved man who was noted for his benevolence, his wit, and his concern for humanity. At the funeral held in Carrizozo, the church was packed with friends from his sixty-six years in Lincoln County who were there to say goodbye to the man who had become a monument to all that was good in old White Oaks. A tribute on the front page of the local newspaper spoke for all Jack's friends the day of his death: "Today, Dave Jackson, White Oaks pioneer . . . went to meet his maker. Time passed swiftly for Jack and even at the ripe old age of 93, he possessed youthful qualities that endeared him to all of

Mary and Dave Jackson's headstone, Cedarvale Cemetery. Photograph by Bob Workhoven. Author's collection.

those who fell within the circle of charm that was his. . . . He kept [the] faith, and countless are they who have sat spellbound as he told of those brighter days. . . . He came seeking riches and found them in the life that was his, a life of fair play, honesty, and charity to all. . . . It wasn't whether he won or lost but how he played the game. . . . Adios, Jack."[23]

NOTES

1. Parker, *White Oaks*, 16–18.

2. David Jackson, "White Oaks Old-Timer Recalls Stirring Days of Mining Boom Town," *El Paso Times*, April 29, 1956, 2-F.

3. Paul Baker, "David L. Jackson," Ramblin' Around Lincoln County, *Lincoln County News*, April 27, 1956; May 4, 1956; May 11, 1956.

4. Amanda Treat, "Glimpses of the life of Mary Jackson," *Lincoln County News*, April 24, 1969, 2.

5. Paul Payton, "In Other Years," *Lincoln County News*, November 11, 1976, 12.

6. Baker, "David L. Jackson."

7. Paul Baker, "A. Lloyd Hulbert," Ramblin' Around Lincoln County, *Lincoln County News*, January 3, 1958, and January 10, 1958.

8. "White Oaks Man Solves Integration Problem," *Southwesterner*, April 1963, B-5.

9. Baker, "David L. Jackson."

10. Glen Ellison, "Paying Tribute To A Great American," *Lincoln County News*, May 11, 1978, 2–3.

11. Baker, "David L. Jackson."

12. Lawrence Queen to John Kelt, August 18, 1971, copy in author's collection.

13. Ibid.

14. Dave Jackson, "Hoyle House and Electricity Come to White Oaks," 74–75.

15. "Alto Power and Light," the story of the White Oaks Power Plant as told to Larue L. Wetzel by Herman E. Kelt, Sr., July 4, 1975, Larue Lane Wetzel's collection.

16. "The Juice Turned On," *Carrizozo News*, January 29, 1915.

17. Baker, "David L. Jackson."

18. Welch, interview.

19. Ibid.

20. Ball, "David Jackson," 34–35.

21. Larue Lane Wetzel, interview by Haldane, June 19, 2001.

22. Ibid.

23. Johnson Stearns, "Our Friend, Dave Jackson," *Lincoln County News*, May 9, 1962, front page.

MAX H. KOCH
Photographer of Note

In November 1888, a German-born furniture maker named Max H. Koch (pronounced "coke") moved from the silver-mining town of Chloride, New Mexico, northwest of present-day Truth or Consequences, to White Oaks. His wife, Alice, and newborn daughter Edna went with him.

Born in 1850, Koch was then in his late thirties.[1] Alice, a former New Yorker of German ancestry, was thirteen years younger than her husband. Baby Edna had been born in Chloride on November 10, 1888; a second daughter, Merle, would be born at White Oaks in 1895.

Chloride, Silver-Mining Boomtown

Chloride sprang to life when eighteen prospectors established their headquarters at the mouth of Chloride Gulch in January 1881.[2] Before long, a band of thirty-five Apaches swooped down on the men, killing two and running off the camp's horses and mules. The remaining prospectors fled then, but they returned in March much better equipped, bearing arms and ammunition. Soon a store was erected and more miners arrived, triggering a meeting to select a townsite. The present site of Chloride was chosen and lots auctioned off by lottery. Tickets were placed into a hat, shaken up, and drawn by one man. The men of Chloride gave a free lot to the first woman to make Chloride her home.

By the summer of 1881, Chloride boasted eight saloons, three general merchandise stores, three restaurants, two butcher shops, a newsstand, a lumber yard, and an assay office. The Pioneer Stage Line deposited newcomers at the livery stable. Before long, Chloride added a hotel, a church, a school, and the *Black Range* newspaper for its five hundred citizens. Chloride reached its peak in the 1880s, though the mines continued to produce after that. Sporadic attacks by the Apaches continued until as late as 1887.

M. H. Koch arrived in Chloride probably early in 1884 as a cabinet builder, and then he opened a furniture store.[3] In June 1884, the *Black Range* ran the notice that "M. H. Kock [*sic*] is increasing his stock of furniture gradually. He can supply the needs of the Range in this line now that the people here are relieved of sending to Socorro or elsewhere if they desire to purchase chairs, bedsteads, tables and the like."[4] A year later, another notice appeared in the *Range*, noting, "M. H. Koch has traded his house and lot at the upper end of Wall Street for the residence and lot of Sam Ferree situated on central Wall Street. This change makes it very convenient for Mr. Koch as his new place of abode is in close proximity to his place of business."[5]

In December 1885, the *Black Range* reported, "M. H. Koch has just received a new stock of household furniture and he also has another lot of goods on the road which will arrive in a few days."[6] By January 1888, Max had earned the respect of Chloride's townsmen, who elected him one of the town's trustees. During this era, a German named Henry Schmidt came to America and worked with a party of surveyors from Colorado to New Mexico.[7] At Chloride, the last place in which the surveyors set up a monument, Henry stayed to become an assayer. Photography was his hobby, and he soon found himself the official photographer of Chloride, shooting thousands of photos of the town and its citizens.[8] These images were preserved on glass plates.

It is not unreasonable to think that fellow Germans Henry and Max became friends or that Henry taught Max the fairly new art of photography. Max chose to become exclusively a portrait photographer. Portrait photographers in towns of that day shot their subjects in a studio, often against painted backdrops and with a wide variety of studio props. They rigged up light sources above the sitters and used mirrors to direct the light.

Edna Koch and Jones Taliaferro. Date unknown. Photograph by M. H. Koch. Courtesy Chris and Jack Harkey.

White Oaks Photographer and Undertaker

In the late 1880s, as Chloride's silver mining began to slow, the newly married Koch cast about for another way to support a family. White Oaks to the northeast sounded promising. Once in White Oaks, Max lost no time in putting his woodworking skills to good use, quickly becoming the town's chief building contractor and carpenter, but his new love was photography.

Over seventeen years, Koch photographed hundreds, perhaps thousands, of Lincoln County citizens, usually in formal poses against elaborate backdrops. He figured prominently among New Mexico's most important early portrait photographers, along with Henry Schmidt of Chloride and J. N. Furlong of Las Vegas.

Within two years, Koch also became for all practical purposes the new town undertaker. John Slack, who had arrived in White Oaks in 1880 and served as its official undertaker for many years, had grown old and needed an experienced helper. Max started out by building coffins and gradually took over the business. His handwriting first supplants Slack's elegant script in the White Oaks Record of Deaths in December 1890.[9] In May 1895, Slack sold the business to Koch.

The Kochs' two daughters, Edna and Merle, were the couple's pride and joy, and "Papa" kept an extensive photographic record of their growing-up years in the Oaks. From the photos, it is readily apparent that Alice Koch was a skilled needlewoman and dressmaker. Her pretty little blondes sported elaborate dresses in the latest styles.

The Move to Tucumcari

In the early 1900s, with White Oaks fading before his eyes as Chloride once had, Max pondered a suitable place to transfer his businesses. The new railroad town of Tucumcari 120 miles to the northeast appeared a promising choice. On November 2, 1905, Koch moved his family into a neat, white cottage south of a photo gallery in Tucumcari.[10] Daughter Edna was now eighteen years old; Merle was seven years younger.

Tucker and Talcot, a team of photographers, already had the town's photo business, and so Koch turned to his third calling: undertaking. In November 1905, he established the first funeral parlor in Tucumcari in a time when horse-drawn wagons were still in use. Specially made wagons were used for funerals. These hearses were completely black and surrounded by glass with heavy cords and tassels hanging from the inside top; the

Merle Koch. Date unknown. Photograph by Max H. Koch. Courtesy Chris and Jack Harkey.

horses drawing the wagon were also black. The casket was prominently displayed.

Services were usually held in the deceased's home and rarely in churches. Pallbearers customarily accompanied the casket on foot to the cemetery no matter how far from the services until the advent of Tucumcari's first gasoline-powered hearse—which was also used as an ambulance—ended this practice.

The *1910 History and Business Review of Tucumcari* published by its chamber of commerce lists "M. H. Koch—Funeral Director and Embalmer." The 1912 edition adds "Furniture and Funerals" to the name.[11]

Leading Citizens of Tucumcari

Max and Alice Koch quickly grew prominent in the community life of Tucumcari. Both were charter members of the Bethel Chapter Order of the Eastern Star.[12] Alice was also Tucumcari's first Worthy Matron of the Eastern Stars and founded Tucumcari's Bay View Club to aid in the "civic, moral, mental and social uplift of the community."[13] The club motto was "Row, not drift." It joined the General Federation of Women's Clubs in 1911. Tucumcari's population was about 350 at the time the Bay View was organized. Over the years, this club helped with many civic needs of the growing town.

Later Years

In 1917, as age began to tell on Koch, he sold the undertaking business to O. G. Reeder. Reeder employed an apprentice named Elton Dunn, who in 1918 bought the business.[14] It is still operated today by the Dunn family.

Max's wife, Alice, died in 1914. Edna kept house for her father for many years, but on January 27, 1926, she left at the age of thirty-eight to marry a handsome Ratón-area rancher named Alfred Davis.[15] Merle had long since married Curt I. Davis (no relation to Alfred). After Edna married, Max went to live with his younger daughter. On April 7, 1926, Edna wrote her dear friend Bertha Mayer Hunter in California of her marriage:

Dear Bertha,

Alfred and I were married the 27th of January. Had a very quiet wedding.... Alfred was a widower and has two girls Ruth age 17 and Wilma age 9.... Both girls come home quite often and spend the week end and of course they are here during the summer vacation.

I love the girls very dearly and always enjoy having them home. Alfred is 6 ft tall and has brown hair and blue eyes and is six years older than I. Believe I told you that Alfred is in the cattle business and we live on a ranch just about ten miles from Raton. We have about 500 acres of land and the greater part of it is irrigated land and we raise lots of hay and alfalfa. Alfred has 4000 acres leased up in the mountains where he has the most of the cattle.

I enjoy living on the ranch very much and we have so much company I really do not have much chance to get lonely. Then too we are on the Highway and there are always cars going by and on another side is Red River and that separates us from the Santa Fe track and there are lots of trains going by so you see this is not such a lonesome place after all.

Merle is at home now, finally got relieved at the office. Marshall Wharton has her place. Alfred has a sister living in Ventura, Cal. and we are expecting her here this summer.

Lovingly, Edna[16]

Edna died the following year after the birth of her son. A local newspaper reported:

MRS. ALFRED T. DAVIS

Edna Alice Koch was born November 10, 1888 at Chloride New Mexico. She passed away at an early hour, Thursday morning, October 6, 1927.

In 1906 [1905] her father moved to Tucumcari, where he engaged in business, being a prominent member of the state embalmers association.

On January 29, 1926 she was united in marriage to Alfred T. Davis, a member of a pioneer family of Colfax county. Since her marriage she has resided at the Davis ranch southeast of the city, making many warm friends here in Raton, being a visiting member of the Order of Eastern Star. Her quiet ways and pleasing manner endeared her to those who knew her best.

Funeral services will... be in charge of Rev. Mayne, pastor at the Presbyterian church of Tucumcari. Burial will be made in Fairmont cemetery.[17]

Max Koch lived on in Tucumcari until his death five years later. He was buried beside Alice on May 9, 1933. Handsome matching headstones mark their graves.[18] Hers bears the emblem of the Eastern Star order, his the Masonic emblem.

NOTES

1. U.S. Bureau of the Census, *United States Census of 1900*, White Oaks.

2. Sherman and Sherman, *Ghost Towns*, 42–43.

3. According to Donna Edmund, curator/docent of the Chloride Museum in Chloride, New Mexico, Koch was not in Chloride in 1881. Edmund looked up his name in several registers kept in the museum.

4. *Chloride Black Range*, June 27, 1884.

5. *Chloride Black Range*, June 19, 1885.

6. *Chloride Black Range*, December 4, 1885.

7. Stanley, *Chloride, New Mexico Story*, 10, 19.

8. McStay, "Beyond a Glimpse," 64.

9. Record of Deaths at White Oaks.

10. Untitled history of the town of Tucumcari, New Mexico, Tucumcari History Museum Collection.

11. Tucumcari Chamber of Commerce, *Tucumcari: 1910 History and Business Review;* Tucumcari Chamber of Commerce, *Tucumcari: 1912 History and Business Review*.

12. "Bethel Chapter Order of the Eastern Star," from untitled history of the town of Tucumcari.

13. "The Bay View Club," from untitled history of the town of Tucumcari.

14. "The Dunn Family," from a history of Tucumcari families, 277, Tucumcari History Museum Collection.

15. Obituary for Mrs. Alfred T. Davis (Edna Alice Koch Davis), no date or newspaper title, from the collection of Ina Hunter Dow, whose mother was "best friends" with Edna when both were girls in White Oaks.

16. Edna Koch Davis to Bertha Mayer Hunter, April 7, 1926, from the collection of Ina Hunter Dow.

17. Obituary for Mrs. Alfred T. Davis.

18. Visit by the author to the Tucumcari Cemetery, January 10, 2006. The graves are in the Masonic Block, Lots five and six of Row one.

DR. ALEXANDER GALLATIN LANE
From Confederate Surgeon to White Oaks Physician and Pharmacist

By the late 1600s, the Lane family, originally from England, had already established a home in New Kent County, Virginia.[1] Alexander's great-grandfather, William Lane, served in the Fourteenth Virginia Regiment during the American Revolution.[2] William moved from Virginia to Elbert County, Georgia, in 1791 and settled on Beaver Dam Creek, where he operated a mill.[3] Son Thomas, born in 1764, was the father of John Allen Lane.

John Allen Lane and Ann Mayfield married in November 1830 and eventually made their way to Vicksburg, Mississippi. In thirteen years of marriage, Ann bore eight children; four of the children, including Alexander, lived to adulthood. Alexander was born in the Phoenix Mansion in Vicksburg on April 30, 1835. He lost his mother when he was only eight years old.[4]

The 1830s through the 1850s were great times to build fortunes in antebellum Dixie, where cotton was king and slave labor was the underpinning of agriculture. By 1834, John A. Lane and his uncle, Rev. John Lane, were partners in a Vicksburg mercantile firm. The younger John's family lived at Oak Grove Plantation in Vicksburg.[5]

Around this time, John Lane also began acquiring extensive land holdings in Mississippi, Arkansas, and Louisiana. Buying over 1,500 acres of land from the Arkansas government from 1838 to 1849 and 12,439 acres in Mississippi from 1839 to 1842 made him a very wealthy man. He used his wealth to provide not only for his own family but also for society in general by personally and financially supporting many religious and educational organizations. In 1858, John purchased Reclusia, a 1,132-acre plantation located on both sides of Bayou Macon in Carroll Parish, Louisiana. He soon transferred this plantation—with its eighty-seven slaves, home, crops, and stock—to his children Alexander, Jane, and Emma Lane.[6]

Like many Southerners, as a young man Alexander had launched a promising career just one step ahead of

Painting of Dr. Alexander Lane, on the occasion of Lane's graduation from Tulane University, New Orleans. Date and artist unknown. Courtesy Larue Lane Wetzel.

the War Between the States. In 1854 and 1856, respectively, he received a bachelor's of arts degree and a master's of arts degree from Centenary College in Jackson, Louisiana,[7] where both his father and his great-uncle, the Reverend John Lane, sat on the board of supervisors.[8] Then he enrolled in medical college, served a year of residency at Charity Hospital in New Orleans, and graduated from the University of Louisiana in 1858.

For a while, Alexander was a successful cotton planter in Carroll Parish, Louisiana, listing his occupations as both planter and physician in 1860. But as war began in April 1861 after the Confederate attack on Fort

Sumter, South Carolina, he joined the Mississippi College Rifles in May, believing "it was in the defense of ... the constitution of the United States." This infantry regiment reorganized not long afterward as Company E, Eighteenth Mississippi Infantry.[9]

Soon Alexander was selected to deliver an address intended to recruit for the Southern cause. His fiery speech before the citizens of Clinton resonated so much with the ladies that, he "walked off the platform, carpeted with showers of bouquets from [the] hands of Mississippi's fairest daughters."[10]

Confederate Surgeon and Hospital Administrator

Lane was commissioned as a surgeon June 6, 1861, after six weeks of boot camp in Corinth, Mississippi. He was then ordered to Virginia with Colonel Burt's Eighteenth Mississippi Regiment. In July 1861, Alexander fought in the First Battle of Bull Run. However, he did not long remain in the infantry and transferred to the medical department in August. Because of his medical and organizational skills, he soon became an officer.

After the battle of Ball's Bluff in October 1861, Lane was made staff surgeon by his commander, General Evans, and placed in charge of Union prisoners at a hospital at Leesburg, Virginia, near the Potomac River. He was then posted to set up and manage Winder Hospital near Richmond, Virginia. Organized in April 1862, Winder was conducted as a Confederate hospital for Union prisoners for three years until its termination ten days before General Lee's surrender. This hospital grew to become the largest in the Confederacy. Winder covered 125 acres of land and had a capacity for 4,800 patients. Here, at the age of twenty-seven, Lane found himself in command of three surgeons and eight hundred hospital attendants caring for a population that ranged from two thousand to five thousand Union prisoners at any given time.

Lane set up the usual dispensary, laundry, kitchens, and corps of nurses and attendants. Prisoners were guarded by 125 men under a commissioned captain. Deeply held beliefs in good hygiene and nutrition caused Lane to go further as an administrator; he installed Russian steam saunas and showers. He had a six-foot ditch cut and had constructed a line of toilets leaning over the ditch. This ditch was flushed out every other day by water from two ten-thousand-gallon tanks pumped full and emptied into the ditch to carry away sewage from the hospital grounds.

With regard to nutrition, attached to the hospital was a bakery that could bake for ten thousand men daily, sixteen acres of garden for fresh vegetables and fruits, a dairy with sixty-nine cows yielding three hundred gallons of milk a day, and an icehouse. Lane had two canal barges constructed to ply up and down the Kanawhac River to the mountains and fetch fresh butter, eggs, chickens, geese, turkeys, honey, and every other available foodstuff for the sick and wounded. He even obtained the best grades of wines and liquors through a personal friendship with Mrs. Snowden, president of the local ladies' hospital association of Charleston, Virginia. The resourceful Mrs. Snowden obtained her spirits from Nassau in the Bahamas with the help of blockade runners.

One of Lane's triumphs as hospital commander was his success in keeping local medical officers from diverting to their own use the wines and liquors intended for patients, a common occurrence at other prison hospitals. Other local women helped the young chief surgeon (suggesting Lane's popularity with women). Emily Mason, niece of Senator Mason, was chief matron of the first division at Winder. Her personal friends, Mrs. Randolph and Mrs. Green, and two daughters of Robert E. Lee often drove out from Richmond in their carriages to visit the sick and wounded. They would sit by the bedsides of patients and write letters for them to loved ones at home. They also fed the prisoners delicacies brought out daily from their own tables.

As chief surgeon, Lane set up a medical examining board of three surgeons and oversaw six division surgeons—each capable of treating some 450 patients—whom he summoned to his office every day except Sunday. Each day Lane inspected one of their divisions, rotating the inspections so that the surgeons never knew which would be their day. With this system, he was able to maintain excellent control of medical standards. Lane set high standards for himself and saw to it that his staff lived up to them as well. Morale ran so high as a result of his challenges that officials and attendants alike competed to excel with the goal of achieving record survival rates.

For the rest of his life, Doctor Lane considered one of his crowning achievements the fact that Winder Hospital had the lowest mortality rate among the more than 150 Confederate military hospitals by the end of the war in spite of having to operate with inadequate funding, supplies, and personnel. Winder's record was unprecedented in the annals of the United States at that time, and even elsewhere in the world.

Conditions in most Civil War prison camps were

deplorable, with confined soldiers suffering terribly from overcrowding, poor sanitation, and malnutrition. Of almost 195,000 Union soldiers held in Confederate prison camps during the war, about 30,000 died. (Union forces held some 220,000 Confederate prisoners in hospitals, with 26,000 deaths.) Mortality rates for most Southern prison camps ran very high; the infamous Andersonville, Georgia, camp led with a 29 percent mortality rate.[11]

Winder was a remarkable exception for general military hospitals. During its years of operation, of almost sixty-five thousand patients treated, only 5.2 % died. The *Richmond Whig* observed that Alexander had "done everything in his power to promote the sanitary condition of the place, and to minister to the relief of the patients." The same newspaper in June 1864 stated that he was "one of the most gentlemanly and energetic surgeons in the Confederacy."[12] Alexander became a major while he was in charge of Winder Hospital.

Marriage

Alexander married Mary Caroline Corbin, probably in 1862, while managing Winder Hospital in Richmond during the war.[13] Born March 1, 1843, at Orange Grove, Georgia, Mary was the daughter of a prosperous Georgia planter. Over the years, the couple had thirteen children. Seven survived until 1900: William, born in Virginia in March 1864; Ida, born in Louisiana in 1865; Anna, born in 1869 at Reclusia Plantation; and the last four children all born in Missouri—Bruce in 1872; Allen Alexander in November 1875; Mary in September 1876; and John in 1879.

After the War

In late March 1865, as the Confederacy crumbled, Alexander was transferred from Winder in Richmond to the Trans-Mississippi Department. After Richmond fell in April 1865, and with the surrender of the Confederate armies over the following summer, the Alexander Lane family made its way across the ravaged South home to Carroll Parish.[14]

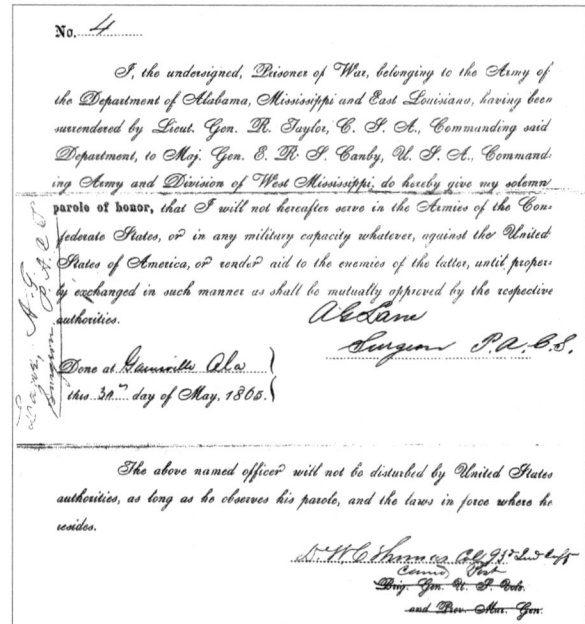

Dr. Lane's parole, May 30, 1865. Lane would have folded this parole and kept it in his coat pocket to present to Union army officers at various points on his journey from Gainesville, Georgia, to New Orleans. Courtesy Larue Lane Wetzel.

Surrender paper for Dr. A. G. Lane, May 30, 1865. Lane surrendered with the rest of Confederate general Richard Taylor's army. Lane's family and servants were with him at the time of his surrender, as he had been transferred from Richmond to the Trans-Mississippi Department. He was provided transportation to New Orleans, where he had several Lane relatives. Courtesy Larue Lane Wetzel.

For two years, Alexander served as sheriff in this parish and struggled to rebuild the family fortunes, but finances continued a downward slide until the couple was finally forced to file for bankruptcy in New Orleans in 1871. Their beloved plantation, Reclusia, was sold to satisfy the creditors.[15]

Like many other young people from the ruined Deep South, Alexander and his wife began looking west for opportunities. They left Mississippi in the spring of 1869, first stopping in Springfield, Missouri, where they lingered until sometime in 1879. One Sunday that year the entire family went to church and returned home to find their house burned down. Only the barn and the horses they had been riding were left. It was then that Lane decided to move farther, this time to the Southwest. In 1879, the railroad had just arrived in Las Vegas in New Mexico Territory, and people started pouring into the country. Alexander took the train to Las Vegas accompanied by son William. The rest of the family joined him in Las Vegas probably in early 1880.[16]

Las Vegas, Territory of New Mexico

In Las Vegas, Lane established a medical practice and drugstore in the same building and did so well that railroad officials soon took notice. They asked him to go to El Paso, Texas, establish an office there, and even offered to give him for his office building the entire block in downtown El Paso where the Mills Building now stands.

In 1880, Doctor Lane left wife and family behind in Las Vegas to run the drugstore while he traveled to El Paso to look things over. He arrived to exciting news about a gold boom in White Oaks, caught the train to San Antonio, New Mexico, and rode the stage on to White Oaks, arriving in May of 1880. Along the way, the stage was held up, but the robbers let him keep his gold watch.

White Oaks

Lane finally landed in White Oaks at the age of forty-five and wrote his wife in Las Vegas to sell the drugstore and bring the family to join him. He had already started to build a home with an office and drugstore in a big room in the middle of the house. Stairs led to the second floor, which was built against a hill. The family living quarters were on the first and second floors around the office and drugstore. Interested as always in nutrition, Lane built a greenhouse on the second floor so that he could grow fresh vegetables all winter.

Dr. Alexander G. Lane. Date unknown. Photograph by J. N. Furlong, Las Vegas, New Mexico. Courtesy Larue Lane Wetzel.

Doctor Lane established a profitable practice and drugstore in White Oaks and spent the rest of his life there, with two exceptions: a brief two years in the little mining town of San Pedro on the Río Grande one mile east of San Antonio near Socorro, and his final year, when because of poor health he and his wife went to California to live near their daughters.

Life in White Oaks

When he first landed in White Oaks in May 1880, the doctor invested in a mine he named after his beloved lost plantation, the Reclusia. The *Las Vegas Daily Optic* called that mine a "well-defined four-foot vein, carrying free gold and gray copper."[17] A week later Doctor Lane himself wrote to the *Optic* to describe the continuing mining boom. He said that the prospecting field now extended north ten miles, south fifteen miles, and west seven miles from town.[18] Lane also reported a scene bustling with energy as log and adobe houses with dirt floors sprang up everywhere. The town site had been surveyed and all lots taken. Town organization followed with the election of a mayor, J. S. Redman; Doctor Lane was named city treasurer.

Mining remained a lifelong interest. In March 1900, the *White Oaks Eagle* reported the return of Doctor Lane from the Oscura mining fields after an absence of six weeks in which he discovered a deposit of 88 percent copper.[19]

Doctor Lane taught a school in the front room of his home for his own children and a few others before a regular school was organized in White Oaks. When townspeople determined to build a schoolhouse, Lane served as treasurer for a group of men who gathered about $10,000. On July 1, 1893, a contract for the new schoolhouse, to be built of brick manufactured in White Oaks, was awarded to the White Oaks Building and Lumber Company. School was held in this building for the first time in 1895 for a student body of thirty-five. The red-brick building still stands today and is used for community events. In the summer, it is open to the public as a museum.

Deaths of Alexander and Mary

Old age and ill health caught up with the elder Lanes. They moved to Oakland, California, to live near their daughters around 1904. Alexander died October 8, 1905. A funeral service was held at his home on the tenth, with mass following at Saint Francis de Sales Church and burial in Saint Mary's Cemetery. Mary lived on with daughter Mary, and later with daughter Ida. She died January 5, 1923, in Oakland, and a mass was held at Saint Jarlath Church. She is also buried in Saint Mary's Cemetery.[20]

The Lane Children

William lived in White Oaks near his parents. In 1893, "Billy" bought the Ozanne Stagecoach Line; the stage mail robberies of 1896 ran him out of business.[21] He also acted as an agent for the Sonora Gold Mining Company of El Paso, Texas, and transferred to that firm his mining holdings in Mexico in exchange for stock and cash. He was killed while drilling a well when he dropped a charge that blew up on the way down into the hole in 1908. He is buried in Cedarvale Cemetery.

Allen attended the school in White Oaks that his father helped build. As one of three partners of the Wild Cat Leasing Company, he operated the Homestake Mines in White Oaks and shipped out over $400,000 worth of gold and tungsten between 1905 and 1930.

Allen A. Lane. Date and photographer unknown. Lane was a partner in the Wild Cat Leasing Company of White Oaks. Courtesy Larue Lane Wetzel.

John, the youngest, ran an assay office his father built for him at the north end of the Lane home. He lived with Allen's family for a time but owned his own home by 1920. John never married.

Daughter Ida came to New Mexico with her parents and in about 1885 married George Sligh, a Nogal merchant. By 1910, the Slighs had moved to Cochise County, Arizona, where George began farming. By

"Snooks," built by Allen A. Lane and his wife. The car was built entirely from spare parts. Lane's wife helped him with the riveting. Date and photographer unknown. Courtesy Larue Lane Wetzel.

1920, after her husband had died, Ida lived in Alameda County, California, where she died in 1949.[22]

Daughter Anna also came to New Mexico with her parents. By 1910, she had moved to Oakland, California, and was living with her sister, Mary Lane Foley. She had married a Mr. Green but was widowed by this time. Daughter Bruce came to New Mexico with her parents as well and married Frank Parker, a mining engineer, in 1891. By 1903, the couple had relocated to Quartzburg, Idaho, where Frank helped dewater a deep, old-time gold mine. They stayed in the Quartzburg-Boise district for five or six years and then finally settled in California and El Paso, Texas.[23]

Daughter Mary moved to White Oaks with her family as a young girl and married John Foley about 1894. Later he worked as a master mechanic for mining companies in California and, briefly, in Mexico. The couple moved to California around 1895. Mary's mother (Mary Lane) and sister Anna were also living with the family in 1920. Mary Foley died in 1957.[24]

NOTES

1. James M. Perrin of Hammond, Louisiana, is a Lane descendant who has done extensive genealogical research on the Lane family in the United States. By letter to Haldane October 29, 2003, he furnished information on the origins of Alexander Lane's family. He obtained these data from the *Layne–Lain–Lane Genealogy*, by Floyd B. Layne, of Los Angeles, 1962, and from the John Lane Bible, Vicksburg and Warren County Historical Society. Mr. Perrin's research is referred to hereafter as JMP.

2. Military service records of William Lane, National Archives, obtained from JMP.

3. Deed Book H, 60, 66; Deed Book K, 91; Deed Book N, 64, Elbert County, Georgia, Elbert County Courthouse, obtained from JMP.

4. U.S. Bureau of the Census, *United States Census of 1840*, Washington County, Mississippi; *United States Census of 1850*, Terrebonne Parish, Louisiana, Bayou Caillou, Fam. No. 514; *United States Census of 1850*, Orleans Parish, Second Ward, Second Municipality, 56; *United States Census of 1860*, Orleans Parish, Louisiana, Second Ward, First District, 185; *United States Census of 1860*, Carroll Parish, Louisiana, Fam. No. 568; *United States Census of 1870*, Orleans Parish, SecondWard, 296; *United States Census of 1880*, Christian County, Missouri, ED 8, 3-A, obtained from JMP.

5. Branton and Wade, *Early Mississippi Records*, obtained from JMP.

6. East Carroll Parish, Louisiana, conveyance records: Book L, 13, 77; Book O, 27, 38, 76, 312, 379; West Carroll Parish, Louisiana: Notarial Book A (Old), 632–35; Notarial Book B (Old), 420–21; Madison Parish, Louisiana: Conveyance Book A, 226–27; Book OR, 360–61; Elbert County, Georgia: Deed Book Five, 218, Book W, 68, all obtained from JMP.

7. "Address of Alexander G. Lane to be delivered before reunion of Surgeons of the Army and Navy. Confederate Veterans, Tulane University of Louisiana, May 1903. Organization and Management of Winder Hospital, Richmond, VA," *White Oaks Eagle* 12, no. 20, May 19, 1903.

8. James M. Perrin to Haldane, January 4, 2004.

9. "Address of Alexander G. Lane." Dr. Lane traced his Civil War service extensively in this article published in the *White Oaks Eagle*.

10. Ibid.

11. National Park Service, *Andersonville National Historic Site*, paragraph on "Civil War Prison Camps."

12. *Richmond Whig*, May 7, 1862, and June 15, 1864.

13. James M. Perrin to Haldane, January 4, 2004.

14. Ibid.

15. Ibid.

16. Larue Lane Wetzel, interviews by Haldane, May 5, 2001, and May 29, 2001. The accounts of Dr. Lane's life in Las Vegas, El Paso, and White Oaks come from the Wetzel interviews.

17. Article by "LEW," *Las Vegas Daily Optic*, May 13, 1880.

18. Letter from Dr. Alexander Lane, *Las Vegas Daily Optic*, May 17, 1880.

19. *White Oaks Eagle*, March 1900.

20. *Oakland Tribune*, October 9, 1905, and January 5, 1923.

21. Wetzel, interviews.

22. Perrin to Haldane.

23. Ibid.

24. Ibid.

JUDGE FRANKLIN HOUSTON LEA
Former Quantrill Guerilla

The American Civil War, not yet four months old, found eighteen local Missouri farm boys milling around in a field on a stifling August day near Lee's Summit in western Missouri in 1861. Among them, brothers Frank and Joe Lea had been captured the previous day while harvesting corn in their father's field. Though all the young men were noncombatants, they were nonetheless closely guarded by a squad of Kansas border militia. The officer in charge began lining up the prisoners in front of a firing squad, moving through the group and barking commands.

As the officer approached the two tall, slender young brothers standing nervously together, he recognized them from a Missouri family known to him. The officer leaned close and whispered, "When I give the command to fire, you both duck and run like hell. Don't stop, either, until you join the Confederates!" Returning to his men, he gave the order to fire. In the chaos that followed, both brothers ran for cover and melted into the countryside. They were the sole survivors of the massacre.[1]

The Seething Hell of the Kansas-Missouri Border

Since 1854, Unionists and secessionists on the Kansas-Missouri border had clashed intermittently as Northern emigrants to Kansas sought to bring that state into the Union as a free state without slaves. Kansas and Missouri soon became locked in a bitter war that foreshadowed the Civil War.

On the Missouri side, a group known as the Border Ruffians mounted a desperate attempt to preserve their slaveholding culture that would turn into a brutal no-quarter-asked, no-quarter-given guerilla war. The Border Ruffians thundered into Kansas bearing brightly colored flags inscribed with mottos reading "Southern Rights" and "The Superiority of the White Race." Their equally brutal opponents, the Kansas Jayhawkers, killed, burned, and pillaged in retaliatory raids across the border into Missouri. Before long, the Jayhawkers and the Border Ruffians had both lost sight of the political reasons for their enmity and took to raiding for the sole purpose of securing booty. The favorite objects of plunder were money, clothing, and Sharps carbines.

The Leas of Lee's Summit, Missouri

Frank and Joe Lea were the sons of Dr. Pleasant J. G. Lea and Lucinda Callaway, and eighteen and twenty years of age at the time of the Lee's Summit massacre. Joseph "Joe" had been born November 8, 1841, and younger brother Franklin "Frank" on July 18, 1843, both in Cleveland, Tennessee.[2]

Their father, Doctor Lea, had attended Jefferson Medical College in Philadelphia in the 1840s, and

Joseph C. Lea. Date unknown. Photograph by B. Schluter, Art Gallery, 516 Main Street, Kansas City, Missouri. Courtesy Historical Center for Southeast New Mexico, Roswell, negative no. 4377.

returned to Cleveland, Tennessee, in the spring of 1847 to practice medicine.

In October 1849, the Leas joined a wagon train to Missouri. They arrived November 22, 1849, at what is now Lee's Summit in Jackson County, Missouri (Lee's Summit may be a corruption of the Lea family name). The doctor and his brother Alfred were among the first to settle on high land away from the swamps and mosquitoes of the Missouri River. Lea prospered as the local doctor and as a successful farmer, allowing him to build a large house on the road to Independence. Lea's wife died in 1857.

Joe and Frank attended school nearby and grew up to become well-educated young men. Joe was six feet, four inches tall, Frank a couple of inches shorter. In 1860, Joe, Frank, and younger brother Alfred, lured by news of gold finds in Colorado, made up a freight-wagon train and crossed the plains overland to Colorado. At the young age of nineteen, Joe was in sole charge of the wagon train. The party encountered Indians on the way and in Colorado as well, where the boys worked as prospectors, lumbermen, and freighters. The brothers were still in Colorado when the Civil War broke out in 1861.

Murder of Doctor Lea

The previous year Doctor Lea had married Fanny, a Yankee woman from Massachusetts. Because Fanny was a Northerner, the doctor became suspect to rabid Southern sympathizers in spite of being a Southerner by birth and owning six slaves. The Kansas-Missouri guerilla war rapidly worsened on the borderlands between the two states, which became known as "The Burnt District" and extended to Lee's Summit. Fanny fled to the safety of her parents' home in Massachusetts with her two young stepdaughters.

In 1861, Doctor Lea was murdered at the hands of a mob of Jayhawkers or Union soldiers (it was often impossible to identify raiders). The mob surrounded Lea's house in Lee's Summit, set it on fire, and shot the doctor through the heart when he came out to confront them.[3]

The constant raids by Kansas Jayhawkers fanned hatred on the part of Missourians, and an equally passionate desire for revenge. To make matters worse, a full-fledged civil war was also now under way within Missouri, with fierce hostility between Unionists and secessionists. The two sides battled viciously for the upper hand throughout the towns and countryside. It was into this maelstrom of violence that the three Lea boys returned from Colorado to find the Lea family home in ruins and their father buried in a nearby grave.

Riding with Quantrill

Lee's Summit has figured prominently in the well-known saga of Jesse James and three of the Younger brothers—Cole, Jim, and Bob. Some researchers have suggested that the Younger brothers and the Leas were cousins. They were certainly close friends.

On July 20, 1862, the Younger boys' father, Henry, met the same fate as Dr. Pleasant Lea: he was ambushed and killed five miles from Independence. Tall, broad-shouldered Cole Younger thereupon joined a guerilla band under the black flag of William Clarke Quantrill, a name that would become feared and despised by thousands in both Kansas and Missouri.[4]

Driven by a desire for revenge, Frank and Joe Lea also joined forces with Quantrill and the Youngers. They quickly learned how to wage guerilla warfare from Quantrill: carefully reconnoiter the enemy and assess the risks, strike rapidly and unpredictably using the element of surprise, never give the enemy a chance to concentrate his forces, retreat when the odds are against you, and, finally, give no quarter, ask no quarter. Quantrill's name came to be linked to some of the most violent and bloody deeds of the Civil War and to some of its most

Franklin H. Lea. Date unknown.
Photograph by B. Schluter, Art Gallery,
516 Main Street, Kansas City, Missouri.
Courtesy Historical Center for Southeast
New Mexico, Roswell, negative no. 652.

daring raids. By the summer of 1862, Quantrill's Raiders had become legendary killers. Cole Younger and Joe Lea were among the leaders of the bushwhackers.[5]

In July and August 1862, Col. Upton Hays recruited a regiment of farm boys for the Confederate army in Jackson County to drive Col. James Buell from Independence, Missouri. Cole Younger helped Hays recruit soldiers into the Second Missouri Cavalry (later part of Joseph Shelby's Sixth Missouri Cavalry Brigade). The winter of 1862 found these recruits with Colonel Shelby's regiment in the Confederate camp at Cross Hollows. This was most likely the outfit that Joe and Frank Lea joined. On August 11, this regiment, led by Quantrill, captured Independence from the Union troops.

It is impossible after Independence to trace the exact movements of Quantrill, the Youngers, or the Leas because of the fluid formation and breakup of guerilla and military units and the complications of the war in this area. The most likely scenario is that the Lea brothers began with Quantrill, rode with Shelby next, then alternated between the two outfits.

The Infamous Lawrence Massacre

On August 18, 1863, after careful planning, Quantrill and his captains completed plans for a massive strike on Lawrence, Kansas, home of known Jayhawkers. His force had now swelled to almost 450 men. The Raiders started from a farm northeast of Lee's Summit and then rode through the night of August 19. They were only four miles from the Kansas line by dawn. They spent the next night riding under cover of darkness toward Lawrence.

At about five o'clock in the morning on August 21, the Quantrill Raiders thundered into Lawrence and executed a massacre that lives on as one of the worst atrocities of the Civil War. They invaded homes, shooting every man in sight and torching the houses. The men of Lawrence ran for cellars, attics, and cornfields; some even crawled under the board sidewalks, but they were relentlessly pursued and systematically slaughtered. The killing continued until about nine o'clock that morning when a lookout warned of the dust of approaching Union forces. The guerillas regrouped into ranks and left as suddenly as they had come.

Lawrence was a smoking pile of ashes, most of its men dead. The exact number of deaths is unknown, but it was at least 150 and probably upward of 200. Joe Lea was wounded in the raid; it is not known whether Frank was wounded.

Quantrill's Raiders Disperse

In the winter of 1863, Cole Younger said that he and Joe Lea were still together with Quantrill at Mineral Creek in north Texas (Frank was likely still with his brother). Quantrill's forces began to splinter, their unity of purpose and enthusiasm evaporating after the Lawrence Massacre. The Lea brothers and others broke off from Quantrill around this time and joined regular Confederate forces. After 1863, Joe Lea served under other officers and became an officer himself, exercising considerable authority. Capt. Jason James states in his memoirs that Joe Lea was his battalion commander during the last two years of the war. He believed that Lea was born to command and had good judgment, and he counted Lea as one of the best men he had ever known.

Reconstruction

After the war, Joe and Frank traveled to Georgia and Louisiana, and then parted ways as Joe moved on to Mississippi to farm cotton.[6] Frank had met and fallen in love with a Louisiana girl, Sue Whetstone. The two were married November 14, 1866, at Auburn, Sue's plantation home near Bastrop in northern Louisiana.[7] The newlyweds lived at Auburn until after the birth of their son Joe and daughter Minnie, after which they eventually moved back to Frank's former home at Lee's Summit. Frank and Sue lived in Lee's Summit for almost a decade, and three more children were born: Carrie, Jennie, and Gertrude.

Joe had shaken off the dust of the Deep South entirely after Mississippi and headed for a brand-new start in New Mexico Territory. He arrived in Colfax County in 1875, and then moved south to the Pecos Valley in the fall of 1877. There Joe would become a leader in the formation and development of Roswell. At the urging of both Joe and brother-in-law Asbury Whetstone, Frank decided to leave Lee's Summit for a new start in New Mexico.

To New Mexico and White Oaks

In 1879, Frank sold his land and traveled with his family by train to Las Vegas, New Mexico. Asbury met the family there, bringing with him a hack and an ox-drawn covered wagon to help move the Leas and their effects to Roswell. The journey across the desert from Las Vegas to Roswell was tiresome. Nothing but barren land could be seen for miles with the exception of a few houses at Fort Sumner. As the Leas approached

Roswell, they crossed North Spring River, so lush, beautiful, and welcoming that they imagined themselves in the Promised Land. Both North and South Spring Rivers ran full in those days.

Roswell turned out to consist of only one store and a hotel/residence of adobe, both owned by Capt. Joe Lea, and a few houses. The Frank Lea family lived in one room of the store while a house and store were constructed for them in the new gold boomtown of White Oaks to the west.

The trip to White Oaks in the early spring of 1881 went without incident, with the family arriving at White Oaks Spring several miles from the town itself a little after dark. After setting up camp, the Leas slept peacefully, but the following night after the Leas' departure in the very same spot, Indians killed and scalped two drummers. They left the bodies on the ground and took the wagon and horses along with all the drummers' clothes and supplies.

A Roistering Mining Town

White Oaks was not quiet and peaceful in the early 1880s; it was like all raw mining towns during that era. Frank must often have thought he was back in the troubled days of his Missouri youth. The Lea home in the Oaks was shot up on one occasion by local drunks who returned later to do more shooting. The second time the drunks were met by a posse of miners gathered to protect the family. The posse fired on the mob, killing and wounding several. Sue Lea, pregnant at the time, was bedridden for several months from the shock. Her last baby, Pearl, was born not long after the shooting.

The *Roswell Daily Record* in February 1905 called Frank Lea an "original settler" of White Oaks.[8] He used a tape measure to lay out the town; he also built and kept the very first hotel in White Oaks, a frame structure with stone facing on both front and sides.

Frank had become a staunch convert to law and order, so much so that he served as justice of the peace in White Oaks for the three years that spanned a determined campaign to catch Billy the Kid and other outlaws. The notorious "Whiskey Jim" Greathouse was arraigned before Judge Lea after the shootout in which Jimmy Carlyle was killed at the Greathouse-Kuch ranch north of the Oaks. Lea heard the evidence and bound Greathouse over to the grand jury under a $3,000 bond.[9]

Drawing of first hotel in White Oaks, believed to have been built by Frank Lea. Date unknown. Traced from a Maurice G. Fulton photocopy by Robert N. Mullin. Courtesy Robert N. Mullin Collection, Haley Memorial Library and History Center, Midland, Texas.

Back to Roswell for Good

In May 1884, the family returned to Roswell, where Judge Lea built one of the first few houses in town and planted a productive peach orchard. He also continued in state and local government service for two more decades.

In Roswell, Lea actively supported many of Joe's projects. Capt. Joe Lea became such a prominent figure in Pecos Valley projects that he was later dubbed "The Father of Roswell." A southeastern New Mexico county (Lea County) is named for him. Joe concentrated on the cattle business, beginning in the summer of 1881, and formed the Lea Cattle Company in 1883. The company grew and was incorporated as El Capitán Land and Cattle Company, headquartered in Fort Stanton in April 1885. By April 1888, Frank Lea was listed as a director of the company, which had by then relocated to Roswell.[10]

One of the Lea brothers' most ambitious projects was the Lea-Cockrell Ditch. Joe, Frank, and others formed a corporation in June 1887 to take water out of the Río Hondo and irrigate the semiarid areas surrounding Roswell. Frank bought acreage along the Río Hondo from its source to its junction with the Pecos River in support of Joe's ongoing irrigation projects. From 1879 through 1885, Joe bought almost 13,400 acres along the Hondo; Frank bought fewer acres. In the 1890s, Joe, Frank, and the Lea Cattle Company became owners of the old Woodlawn Ditch on the South Spring River, another step in Joe's plan for turning Roswell into a desert oasis.[11]

Frank attended the first meeting to organize a Masonic lodge in Roswell on October 28, 1888, and served as its first secretary in 1889. Roswell Lodge No. 18 was granted its charter on January 20, 1890.[12] In November 1890, Cháves County voters elected Frank Lea as county clerk.

Pecos Valley was still without a railroad. Finally, on

Judge Frank H. Lea family and relatives, July 1888. Standing, *left to right*: Gertrude Lea, Sam Parsons, Mary Lea, Asbury Whetstone. Sitting, top row, *left to right*: Carrie Lea, Frank H. Lea, Susan Caroline Whetstone Lea. Sitting, bottom row, *left to right*: Pearl Lea, Park Lea, Doss Whetstone, Joseph D. Lee, Ralph Parsons, Mack Minter, Jennie Lea. Photographer unknown. Courtesy Historical Center for Southeast New Mexico, Roswell, negative no. 462-C.

October 12, 1894, the locomotive whistle of the Pecos Valley Railway announced the arrival of Roswell's first passenger train. When news spread that the special car of general manager Jeff Miller had reached the hill by the schoolhouse, Capt. Joe Lea and Judge Frank Lea headed a stampede for the scene, with Frank urging his horse "Old Dick" on as fast as he could.[13]

Joe Lea died a decade later on February 4, 1904. Frank became ill about the time of Joe's death in 1904 and died on February 10, 1905, just over a year after Joe died. (Frank's wife, Sue, had died in 1902.) In Roswell, Judge Frank Lea had come to be known as a man of marked individuality, with strong convictions and opinions but first and foremost a loyal, law-abiding citizen and a law-enforcing official. Back in White Oaks, the *Outlook* observed that Judge Lea "held the respect of all, and his many friends all over the southeastern part of New Mexico will be saddened to learn of his death."[14]

NOTES

1. Two sources report this incident: Poe, *Buckboard Days*, 120–21; and Fleming, *Captain Joseph C. Lea*, 15. Sophie Poe lived with J. C. Lea's family when she first came to New Mexico in the early 1880s and was a distant relative of the Leas through her uncle Stephen A. Lea. She no doubt heard firsthand accounts of the Lea brothers' past.

2. Information about the early life of Joe and Frank Lea came from Fleming, *Captain Lea*.

3. Ibid., 13–14; Poe, *Buckboard Days*, 121.

4. Fleming, *Captain Lea*, 17–20.

5. Castel, *William Clarke Quantrill*. This book covers the short and bloody career of Quantrill in the Civil War from winter 1861–1862 to his death on June 6, 1865. His worst atrocity was the Lawrence Massacre of August 21, 1863. In *Captain Lea* (p. 24), Fleming places J. C. Lea as a participant in the Lawrence raid and quotes Donald R. Hale and Joanne Eakin from their book *Branded as Rebels* regarding J. C. Lea: "With Quantrill . . . Wounded at Lawrence on August 21, 1862."

6. Poe, *Buckboard Days*, 122; Fleming, *Captain Lea*, 45.

7. Frank H. Lea, interview by Gertrude Lea Dills, February 26, 1937. Dills follows the Frank Lea family from the time they left Lea's Summit in Missouri through the White Oaks period.

8. *Roswell Daily Record*, February 1905.

9. Keleher, *Violence in Lincoln County*, 286; Rasch, "Alias 'Whiskey Jim,'" 92.

10. Fleming, *Captain Lea*, 133.

11. Ibid., 138, 124, 143. Fleming quotes the acreage bought from an article by Victor Westphall in the *New Mexico Historical Review* of January 1958.

12. Ibid., 167.

13. Ibid., 197.

14. *White Oaks Outlook*, February 16, 1905.

NEW ENGLAND SEA CAPTAIN JOHN LEE
Around the World aboard the Yankee Clipper

Though he inhabited a small body, Capt. John Lee possessed the heart of a lion.[1] He feared no man, and while not inclined to look for trouble, if challenged he never failed to attack, joyfully engaging in battle with fists, axe handles, or broken bottles. He was never known to lose a fight, and toughs who dared test him quickly learned to give him a wide berth. He seemed to actually prefer the intimacy of hand-to-hand combat, but he was also proficient in the use of firearms.

The bearded captain always wore a black bow tie and to the end of his days played the martinet with his large family of children. The children would be sent from the table if they even laughed. Everyone had to have permission to do anything, just as it had once been with his seamen.

Born for the Sea

John Lee began life as Jon Lees, firstborn of James and May Pearson Lees, on November 27, 1838, in Scotland. His father, James, came from Abdie in Fife County, Scotland, and his mother, May Pearson, from Kilmany in the same county. James and May Lees emigrated to the United States in 1840, when their son was eighteen months old. Once in this country, the couple dropped the "s" from their last name, and Jon became John. Thereafter he was known as "John Lee."

The Lees established a home in Moodus, Connecticut, not far from the Connecticut River and a scant fifteen or sixteen miles from Long Island Sound and the Atlantic Ocean. Four more children were born to the elder Lees: Mary Ann in 1840, Jeannette in 1845, Robert in 1848, and James in 1859.

The salt air blowing inland from the Atlantic stirred a powerful longing for adventure in John, compelling the headstrong youth to run away to sea at the age of fourteen and ship out as a cabin boy. He made three trips around the world and two trips around Cape Horn, putting into every port. He also sailed into both the Arctic and Antarctic Oceans. By the age of twenty, Lee managed to buy his own trading vessel, a sailing ship called the *Yankee Clipper*. It speaks well for the leadership abilities of the New Englander that he could command his own ship at such a young age. On one of his trips, Captain John even discovered a small island, which he named after himself, Lee's Island. (This island appeared on the world globe kept in the White Oaks Schoolhouse, greatly boosting the stature of little May and Nettie Lee in the eyes of their classmates.)

Samoa

At the helm of his ship, Captain John sailed the Pacific Islands in the late 1800s, trading with the natives for copra (dried coconut meat) and hemp, which he then sold in Sidney, Australia. As part of his usual routes, he visited the many ports of the Samoan islands of Upolu, Savaii, and Tutuila (the largest island of American Samoa and site of the important port of Pago Pago).

Before long, as a natural outgrowth of the trade in copra, Lee bought his own coconut plantation and opened a store at Apia on the gateway island of Upolu, one of nine islands in the Polynesian island kingdom of Samoa. At Pago Pago, Captain John also bought a large piece of land (which today houses a naval base) and erected another store and stocked it with trade goods.

Captain John always said that Samoa was the most beautiful place he had ever seen in his travels and one of Robert Louis Stevenson's poems was a favorite with him:

> Fair Isle at Sea—thy lovely name
> Soft in my ear like music came.
> That sea I loved, and once or twice
> I touched at isles of Paradise.[2]

In those days, the Samoan islands were administered jointly by Germany, England, and the United States.

Under the American flag, Captain John sailed freely throughout the islands.

Mary Magdaline Purcell, Samoan Princess

A striking half-Polynesian girl named Mary Magdaline Purcell lived with her parents, Ned William Purcell and Salaevalu Si'ilatamai (or Ruth Hanna Mata'afa, her English name), on the Samoan island of Upolu. Mary was born August 18, 1843.

Her father, the Englishman Ned Purcell, born in 1813 and a graduate of Oxford, had found himself on Samoa because of a shipwreck. It is thought that he was a missionary sent from the Church of England to help convert the Polynesian natives of the Pacific. Missionaries dispatched from England had been in Samoa since the 1700s and had already converted a few of the natives. Ned fell in love with Ruth Hanna Mata'afa, married her, and stayed to raise a family at Apia in Western Samoa. Hanna's father was Luteru Si'ilatamai, a Polynesian who had been king of Samoa for many years. As king, his honorific title was "Mata'afa."

It was on the island of Upolu that Captain John first laid eyes on Mary Purcell. She was a tall, beautiful, big-boned Polynesian princess (granddaughter of the king of Samoa). The couple married in 1861, when Captain John was twenty-three and Mary was eighteen. Her wedding photo shows a serious young woman with long

Ruth Hanna Mata'afa Purcell, mother of Mary Magdaline Purcell Lee. Date and photographer unknown. Courtesy Marilyn Lemon Ingley.

Mata'afa, King of Samoa. Date and photographer unknown. From *The Works of Robert Louis Stevenson*, ed. Charles Curtis Bigelow and Temple Scott (New York: Bigelow, Smith, 1906), 7:175.

Samoan war club. This club is 19 inches long and 4 3/4 inches at its widest. At one time the point was sharper but has since broken off. Date and photographer unknown. Courtesy Carol Queen Watt.

Mary Magdaline Purcell Lee in Samoan costume. Believed to be Mary's wedding photo. Date and photographer unknown. Courtesy White Oaks Historical Association.

flowing tresses underneath a crown of braids across her head and a circlet of flowers tucked above her right ear.

Eleven children would be born to the couple, nine in Samoa and two in the United States. The first three children born in Samoa were Edward (November 2, 1862, at Apia on Upolu); John (January 18, 1865, on Savaii); and Robert (April 23, 1867, also on Savaii). The next six children were born at Aleipata on the eastern end of the island of Upolu: Mary Annie (April 13, 1869); George (December 23, 1870); Isabella (April 29, 1873); Elizabeth (December 18, 1875); Elen (March 22, 1878); and James (August 30, 1879). (The last three girls all died young and are buried on Upolu.)[3]

Two more girls were added after the family went to the United States: May in Kinney County, southwest Texas, June 1, 1882; and Jeanette (always called "Nettie") in Kinney County on January 19, 1884. Two or three years after John and Mary were married, the Germans moved to take over Savaii in Upolu. They sent warships to the island, landed, and hoisted the German flag above the English flag. After a night of partying involving too much of the local brew, a brash Captain John climbed the flagpole and tore down the German flag. The Germans threw him in jail, and his missionary father-in-law had to extricate him. As luck would have it, a typhoon struck the island that same night and sank the German warships.

The couple prospered in the Pacific from their plantation, stores, and the sea trade through the islands.

Back to the United States

In 1879, after twenty-seven years of life at sea, Captain John longed to return to America to visit his aging parents. He also wanted his younger children raised in the United States. He sold everything he owned except the coconut plantation in Apia, which he left with Mary's three brothers—William, Edward, and Sam Purcell—to manage. Lee thought that he might some day return, but that was not to be. The *Yankee Clipper* also went on the block. Many of his wife's people, the Purcells, left Samoa at this same time to settle in Sidney, Australia.

Captain John and his family boarded an old commercial sailing clipper and began the three-month trip

Apia Harbor in Samoa, where Capt. John Lee's coconut plantation was located. Date unknown. Photograph by John Davis, Samoa. Courtesy Marilyn Lemon Ingley.

Capt. John Lee and his wife, Mary. Date and photographer unknown. Courtesy Donald M. Queen.

across the Pacific to the United States. Mary had sewn $50,000 in gold certificates into a chamois skin belt that she always wore under her clothes for safekeeping. She also packed several keepsakes of her Samoan heritage. One, a rush basket, had been woven for her by her mother, Hanna. Another was a Samoan war club called a *talavalu* or *lapalapa*. There was also a cloth many feet long and folded accordion-fashion made of beaten plant fiber from the inner bark of the paper mulberry and painted with beautiful designs. Mary later chose to keep the cloth on the covered-wagon trip from Texas to New Mexico in preference to her treasured china.

At sea, the passengers on the clipper were becalmed for thirty days in the horse latitudes, also called the "doldrums." No winds blew and the sea was absolutely still. Water and food grew scarce and had to be rationed. The captain began acting as though he were deranged, and the near mutiny of the crew forced Captain John to assume command of the clipper. Just before the food ran out, the ship made the port of Honolulu. After restocking, Captain John brought the ship safely on to San Francisco. It was a week's trip by rail from San Francisco to New York City on the Union Pacific. The Lees were caught on the train in a severe snowstorm at Cheyenne, Wyoming, and were stalled for three days, during which Mary shared her food and blankets with others on the train.

The Lee family visited John's family in Connecticut for some time. His father, James, was now seventy years old and his mother, May, was fifty-six (she would live until June 1888). After visiting in Connecticut and in other places in the East, Captain John moved his family to Richmond, Virginia, bought a farm, and stayed there for a year and a half.

Ranching in Texas

Lee soon became bored with farming and decided that his next venture would be ranching in Texas. He had wanted a ranch all his life. The Lees traveled by train to Kinney County on the Nueces River about twenty miles from Brackettville, Texas, and bought two ranches, stocking them with horses and cattle. The ranches were not far from the Río Grande. Five years of ranch life proved enough for the always-restless captain. Also, he was fast losing money in his latest venture. A neighbor named McBee had a ranch at White Oaks in the territory of New Mexico, and McBee was always telling Captain John what a great cattle country New Mexico was. In 1886, Captain John packed up his family once again and got on the trail for New Mexico in a cavalcade of five covered wagons drawn by horses. The men drove two hundred head of cattle and sixty horses; the women did the cooking.

Besides John and Mary, there were eldest son Edward, his wife, Addline, and their first son, William, two years old; John and his wife, Hester Cox; son Robert; daughter Mary Annie, her husband, Fred Cox, and their first child; sons George and James; and the two smallest Lee daughters, May and Nettie. The women and children very much feared Indian attack as they had heard of the terrible things done to wagon trains. Their luck held; the Lees were not attacked, though they saw Indians several times.

Disaster on the Pecos

All went well on the trip until the wagon train reached the Pecos River near the town of Seven Rivers in southeast New Mexico. One morning four-year-old May woke up to hear her mother crying. Looking out from the wagon, she saw piles and piles of dead stock all around, poisoned from drinking alkaline water.[4] Now without enough horses to pull all five of their wagons, the family broke most of the good china to lighten the load. They took what remained of the herd up the Peñasco

River to the little settlement of Weed in the Sacramento Mountains. Captain John bought a farm, and the family lived there a year. They raised potatoes, which the boys peddled as far south as the Pecos River.

The captain decided to move on to the ranch of his Texas neighbor, McBee, two miles outside White Oaks. He sold the Sacramento farm and some of the cattle they had left in 1888, then traveled to White Oaks by way of Tularosa and back up north.

A New Start in White Oaks

The Lees lived on the McBee place for a year and ran a dairy there. The captain, often taking daughter Nettie along in his wagon, delivered milk to homes and also to the mines. Milk was delivered in ten-gallon cans, with pints and quarts dipped out from the cans using long-handled dippers. Customers paid ten cents a quart. Captain John and an Irishman named Shattuck partnered in a slaughterhouse and meat market set up in an old log building at White Oaks. The captain built a smokehouse nearby, where he smoked sausages and bolognas and made hogshead cheese and liverwurst. There was such a big family of the Lees, with the married children all settling nearby, that neighbors soon dubbed the settlement "Leesville." (They lived east of Cedarvale Cemetery.)

Captain John also worked as one of the stage drivers from White Oaks to San Antonio and back. On one occasion, he did not return when the stage was due, making Mary very uneasy. The stage had been held up in the past, and she feared that it had happened again and that her husband was dead. Lee was an entire night and day late. Just as the Lee sons and neighbors were saddling their horses to go looking, the stage appeared over the hill. The travelers had run into a terrible blizzard, and the horses could not pull on through the blinding snow. It was very cold, and Captain John and the passengers were nearly frozen. He stopped the stage at his house, and the passengers alighted to thaw out with hot coffee before going on into town. The mustache and beard that the captain always wore were completely covered with ice and snow so that only his eyes could be seen.

After the mines began to fail, the meat market and dairy business fell off. Shattuck left for Bisbee, Arizona, where he discovered the Shattuck Den Mine, started the Bank of Bisbee, and became a multimillionaire. The captain stayed on in White Oaks and eventually bought the Little Casino Saloon, which he operated until moving the saloon a few miles south to the new railroad town of Carrizozo. There he built the Headlight Saloon and ran it until he retired.

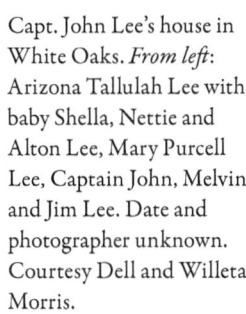

Capt. John Lee's house in White Oaks. *From left*: Arizona Tallulah Lee with baby Shella, Nettie and Alton Lee, Mary Purcell Lee, Captain John, Melvin and Jim Lee. Date and photographer unknown. Courtesy Dell and Willeta Morris.

The Last Years

On Captain John's seventieth birthday, the *White Oaks Outlook* recalled some of his early wanderings:

> Capt. John Lee permitted his friends to smile on him for the seventieth time last Monday.... He became a Nantucket sailor, and as such has sampled salt water in about every port the world over. In one of the Arctic expeditions he came as near climbing the north pole and nailing the flag to that elusive sapling as any man living. He has frequently skimmed along the borders of the Antarctic . . . and counted the innumerable islands of the South Pacific seas. . . . For many years, however, he has been a dry land duck . . . and now wears number nine brogans, at a rough guess. From being a tar he has been tarred with a different kind of stick, and is now the patriarch of a score or more of children and grandchildren.[5]

Because son Jim's family was now in the Arizona mining town of Douglas on the Mexican border, the elder Lees and the Ned Lee family moved to Douglas in 1916. On a hot August night in 1919, Captain John lay semiconscious and dying in the bedroom of a small house in the dusty border town of Douglas. As family watched spellbound, his imagination carried him back in time to the deck of his old sailing ship, the *Yankee Clipper*. All the night long the captain barked orders through his white beard and talked and joked once again with his sailors, the words carrying their accustomed air of authority.

With the morning came the end of a life of high adventure on both land and sea. Lee died at the age of eighty-one. Mary returned to Carrizozo to live with family and died in 1925 at the age of eighty-two. Both Captain John and Mary are buried in Douglas, Arizona.

The Children of John and Mary Lee

EDWARD

Edward, or "Ned," was the smallest of the Lee boys. Captain John always took Ned along on his many sea voyages to act as cook. The boy developed his sea legs early, along with a lasting love for the sea. On one trip, they were caught in a terrible storm, and Captain John had to tie Ned to the mast to keep him from washing overboard while he and the seamen toiled to keep the vessel afloat. He afterward joked that Ned never managed to grow another inch.

In later years, Ned told his family that on their travels through the Pacific islands, they found the natives to be uniformly loving and hospitable. The Polynesians liked hearing even one word spoken in their own language, such as *malo* ("hello"). Ned also described how the natives made their homes using whale bones for the framing, which strengthened the structures enough to withstand the occasional storms that would rake the islands.

By the time Ned was sailing with his father, the Samoans had changed some of their practices from when Captain John first began sailing in the islands. On one occasion, the natives prepared a feast for their distinguished visitors. Still practicing cannibalism, they served a baby as one would a suckling pig. Captain John did not share in that part of the dinner.

Sometimes father and son watched from the deck of

Edward "Ned" Lee family. Back row, *left to right*: Isabella, John, Ned, William, Addline, Lucinda, Fred. Bottom row, *left to right*: Robert, Beulah, Lillie. Date and photographer unknown. Courtesy Winnie Whitaker Eoff.

Emma Amy Lee (Ned and Addline Lee's daughter) with pet Angora goat, 1912. Photographer unknown. Courtesy Winnie Whitaker Eoff.

the ship as islands rose and sank in the ocean owing to volcanic action under the sea. Ned also accompanied Captain John on a memorable sea voyage to Scotland to visit his father's relatives, probably in 1874 when Ned was twelve. The long trip would have taken them around Cape Horn (the southernmost tip of South America) and from there into the Atlantic. They would have sailed along the northwest coast of Africa and along the western coasts of Portugal and France, thence to the English Channel, the North Sea, and finally the Scottish coast. The trip would have been long and formidable.

Ned arrived in the United States when he was eighteen years old and went on with his parents to Texas. In Texas, he married Addline Cox of Kinney County on January 9, 1886, when he was twenty-three; she was three years younger. They had ten children.

Their first son, William, was only a baby when his parents joined the Lee wagon trek to New Mexico late in 1886. The family returned to Brackettville for a time, where son Robert was born. The 1900 census shows the Edward Lee family back in White Oaks, where the last three children were born. In 1914, Ned, at the age of fifty-two, was working at the North Homestake Mine in White Oaks. Son John "Bud" was shift boss in the same mine. Two years later the Ned Lee family followed his parents to Douglas, Arizona. The year 1918 brought great tragedy; oldest son, William, died in October and John in December, victims of the influenza epidemic then raging throughout the United States.

By 1922, Ned had become superintendent of the Douglas Water Department. Ned, Addline, son Fred, and two daughters all moved to Camas in Washington State in 1927. Six years later, on September 26, 1933, Ned died at home. He is buried in the Camas cemetery. Addline lived on until March 9, 1937; she is buried next to her husband.

JOHN

John was a boy of fourteen when he left Samoa for the United States and twenty-three when he arrived with the Lees in White Oaks in 1888. By March 1886, he had married Hester Cox, daughter of a south Texas family. Hester was a feisty woman who hardly ever wore shoes and was as comfortable living outside in a tent on a dirt floor as in a house. She could cook over a campfire as well as other women could on a stove. The couple joined the Lee caravan from Texas to New Mexico, but they seemed constantly to be traveling back and forth between Texas and White Oaks, probably owing to the pull from two

John Lee. Date ca. 1910. Photographer unknown. Courtesy White Oaks Historical Association.

families. John and Hester would have seven children in twelve years. By November 1910, the family was living near Bennett, Texas, and John was working at the Acme Brick Yard. He and eighteen-year-old son George were digging out the clay needed to make the bricks when the roof caved in, killing John instantly. He is buried in a cemetery near Weatherford, Texas, close to Fort Worth.

Hester remembered that day vividly until her death. John was teasing her that morning before he left for work, and she told him to be gone and leave her in peace. John laughed, slapped her affectionately on the bottom, and kissed her goodbye. Shortly after John died, Hester took her children and moved to East Texas.

ROBERT

Robert was twelve when he came to the United States with his family and twenty-one when the family arrived in White Oaks in 1888. He found a wife in the Oaks. Her name was Zaney Paulina Coats, born June 18, 1875, in Cedar Valley, Missouri. In 1884, when she was nine years old, she accompanied her father, William, to New Mexico as part of a twenty-one-wagon Missouri wagon train. William split off from the main group and went to Denver, where the family stayed awhile before turning south again to White Oaks. For some unknown reason, he then took most of his family back to Missouri, leaving Zaney and her sister behind with relatives.

Robert and Zaney married in 1892 when Robert was twenty-five and Zaney seventeen. Daughter Jeannette

Robert Lee family. Wife Zaney Coats holds baby Jeannette. Taken only months before Robert's death in a mining accident. Date and photographer unknown. Courtesy White Oaks Historical Association.

Elizabeth was born March 2, 1893. At the time, Robert was working as a miner in the South Homestake Mine.

On May 13, 1893, Robert was killed when he drilled into an unexploded percussion cap (a "misfire" in mining terms), and it went off. He suffered terrible injuries but lived until the early morning hours before he died. He was only twenty-six years old. Baby Jeannette was two months old and his widow pregnant with their second child, a boy. The boy was named Robert after his father when he was born the following year.

MARY ANNIE

Mary Annie was a child of ten when the Lee family left Samoa. By the time the Lees had sold their ranches in Texas and started by covered wagon to New Mexico, Mary was seventeen and already married (on May 21, 1885) to Fred Nathan Cox. Her brother Ned would marry Fred's sister Addline Cox on January 9, 1886. Little is known about Mary's movements thereafter, except that the Coxes soon returned to Texas.

A descendant of John and Hester maintains that the Cox family came from far down in south Texas, where they were famous for fighting Comanches.[6] Indians continued to be a problem in that area even in the early 1900s. This descendant claims that Fred Cox was scalped by the Indians and lived almost twenty-four hours afterward, but it is not known where or when this happened.

Mary died August 17, 1897, at the age of twenty-eight. She may have perished in the same attack that killed her husband. She is not buried in White Oaks. At the time of Fred's death, the couple had four children. Three of the children died in a house fire a few years after Mary's death.

GEORGE

George was eight years old when he came to the United States with his family and eighteen when they arrived in White Oaks in 1888. For a time, he joined his brothers as a miner. In 1894, he married twenty-four-year-old Cynthia Coats, elder sister of Zaney. Cynthia had also come to White Oaks as a member of the Missouri wagon train and had previously been married to a Frank Wilson. The couple had five children over the years.

A 1907 advertisement in a Carrizozo newspaper reported on George operating a dairy there. He was still in Carrizozo in 1912, according to this newspaper account: "I have been a resident of Lincoln County for 26 years. I own a homestead of 160 acres 2 miles east of Carrizozo and have lived on it for six years. My principal

George Lee family. Top row, *left to right*: Elsie, Rena, Frank Hayward. Bottom row, *left to right*: Robert Dewey, George, Cynthia, Jim. Date and photographer unknown. Courtesy White Oaks Historical Association.

crops have been milo maize and cane, with some corn.... I have also raised the very best of garden truck, irrigating from my well 22 feet deep. I have half an acre in dry farm alfalfa and got one ton this year from one cutting."[7]

Not long afterward, George moved his family to Weiser, Idaho, and operated a meat market there. The move was probably made because Cynthia's father, William Coats, had left Missouri for Idaho and Cynthia wanted to be closer to family.

George lost his eyesight in 1923 following eye surgery. He continued to perform basic duties such as shaving with a straight-edged razor and walking down a plank walkway to a woodpile where he chopped firewood and kindling. He also enjoyed listening to baseball games on the radio. The couple stayed in Idaho until 1924, when George moved the family to Camas, Washington. They lived there until George's death at eighty-four years of age in Vancouver on November 27, 1955. Cynthia lived two more years until her death in 1957 at age eighty-six. Both are buried in the Camas cemetery.

JAMES

James "Jim" was a baby when his parents left Samoa for the United States in 1879 and seven years old when the Lees left Texas for New Mexico. This handsome boy, the last son, would live to be his parents' rock in their old age. He grew up in White Oaks and attended school there. As a boy, he and friends were playing with blasting powder in a glass bottle when the bottle exploded, severely cutting his face and leaving a deep scar. Years

James Lee family. Top row: Alton and Melvin. Bottom row, *left to right*: James, Shella Doyle, Arizona Tallulah. Date and photographer unknown. Courtesy White Oaks Historical Association.

White Oaks baseball team, sometime between 1891 and 1895. Back row, *left to right*: Jess Vandever (manager/umpire), Joseph A. Gumm, Jim Tate, John Gumm, Frank Parker. Second row, *left to right*: Jim Lee (pitcher), Pete Gumm (second base), unknown, Paul McCourt. Front row, *left to right*: Gene Stewart (shortstop), Bob Ransom (third base), John McCourt (fielder). Photographer unknown. Courtesy White Oaks Historical Association.

later as a grown man Jim would have to cross the border from Douglas, Arizona, where he lived and worked, into Agua Prieta, Mexico, to find a barber to shave him; the American barbers did not want to bother having to shave around the scar. He spoke fluent Spanish and had no trouble communicating with the Mexicans.

On February 6, 1902, in White Oaks, Jim married Arizona Tallulah Smith, a pretty nineteen-year-old who was always called "Tallulah." Jim was twenty-three. Their five children were all born in White Oaks. Years later the couple also adopted grandson James Melvin "Mel" Lee, born April 6, 1930, and raised him as their own from the age of three months.

As a young man, Jim loved baseball; he was a champion pitcher on the wildly popular White Oaks baseball team. On one occasion when the team was playing near a railway station, the train stopped close to the field where Jim was pitching. A scout from one of the major league ball clubs happened to be aboard and saw the standout pitcher in action. After the game, he collared Jim and offered to sign him to a contract immediately if Jim would board the train right away. Jim said he would have to go home and talk it over with his wife.

Tallulah delivered an emphatic "no." She thought that ballplayers were worthless playboys who never amounted to anything. Later she deeply regretted her words, realizing that this could have been the big chance of Jim's life. He would spend most of the rest of his life as a miner in that tough and unforgiving occupation; it ruined his strong body. He developed silicosis (the miners' disease), which destroyed one lung and part of the other.

In 1913, with the mines played out, Jim tended bar at his father's saloon in White Oaks and in 1916 moved his family to Douglas, Arizona, along with the elder Lees and many other relatives. In Douglas, Jim first went to work at the old C&A Smelter. In those days, there was no heavy equipment such as bulldozers, just "pine handles" (the name all miners who lived the life gave the shovel). At the smelter, he was also a champion weightlifter; at lunchtime Jim would lift train wheels on a big steel bar in contests of strength. Adopted son Mel said of Jim, whom he always called "Pa":

Pa was an expert powder man, as other members of our families were, and I saw many times . . . a new truck with Mexicans pull up to our house . . . and [get] Pa to go to Mexico on one of the large ranches. [He would] spend a few days to "shoot" some deep water wells they had sealed off. This is a very specialized operation. Not quite enough [explosive]—no improvement; too much, and the whole well is ruined . . . just the right amount and the well was renewed. Pa had total success and went down there numerous times. The big ranchers treated him like a king. Best food, best quarters, fair pay in American money.[8]

Mel also recalled that Jim was expert in double jacking, a mining operation in which one man holds the steel drill and the other swings an eight-pound hammer to strike the steel and dislodge ore. The operation is very dangerous for the man holding the steel. Jim never missed with the hammer.

After twenty-three years at the mining firm of Phelps Dodge Corporation, Jim came home one day to deliver the worst of news. His foreman had told him that he was through after he finished out the day. He said that Jim had done nothing wrong, but "high officials" felt he was too old to keep up such hard work. He should have received a pension, but management counted every single one of his days off and sick days and whittled the total down to just weeks under the twenty years required. He would not get one cent. He was given a cheap yellow-handled pocketknife as a memento of his twenty-three years of hard labor.

Jim died in Douglas on January 9, 1962. Tallulah followed four years later, on January 8, 1966. Both are buried in Calvary Cemetery, Douglas.

MAY

May was born in the United States; she was four years old when the family left Texas and six when they reached White Oaks. The friendly, outgoing girl with the big brown eyes grew up in the Oaks and attended school in the red-brick schoolhouse. She spoke fluent Spanish. On January 1, 1902, she married Edward Queen, a handsome mining man with striking blue eyes.

Edward and May Lee Queen on wedding day, January 1, 1902. Photographer unknown. Courtesy Donald M. Queen.

On January 9, the *White Oaks Eagle* reported, "One of the prettiest weddings that has taken place in White Oaks for a long time was celebrated at the Methodist church New Year's night.... The bride's dress was a pearl white china silk, with yoke of over lace, and trimmed with bertha silk, finished with lace. The skirt was walking length, trimmed with tucked ruffle. Her hat was white velvet, trimmed with white china silk, and a cluster of white tips. Shoes and gloves were white to match [the] rest of costume."[9]

Ed and May remained devoted to each other throughout life. Three children were born: Ellyn on September 21,

Edward Queen family. Top row, *left to right*: Donald John, Ellyn Allie, Lawrence. Bottom row: Edward and May. Date and photographer unknown. Courtesy Donald M. Queen.

1902; Lawrence on March 15, 1905, and Donald John on December 6, 1909. Lawrence was born at 6:00 A.M. just as the whistles began blowing for the men to go to work at the mill. This was also the day the Wild Cat Leasing Company, in which his father was a partner, began making its first cleanup at the mill.

Before long, Ed had prospered enough to buy the Gumm House, one of White Oaks' finest. May was a model wife and mother. She loved to cook and sew and was a talented musician, playing the organ and piano. Her Methodist faith was an important part of her life, and she played the piano and sang for the Methodist Church. She also liked an occasional drink and her cigarettes, though in her time women pretended they did not smoke and their husbands went along with the game.

In 1917, the Queens left White Oaks, Ed's mining interests taking them to Douglas, Arizona; Zacatecas, Mexico; Los Angeles; and Nevada. During the Great Depression, they mined in California, Arizona, and Nevada. In 1933, they returned to White Oaks, bought the Lane house, and survived those dark times by mining, farming, and hunting. With the coming of World War II, the family moved to San Diego to work in the war plants. They also managed the Hotel Ramona in Ramona, California, until 1948. Two pictures hung on the wall of their kitchen at Ramona. One was of Jesus, the other of Franklin Delano Roosevelt.

In 1949, age and ill health (May had cancer and Ed suffered from silicosis) brought them home to White Oaks for the final time. Ed died April 22, 1949, of a heart attack during a family gathering. May followed him on November 1 of the same year. Both members of this devoted couple are buried in Cedarvale Cemetery.

JEANETTE

Jeanette "Nettie," the last child of the Capt. John Lee family, was born in 1884. She was almost three years old when the family left Texas for New Mexico and four when they arrived in White Oaks. Like her sister May, she attended school in the Oaks. In 1906, she moved to Carrizozo to teach. After nine months, she switched jobs and went to work at the old Exchange Bank, Carrizozo's first bank. She worked there until her marriage to Raymond E. Lemon.

Ray, a Kansan born in May 1883, had come to Carrizozo in 1908 as a cashier for the El Paso Southwestern Railroad but returned to work for the Rock Island

Nettie Lee Lemon, ca. 1918. Photographer unknown. Courtesy Donald M. Queen.

Railroad at Pratt, Kansas. When the two decided to marry, they chose Dalhart, Texas, as their meeting place. Sister May and brother-in-law Ed Queen accompanied Nettie to Dalhart, Texas, where Ray met the party. The couple was married in the parlor of the De Soto Hotel at Dalhart on February 16, 1911.

The Lemons lived in Pratt for nearly a year before returning to White Oaks. Ray kept books for the Wild Cat Leasing Company and the White Oaks Mercantile Company.

The couple had one son, Maurice, born in 1912. Nettie and May and their husbands remained best friends throughout life, despite different political views. Their paths diverged when, with the "ghosting" of White Oaks, the Queens left for mining interests elsewhere and the Lemons moved south of White Oaks to the railroad town of Carrizozo. Eventually, Ray became a banker and managed the Citizens State Bank in Carrizozo for several decades.

Nettie was a warm, intelligent woman with enormous energy. Like her sister May, she played the piano and organ, and religion was an important part of her life. She stayed active in the Carrizozo Women's Club and Eastern Star to the end of her life. She enjoyed the domestic life and entertaining—and (like her sister)

Raymond Edward Lemon, ca. 1918. Photographer unknown. Courtesy Donald M. Queen.

her cigarettes and an occasional drink.

Ray and Nettie personified the best of American middle-class values: they were conservative, compassionate, honest, industrious, optimistic, and tolerant. After long and useful lives, Ray died in August 1972 at age eighty-nine and Nettie in 1977 at ninety-four. They are buried in Cedarvale Cemetery in White Oaks.

NOTES

1. In compiling this piece, I had the invaluable help of a host of Lee descendants. Correspondence, telephone calls, and a series of in-depth interviews over a period of eight years produced many reminiscences told to Lee children and grandchildren by the sons and daughters of Capt. John and Mary Lee. These descendants also shared a wealth of photographs, memoirs, genealogical records, and maps. Those who provided helpful information include Judith A. Carey, Bellevue, Washington; Winnie Whitaker Eoff, Roswell; Marilyn Lemon Ingley, Bend, Oregon; Brenda Lee-Sprouse, Arlington, Texas; Melvin James "Clem" Lee, Thatcher, Arizona; Pat and Dee Ann Lee, Appleton, Washington; Dell and Willetta Morris, Fort Thomas, Arizona; Jeannette Lemon Olson, Gales Ferry, Connecticut; Donald M. Queen, Carrizozo; Tallulah Lee Sherer, Payson, Arizona; and Carol Queen Watt, Port Charlotte, Florida.

2. Stevenson, *New Poems*, 82.

3. John Lee family Bible, "Birth" page listing births and deaths of the eleven Lee children, copy in author's collection.

4. May Lee Queen, interview by Edith Crawford, June 13, 1938.

5. *White Oaks Outlook*, November 30, 1905.

6. Brenda Lee-Sprouse to Haldane, September 25, 2000.

7. *Carrizozo News*, August 2, 1912.

8. Carol Queen Watt to Haldane, March 28, 2000.

9. *White Oaks Eagle*, January 9, 1902.

John Canning house, built in the late 1880s. Raymond Lemon, his wife, Nettie, and son Maurice are in front. Date and photographer unknown. Courtesy Marilyn Lemon Ingley.

THE LESLIES
Settlers at Texas Park

In the late 1880s and early 1890s, several large families came into Lincoln County and founded a settlement just north of Tucson Mountain and southeast of White Oaks. Most hailed from Texas, hence the settlement came to be called Texas Park. The Leslies were one of those vigorous frontier families destined to grow and thrive in Lincoln County.

The First Robert Leslie

Robert Leslie came to Erath County in central Texas from Georgia in 1855 by ox wagon train.[1] Three children had been born to Robert and his wife in Georgia; six more would be added in the hamlet of Old Doublin southwest of present-day Fort Worth.

Leslie operated a sawmill, a cotton gin, and corn and flour mills in Old Doublin (later simply Dublin) near the joint county lines of Erath and Comanche Counties. Old Doublin got its name from the practice of "doublin' up" by frontier families to protect themselves from raiding Comanches.

The Leslies in Erath and Comanche Counties

Robert Leslie and his family lived a frontier life, working the land and raising hogs. Leslie's wife raised large flocks of different kinds of birds and every year plucked enough duck and goose feathers to make a big feather bed. Like other frontier women, she spun and wove the cloth for all the family's clothes. The children walked two and a half miles to school while carrying bottles of milk and tin lunch buckets of buttered biscuits sweetened with molasses. On the way to school as often as not they met rattlesnakes, centipedes, tarantulas—and sometimes even wolves, bears, or wild cattle. The school had only one book, the *Blue Back Speller*, and the children wrote and ciphered using a slate and chalk.

The Leslie clan lived in Dublin until after the Civil War, when Robert moved his family to Rusk Creek and preempted 160 acres of public land. On this land he raised hogs. He and two of his sons periodically traveled the ninety miles to Fort Worth. They drove their herd of hogs by foot and set up camp on the banks of the Trinity River, where they killed the hogs and prepared them for market. A big ox wagon accompanied them to hold equipment for processing the hogs, including a huge wash pot for scalding the animals after slaughter. It took a month to prepare ninety hogs for sale.

On the return trip from Fort Worth, the elder Robert always hauled one thousand pounds of flour back home. On one such trip, he hauled back a cook stove—the first in Erath County, since most people cooked on open fireplaces. Depending on the time of year, he would also haul back loads of apples.

In Comanche and Erath Counties, the elder Robert raised so much corn, originally intended for the hog-raising business, that he had to buy a small mill to grind the corn. Once in business, he ground corn as well as flour for all the people in that part of the country.

Unsolved Murder of Robert Leslie

One day, the Leslies' daughter Arizona Vera who was the first child born in Old Doublin in 1862 found her father unconscious from a bullet wound and half-frozen in a lonely thicket. Hours later she watched him die without having been able to positively name his attackers. The mystery of Robert's death was never solved. The Hardin Gang could have been responsible. The Hardin Gang was suggested by the fact that Robert mumbled, "Hardin, Hardin" several times after his family had moved him to a fire and tried to revive him. The Hardins were operating nearby at this time and in May 1874 killed Brown County deputy Charles Webb on the streets of Comanche, Texas.

The murderers could as well have been Comanches. During the late 1860s and early 1870s, Comanches still raided that part of Texas. One moonlit night a neighbor

Robert Leslie, Sr. Date and photographer unknown. Courtesy Robert and Dorothy Leslie.

of the Leslies heard a noise and, looking out his cabin door, saw an Indian ready to steal his horse. He shot and killed the Comanche. Another night this same neighbor's dog sounded an alarm down near the river. Taking his gun, the man set out to investigate and crept along the trail to the river when he was jumped by an Indian, who pinned him in such a way that he could not use his gun. Desperate, the man managed to reach his bowie knife and kill his attacker.

The Second Robert Leslie, "Robert Senior"

Robert Leslie, son of the elder Robert, was a native of Fulton County, Georgia, and was born May 17, 1853. His wife, Elizabeth Ward, was born January 6, 1857. The couple married in Dublin in 1871 and became the parents of ten children including Elisha "Lish" and Robert, the third Robert in this line.

The younger Robert, who came to be known as Robert Leslie Senior, owned several big, heavy teams and sometimes used them in construction work. At one time, he had four teams at work for a railroad company.

From Texas to the New Mexico Territory

In 1883, a man named Jack Farr came to Dublin from Lincoln County in New Mexico Territory.[2] He spread the word about a beautiful land and friendly people in Texas Park near White Oaks. Robert Leslie was ready to make a move; he thought Texas was becoming too crowded with people from the East. No doubt the murder of his father was another incentive. He decided to strike out for this new land. Joining him would be the Yorks, the Arthurs, and two brothers of the Carter family, who would care for the wagons and stock on the move. In this Leslie family, there were the parents, Robert and Elizabeth, and children Lish, Lura, Jim, and Callie. Eldest son Lish, a lad of only ten years, was recruited as a third hand to help with the two hundred head of cattle. Three covered wagons made the trip—one for each family with children. Each wagon had its own provisions, chuck box, and water kegs, and each family did its own cooking over a campfire. Women and children slept in the wagons, the men on the ground. They grazed the cattle along the way. If there was good grass and water, the travelers might stay a week in a camp. For fresh meat, the men shot prairie chickens and antelope. Fortunately, they saw no Indians and arrived in New Mexico without any serious trouble.

At Texas Park and White Oaks

The little wagon train crossed the Pecos River at Fort Sumner. Billy the Kid had been dead only a couple of years, and the families rode out to see the grave of the notorious outlaw. (Robert Senior had once met Billy at a railroad construction camp but did not know him well.) From Fort Sumner, the wagon train traveled to Roswell and then up the Río Hondo in Lincoln County, through the Mescalero Indian reservation, and finally to Weed, New Mexico.

The travelers arrived at Weed in the fall of 1883, after being on the road for three months. There the two Carter drovers left, and the Arthurs stayed only a year or so before returning to Texas. The Yorks settled around Weed in Peñasco country. After only a month in Weed, Robert Senior decided to scout out the Farr Ranch, which was some eight miles from White Oaks. (This same ranch was later known up to the 1930s as the Felix Guebara Ranch.)

Robert must have liked what he saw, because after he returned to Weed the Leslies gathered their cattle and drove them through the mountains to the Farr Ranch, where they stayed all winter. In the spring of 1884,

Robert Leslie, Sr., house. Photograph by Bob Workhoven. Author's collection.

Robert Senior filed on a homestead in Texas Park at the foot of Tucson Mountain. After living on the homestead for several years, Robert bought the Paden place in White Oaks as a base for sending his children to school. When school was out, everyone returned to the Texas Park homestead. The White Oaks house was near enough that Robert could travel back and forth and keep the homestead going.

In 1916 the *Carrizozo News* reported, "Bob Leslie, Sr. is erecting a fine adobe residence . . . on the site of the old Paden home. It is to be hoped this will be a starter for a new building revival in White Oaks."[3]

Robert built a big barn on the place in White Oaks where he could drive a team in, stop, and have a place to hang the harness. Robert Senior was father and grandfather to many. Grandson Robert Ward Leslie and his two sisters grew up with these grandparents in White Oaks after little Robert's mother died. Robert Ward wrote a poem about that house:

> My grandfather's house still stands
> Sturdy, hearty, and hale
> Keeping watch over its people
> While they rest in Cedarvale.
>
> Its foundation extends above the ground
> Exactly two feet high;
> It is made of sandstone that
> Came from the hills nearby.
>
> Its floors are made from hardwood;
> They were shiny and slick,
> And its sturdy walls are built
> Out of adobe brick.
>
> In its warm fragrant kitchen, Grandmother
> Kneaded a mountain of bread.
> And each night I climbed its stairs
> Up to my frigid bed.[4]

Robert Senior died in February 1932, and his wife, Elizabeth, followed just one month later in White Oaks. Both are buried in Cedarvale Cemetery.

Lish Leslie, One of the Great Ranch Hands

One of the most colorful members of Robert Senior's family was son Lish, born February 14, 1873, in Dublin, Texas.[5] He was an energetic, restless boy who quickly tired of school (though he did enjoy playing on the White Oaks baseball team) and bucked to get out on his own. At sixteen or seventeen, he finally ran away from home and rode to the nearby Block ranch to hire out as a bronc buster. A superb rider, he was unafraid of any horse and soon became one of Lincoln County's all-time great cowboys. For about five years, he worked for the sprawling Block, then owned by Andy and Mel Richardson.

Lish Leslie Bucks out of a Bad Situation

Sometime during Lish's early years, New Mexico Territory swore out a warrant for his arrest for larceny and unlawful branding of cattle—a misdemeanor—and he was convicted."[6] Two deputies were assigned to escort Lish on horseback to Santa Fe and prison. When they stopped to water and rest their horses at shallow Red Lake, Lish put the spurs to his horse and staged a first-class rodeo show, making his horse buck from one end of the lake to the other. The deputies, who had dismounted, stood and watched in amazement. Once Lish reached the other side of the shallow lake, he halted his pony, waved his hat at the deputies, and took off at a dead gallop. The lawmen fired a volley of shots, but Lish was long gone.

Lish landed in Arizona but pined for his folks and friends back in Texas Park. He was not seen again in New Mexico until a couple of years later when he vol-

Lish Leslie. Date unknown. Photograph by Max H. Koch. Courtesy Robert and Dorothy Leslie.

untarily returned and gave himself up. After serving a short prison term at the New Mexico state prison, he came back home, eventually becoming one of Lincoln County's most upstanding citizens.

In 1894, Lish opened a meat market in Springerville, Arizona. A year later he met and married Minnie English and went to work for Harris Miller, who owned a ranch near Springerville. While breaking wild horses on this ranch, his horse fell on him and crushed him badly. It took more than six weeks in a Saint Johns Hospital in Arizona before he could hobble out the front door. His bronc-busting days over, Lish returned to his father's place in the Tucsons and filed on his own homestead nearby. There he raised a few cattle and horses and did a bit of dry farming. As a teamster, he hauled many a load of high-grade gold ore from the North Homestake and Old Abe Mines in White Oaks. Lish and his brothers also took teams of horses and hired out to work on roads in Lincoln County.

Neighbors with the Block Ranch

William Blanchard, another settler in Texas Park and friend to Lish, mentions Lish several times in his memoirs.[7] Around 1896, the Block ranch north of present-day Capitán hired a few "bold, bad men" to get rid of the small-scale cattlemen they called "nesters" and saw as threats to their ranching business—either by driving them off or convicting them of cattle stealing and sending them to the state pen. (The Block was losing hundreds of their cattle to rustlers in these years; many nesters in the area were not eating their own beef.)

Once, when the far-flung Block roundup was due to come to Tucson Mountain, the small cattle outfits around Texas Park and Tucson Mountain banded together and asked Lish to go to the roundup and look after their cattle. This was to ensure that the Block did not claim as its own the cattle from the small ranches. These cattlemen put together a remuda of a few horses for Lish; he would do the work, and the cattlemen would pay him. The day the roundup started, Lish took his remuda and went down to the roundup. Block boss Tom Pridemore was away; he had told his second in command to turn his horses into the remuda and stay to watch the roundup. Lish told this second boss that he had come to represent Texas Park and Tucson men. The boss instructed him to turn *his* horses into the remuda and stay to watch.

Around noon Pridemore showed up, rode over to Lish, and demanded to know what he was doing there. Lish told him that he had come down to represent the upper country in the roundup. Pridemore ordered him to cut out his horses and ride off. Lish behaved himself and returned home this time. There was no shooting.

Not long after, Lish and the Blanchard brothers heard that one or two of their steers had become mixed in with the Block steers at a Block corral. They started down to the Block to get their cattle and went to the ranch store, but, seeing nobody around, they tied their horses to the hitching rack in front and went in. Andy Richardson, principal manager of the Block, was inside with three clerks—and lots of firepower. Lish and the Blanchards told Richardson that he had better turn their calves out and do it quick.

Everybody tramped down to the corral, and Richardson tried to send the Texas Park boys in to retrieve their yearlings. The boys told him that they had not put the yearlings in that corral and were not going to turn them out—*he* was. A decidedly large, fat man and short of breath, Richardson tried to turn out the half-wild yearlings, who were thin and long of breath. He could not, and asked for help; Lish's party refused. Things went from bad to worse. The Texas Park boys decided that three of them needed to spread out and get on three sides of the corral, get off their horses and hobble them, then lie down on the ground with their rifles while the fourth man went in to bring out the calves. The plan worked; there was no trouble that day, either.

Although the episode may not sound serious, during those times three or four men in Block country *had* been bushwhacked. And there would have been more except that the Texas Park boys early on heeded the advice of friend Burleson. A man with much experience in cattle troubles, Burleson advised, "If you want to keep Old Man Trouble out of your house you'd best meet him at the door."

Meeting Old Man Trouble

Lish and the other Texas Park boys thought that Burleson's advice was sound. On one occasion in the early 1890s, the "X" outfit (Blanchard names no names in his memoirs) came into the Tucson Mountain area with three wagons and twenty-four men who intended to get rid of the nesters by raiding all the small ranches—and one ranch in particular, the one owned by Lish.

Somehow the small ranchers got word and Lish made his preparations. Four men were to stay at the home camp; three would hide behind rocks and trees on three sides of the camp, and one man would be sitting in front of the tent. When the X Ranch men showed up, they were to be warned to leave. If they made any hostile moves, they would be shot.

The evening before the X Ranch men were to arrive, Lish sent word to all his friends and relatives that he could use more help. The next morning twenty-seven men and boys arrived at his camp and stayed all day. Among them was one slow-spoken, drawling old fellow who carried an ancient single-shot rifle with a thirty-two-inch barrel. Blanchard mentioned to him that the barrel was crooked. The old man replied, "That's all right. All you have to do is aim old Betsy about four feet high and three feet to the left and you'll hit dead center every time."

A seventeen-year-old, one-eyed boy named Allie Lee was there too. He was from Oklahoma, where he was rumored to have killed a man and been run out of Indian Territory; he was already one of the toughest citizens to be found in Tucson country. Lish had to restrain him from drawing a bead on several of the X Ranch men. Allie explained that he figured the woods were full of X men and he needed to whittle the crowd down *before* he had to fight. When the X men climbed to the top of a ridge eight hundred yards away and looked over the crowd with a glass, they rode off.

The Texas Park Boys in Court

In another case of meeting Old Man Trouble head-on, the Block ranch had a man named Will Yates arrested for allegedly stealing a calf from a Block cow and putting his brand on the calf. Blanchard claims that the charges were trumped up and that there was evidence to prove it. Nonetheless, Yates was served a warrant by a justice of the peace who was working for the Block at the time. Yates was to sit at a preliminary trial.

Six or eight of the Texas Park boys decided to go along to see that justice was served. When the contingent arrived at court, the official quickly decided to shift the case over to another justice of the peace at Salado (now Capitán).

On trial day four, the Texas Park boys rode to court with Will Yates. Since it was against the law to carry weapons into the courtroom, everyone had to turn them in (or, as Blanchard explains, hide them carefully). The Texas Park boys found four men already sitting on one of two benches inside the cramped courtroom. The deputy sheriff sat on the end of the bench with the four men in line down from him. The Texas Park boys expected trouble, so the four of them sat down on another wooden bench exactly opposite the first four men, with about two feet between. Then each Texas Park boy took out his long-bladed pocket knife and carefully began to carve some urgently needed toothpicks. In a very short time, the case was dismissed. Afterward, the Yates case was sent to a grand jury, which promptly dismissed the charge.

Lish's Later Years

In about 1923, Lish moved to Carrizozo south of White Oaks, again opening a meat market.[8] He brought his three older children with him and sent two to school while the oldest girl, Ruby, kept house for her father after her mother's death. In 1933, he married for the second time to Ruby Wright of Albuquerque. The next year the couple moved back to White Oaks and bought the place that Robert Senior had owned. In his old age, Lish went to El Paso, Texas, to live with daughter Ruby. He died there of food poisoning in 1950 or 1951 at around seventy-seven years of age and is buried at El Paso.[9]

Robert Ward Leslie

One of Robert Senior's grandsons was Robert Ward Leslie, born in 1912. When Robert Ward's mother died, he and his two sisters went to live with their grandparents in White Oaks. Robert grew to manhood at his grandfather's house. After a long and interesting life that included many years in Alaska, Robert and his wife, Dorothy Pratt Leslie, returned to Carrizozo where they built their own house and became very active members of the White Oaks Historical Association. The couple worked hundreds of hours helping to preserve Cedarvale Cemetery and all the historical treasures around White Oaks.

In recent years, on summer and fall weekends Robert could generally be found at the White Oaks Schoolhouse Museum, where he welcomed all comers, showing them around and regaling them with tales of a bygone era. Robert Ward Leslie died in January 2003 at the age of ninety-one and was buried in Cedarvale Cemetery as family and friends sang several old hymns to the accompaniment of a guitar.

Robert Ward Leslie at the White Oaks Schoolhouse, June 1993. Photograph by J. Pierre Revel of Switzerland. Courtesy Robert and Dorothy Leslie.

NOTES

1. Early history of the first Robert Leslie family is from undated memoirs by Mrs. Arizona Vera Leslie Gillett of Fort Worth, Texas, reprinted in the *Dublin Progress*, 55th year, No. 51, 1–3.

2. Elisha Leslie, interview by Edith Crawford, October 17, 1938.

3. *Carrizozo News*, May 16, 1916.

4. Excerpts from Robert Leslie's poem "My Grandfather's House," in Birdsong, "Leslie House."

5. Leslie, interview.

6. Stearns, *Carrizozo Story*, 54–55.

7. All accounts of Lish's experiences with the Texas Park boys and the Block ranch are from Blanchard, William E., untitled and undated memoirs of Lincoln County, about 143 pages (the pages are not numbered, and it is not known if all pages have been included), from the collection of Robert and Dorothy Leslie of Carrizozo. A copy also exists in the archives of the Lincoln County Historical Society in Lincoln.

8. Leslie, interview.

9. Dorothy Pratt Leslie, telephone conversation with Haldane, December 29, 2006.

THE ENTERPRISING MAYER BROTHERS
Frederick, Paul, and Charles

The three Mayer brothers of White Oaks came from a German family. Their father, Pelag (or Pilag or Peleg), was born in 1807 in Dunningen, Würtemburg, and died in 1861 in Geneva, Illinois. He married his wife, Anna, in 1841. Anna was born in 1819 in Wilflingen, Hohenzollern, and died in Wapakoneta, Ohio, in 1900.[1] Pelag and Anna arrived in the United States in 1854, after forty-three days at sea with seven children: Felix, Frances, Frederick, William, Mary, and twins Robert and Augusta. Three more children—Paul, Meinred, and Charles—would be born in the United States.[2] Frederick, Paul, and Charles would later go west and settle in White Oaks. Frederick was born February 22, 1846, in Deisslingen, Würtemburg, Germany; Paul on May 24, 1855, in Geneva, Illinois; and Charles on April 13, 1860, in Oswego, New York.[3]

Frederick: The Trailblazer

Frederick, or Fred, was oldest of the three brothers who came to White Oaks. He enlisted in the Union army during the Civil War on May 6, 1864, as a private in Company G of the 141st Illinois Infantry Volunteers, a one-hundred-day unit garrisoned at Columbus, Kentucky. A month after Fred enlisted, he was wounded in the back during a sortie defending a supply depot.[4] After being hit, Fred lay all night where he had fallen, fearing he would die. The next day he was found and taken to a hospital where gangrene set in, forcing him to stay in the hospital lying on a rough bed of pine boughs for a long time before being discharged with a "permanent injury from fever sores."[5]

Fred's doctor ordered him to go west and regain his strength. Fred went as far as the Mississippi River, where he joined a bridge-building gang. While working on the bridge, he fell into the river and injured himself, resulting in more time in the hospital.[6] On his release, he traveled for perhaps thirteen to fifteen years—Ohio for three years; Guthrie Center, Iowa, for a year; Fort Scott, Kansas, for two years; and Texas, for seven years. During the later years, he spent a good deal of time with a party of surveyors in Texas and New Mexico.[7]

Fred worked a summer in southwest Colorado and later went to Taos, New Mexico, before finally arriving in Las Vegas, perhaps in 1879. He freighted goods from Las Vegas and Santa Fe to and from Mexico. Seeing opportunities at every hand, he encouraged several of his younger brothers to come join him. Fred appears in the 1880 Las Vegas census living in a hotel on the plaza of Old Town along with brother Meinred (Menry).[8] Menry did not stay in New Mexico long, traveling instead to Washington State.[9]

When local papers began trumpeting the White Oaks gold strike, Fred joined the gold seekers, arriving at the Oaks late in 1880. Fred became a teamster, sometimes a stagecoach driver (perhaps for brother Paul across the malpais to San Antonio), and owner of two mines at White Oaks—the Flora Temple and the Belle of Tucson.

As a teamster, Fred had a close brush with Indians in 1883.[10] The *Golden Era* reported Fred's return to White Oaks from Red Cloud nearby with a load of ore for a smelter. The load became too heavy for the team, and he stopped to unload a few sacks of ore. As he unloaded the sacks, a party of fifteen renegade Jicarilla Indians suddenly appeared. They climbed atop the wagon and dumped several sacks of ore on the ground to see what they contained. Then they seized his provisions and rode off, saying they had left the reservation because no supplies had been issued to them and they were hungry. Fred escaped unharmed to continue on to White Oaks. Whenever he had a chance, Fred also began to buy up real estate; he eventually owned several lots in White Oaks and applied for a homestead on 160 acres near Willard in January 1906.

Frederick Mayer. Date and photographer unknown. Courtesy Mary Jo Jamison.

Maria Louise Schultz Mayer. Date and photographer unknown. Courtesy Mary Jo Jamison.

In the spring of 1891, a pretty young girl of eighteen named Maria Louise Schultz ("Hattie" to her friends) came to White Oaks from Lakin, Kansas, with her father, Charles.[11] Charles, another German immigrant, had been the blacksmith in Lakin and planned to continue this trade in White Oaks. His daughter came along to keep house for him and to give music lessons. Charles soon returned to Lakin, but Hattie stayed on, living with the Biggs and Hewitt families and making her living teaching piano.

Hattie met Fred Mayer, who lost no time in sweeping the newcomer off her feet. During their courtship, he gave Hattie a necklace and bracelet set of Mexican gold coins brought from Mexico City.[12] When they became engaged, Fred also gave her an emerald engagement ring surrounded with tiny diamonds. The couple married on January 2, 1892; she was nineteen years old, Fred much older at forty-five. Brother Paul also courted Hattie, but Fred won out.

The *Lincoln County Times* reported, "Last Tuesday night Rev. Hurd pronounced Fred Mayer and Maria Louise Schultz husband and wife. While our congratulations go forth to Fred, our sympathies reach out toward Paul. He must feel very much like an orphan. But there are fish yet left, and Paul could hook a good one if he'd only go a'fishing."[13]

Four children—three girls and one boy—were born in ten years to Fred and Hattie: Mary Gertrude, Grace Caroline, Anna Laurie, and Frederick Charles. The family lived in a house across the street from the Biggs family, where Hattie had boarded before she married.[14] Biggs was a Southerner, and since Fred had fought as a Union soldier, the two men were neighborly but never grew close, unlike their wives.

Ten years after her marriage, Hattie died of a ruptured appendix at the age of twenty-nine. Her youngest, Charles, was only four years old. The *White Oaks Eagle* reported that the funeral took place at the Congregational Church, with burial in Cedarvale Cemetery.[15] For some months after their mother's death, the children were cared for by Ina Mayer, brother Charles's wife. Little Charles would sit in Ina's lap and cry. Ina would cook dinner, get it on the table, and then hold him afterward for comfort.[16] After Hattie's death, Fred moved in 1905 to his 160 acres near Willard, New Mexico, and tried to keep his family together with the help of housekeepers. For years, besides his Civil War injuries, he had lived with a cancer on his face. He died of this cancer February 15, 1910, and is also buried in Cedarvale Cemetery.[17]

Fred's brother, Paul, was appointed guardian of the four children. Until the children came of age, they were sent to live with relatives, and Paul later sponsored one of Fred's children (Gertrude) at Ottawa University in Ottawa, Kansas.

Paul: The Entrepreneur Brother Who Lived the American Dream

Paul joined Fred in White Oaks on March 15, 1881, having first landed by rail in Las Vegas on March 3.[18] Departing from his former trade of cigar maker in Covington, Kentucky, he started up a livery stable and feedstore in the bustling boomtown of White Oaks. According to the *Golden Era*, he worked very hard: "Paul Mayer can be found at his feed stable, on White Oaks Avenue, at any hour, day or night. He attends to business strictly, and will not starve your stock."[19]

That same year he successfully drilled a water well northwest of Young and Butler's general merchandise store. Besides his livery stable, he had a dangerous side business. After Paul gained the complete trust of mine foremen at the Old Abe Mine, he was hired to deliver bullion and gold dust to the Santa Fe Railway depot at San Antonio.[20] The mine foreman would pour solid gold ingots at White Oaks, which Paul then couriered to San Antonio, put on a train, and shipped east to the mine owners. Paul invented several ways to hide the gold and slip out of town undetected.[21] A man who worked for Paul named Felix Guevara would sit around White Oaks on a bench pretending to be half asleep. Felix would listen and report any newcomers or suspicious characters.

Supply wagon on White Oaks Avenue. Date and photographer unknown. Courtesy John and Wesley Mottinger.

Bearded man driving a supply wagon. Date and photographer unknown. Courtesy John and Wesley Mottinger.

Taliaferro store and Paul Mayer's livery. Date and photographer unknown. Courtesy John and Wesley Mottinger.

Paul Mayer's ad for his passenger line in the *White Oaks Eagle*, May 23, 1901. Author's collection.

Paul Mayer's ad for buggies and spring wagons in the *White Oaks Outlook*, February 16, 1905. Author's collection.

Paul never told anyone when he would leave. He never left during the day, but always at night, at different hours of the night, and in different directions. He hobbled the horses for grazing and sought out trees or other secluded places to sleep. Once he packed and rode *east* from White Oaks, hiding out a week before setting out again to San Antonio. Paul had a packhorse with a bedroll and camping gear so that he could be gone for a week or ten days. He led another packhorse with the bullion. At times, he took Felix with him on these trips.

Paul also got in the stage line business, placing an ad in the *White Oaks Eagle* that read, "Passengers carried to White Oaks and any part of the country on the shortest notice. Good rigs, careful drivers. White Oaks Passenger Line. Paul Mayer, Proprietor."[22] The *White Oaks Eagle* reported in the spring of 1900 that Paul built a new stable and carriage house fifty feet square and ten feet high of adobe, with a foundation of stone.[23] Paul ran the stage line from White Oaks to Carrizozo after Urbain Ozanne got out of that business. In 1902, Paul added a new Studebaker hack to his stage line.[24] Besides the usual renting of horses, carriages, and hacks, he also rented out sleighs so the locals could enjoy rides in the winter snow.

After failing to catch his first "fish"—Hattie, who had married brother Fred—in the late 1890s Paul began courting Ula Gilmore, who had come to White Oaks to teach fifth grade. For many years Miss Ula played a game of cat and mouse with Paul, finally capitulating to marriage in 1914 when Paul was about fifty-nine and Ula was thirty-six. The couple had no children.[25]

From 1890 to 1894, Paul served as a member of the Lincoln County Board of Commissioners. In 1913, still reluctant to leave White Oaks, he became secretary of the Lincoln County Road Commission.[26] In about 1918, Paul moved to Carrizozo and became a successful realtor and financier. He and Ula built a house of adobe, plastered inside and adorned with beautiful woodwork. The outside was stuccoed, as were many of the houses in Carrizozo. Stunning Navajo rugs, blankets, and elegant oil paintings were scattered all over the house.[27]

Paul Mayer. Date and photographer unknown. Courtesy Mary Jo Jamison.

Ula Gilmore Mayer. Date and photographer unknown. Courtesy Mary Jo Jamison.

Lawyer John Hall recalls Paul's Carrizozo days: "In the 1930s Mr. Mayer had a small office about one-half block east of the old First National Bank. His only business [then] was making loans of his own money to ranchers in the county—all to well-known people with four or five sections of land and some livestock, all on first mortgages and at 10%. One exception—he loaned me a few hundred dollars to make a down payment to Mr. Hudspeth for a few hundred lawbooks—all without a mortgage. He took the risk. I visited with him, did some legal work and probated his estate."[28]

Paul stayed active in the Carrizozo Businessmen's Club for many years. He died in 1945 at the age of ninety after a long bout with Alzheimer's disease and is buried in Cedarvale Cemetery in White Oaks.[29] At the time of his death, he owned deeds to many of the old White Oaks mines. Ula died in 1956 and is also buried at Cedarvale.

Charles: Blacksmith, Deputy Sheriff, Mercantile Owner

Born in New York State, Charles grew up in Ohio and left home at age twenty-one. Later he traveled by train to Las Vegas, arriving in 1883.[30] After some time there, he traveled by rail from Las Vegas to San Antonio and on to White Oaks by freight wagon to join his brothers, arriving during very cold weather in January 1884. There were not yet many buildings in town, and what few there were, were built of logs. By March, Charles

Charles D. Mayer. Date and photographer unknown. Courtesy Ina Hunter Dow.

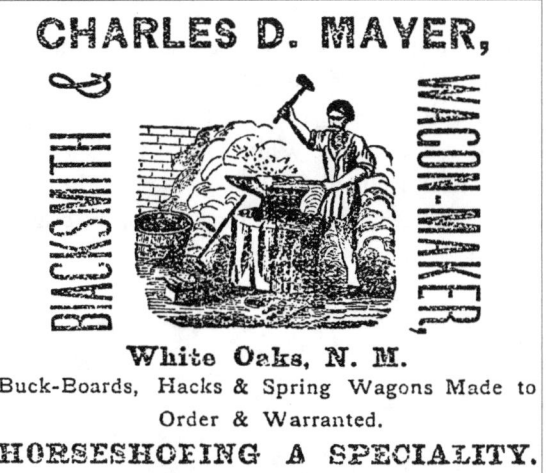

Charles D. Mayer's ad as a blacksmith and wagon maker in the *Golden Era*, September 24, 1885. Author's collection.

Charles D. Mayer's ad for horseshoeing in the *Golden Era*, October 8, 1885. Author's collection.

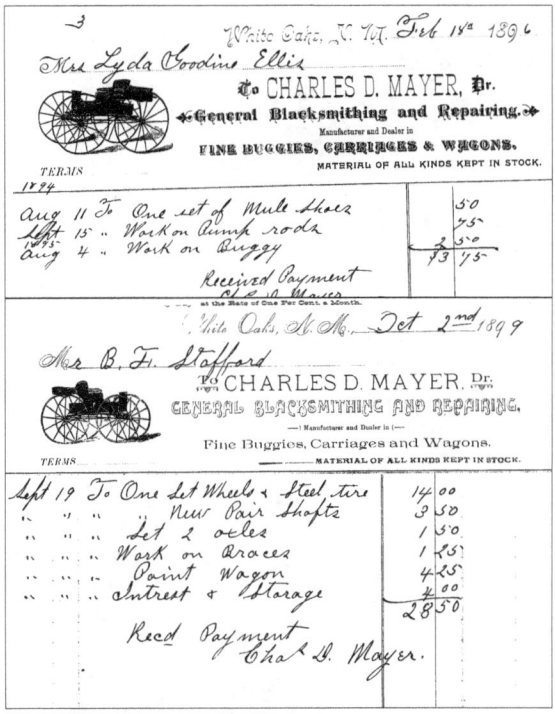

Sample bills, Charles D. Mayer General Blacksmithing. Author's collection.

had opened up a blacksmith shop and did work for the miners. He also shod horses, mules, and oxen for local farmers and freighters. A master ironworker, Charles maintained tools for workers such as the stonemasons from Saint Louis who cut stone for more than a year during construction of the Hoyle House.

Some excitement came Charles's way in 1886, after he was appointed deputy sheriff for White Oaks under Sheriff George Curry. He happened to be in Roswell at the time a man was gunned down at a roundup camp on a mesa above Picacho. A man named George Musgrave was settling a long-held grudge against his former partner, George Parker, who he claimed had turned on him and had him indicted for illegally branding cattle that both men had stolen. Musgrave was tipped off in Arizona about his ex-partner's whereabouts in New Mexico and rode all the way to Picacho to find Parker with the sole purpose of killing him. Musgrave rode into the roundup camp and waited for Parker to arrive. As soon as Parker dismounted his horse, Musgrave fired his pistol four times and killed him. He and his companion, Bob Hayes, then rode off toward the Diamond A Ranch near Roswell. Both before and after the Parker killing, Musgrave and Hayes had taken the opportunity to rob the White Oaks stagecoach at the head of the malpais. The stagecoach was robbed twice, both coming and going. Sheriff Curry asked the twenty-six-year-old Charles to form a posse and follow Musgrave and Hayes. Charles immediately returned to White Oaks and formed a five-man posse to trail the outlaws.

The posse crossed the San Andrés Mountains, came out on the Jornada Flats, and rode on to the Río Grande. When they got to the banks of the river, it was running out of control, full of whirlpooling muddy water.

Finally, the men crossed safely and traveled for two more days until they neared Fairview (now Winston). Charles rode alone into Fairview to the post office to talk to the postmaster, also a deputy sheriff. This man advised him strongly to turn back as he was heading straight into a nest of dangerous men in rough country where the posse would be at great disadvantage. The posse concurred and returned by way of Magdalena and San Antonio, having failed to get the outlaws.

About this time Charles also found the love of his life right at home in White Oaks—Ina Jane Wauchope, born January 18, 1868, in Illinois. She was a granddaughter of John E. Wilson, one of the original locaters of the famous South Homestake Mine in 1879.[31] On January 14, 1888, Charles and Ina were married in White Oaks. Daughter Bertha was born in 1889 and son Paul in 1891.

After twelve years running the blacksmith shop, Charles sold it in 1896, and he and Ina bought a general merchandise store from former owner Jones Taliaferro.[32] He also sold groceries.

To advertise his business throughout the region, he painted "C. D. Mayer for Dry Goods and Groceries" on a huge rock on the left side of the road going up to White Oaks, creating an early-day ad. During World War I, grocery stores were required by the government to sell a certain amount of cornmeal with a certain amount of flour (flour could not be sold without cornmeal because there was a serious rationing of flour), and sometimes it was hard to convince customers of this governmental requirement. Ina had an especially hard time with a little Hispanic boy whose mother would send him to the store with instructions to get five pounds of flour but *no cornmeal*. The boy did not understand the rules and definitely wanted no part of cornmeal.[33]

The Charles Mayers occupied the second floor of their store from 1896 to the fall of 1921; before that they lived in their much-loved little white house bought from the Cannings.

The couple moved their business to Carrizozo in September 1921 to operate a general merchandise store there. When his health failed in 1929 or 1930, Charles retired.

Charles D. Mayer's ad for dry goods store painted on a rock west of White Oaks, 1914. Ethel Graham is the woman on the donkey. Photographer unknown. Courtesy White Oaks Historical Association.

Charles D. Mayer, General Merchandise. This mercantile was first the Young & Taliaferro Mercantile, before it became the Charles D. Mayer, General Merchandise, and finally the Wayne Van Schoyck, Sr., store. Date and photographer unknown. Courtesy of Ina Hunter Dow.

Interior, Charles D. Mayer, General Merchandise. Date and photographer unknown. Courtesy Ina Hunter Dow.

Charles suffered from "pernicious anemia" and became an invalid confined to his bed for some eight years, cared for faithfully by his wife. He died January 29, 1944, and is buried in Evergreen Cemetery in Carrizozo. Ina stayed active in church and community affairs in Carrizozo before moving to Modesto, California, in 1950 to live with her daughter. She died in September 1958 and is buried beside her husband in Carrizozo.[34]

Ina Mayer and daughter Bertha. Ina made the elaborate hats both wore. Date and photographer unknown. Courtesy Ina Hunter Dow.

Charles D. Mayer family. Charles, wife Ina Wauchope, daughter Bertha, son Paul. Date and photographer unknown. Courtesy Ina Hunter Dow.

Bertha and Paul Mayer. Date unknown. Photograph by Max H. Koch. Courtesy Ina Hunter Dow.

C. D. Mayer and horse. Date and photographer unknown. Courtesy Ina Hunter Dow.

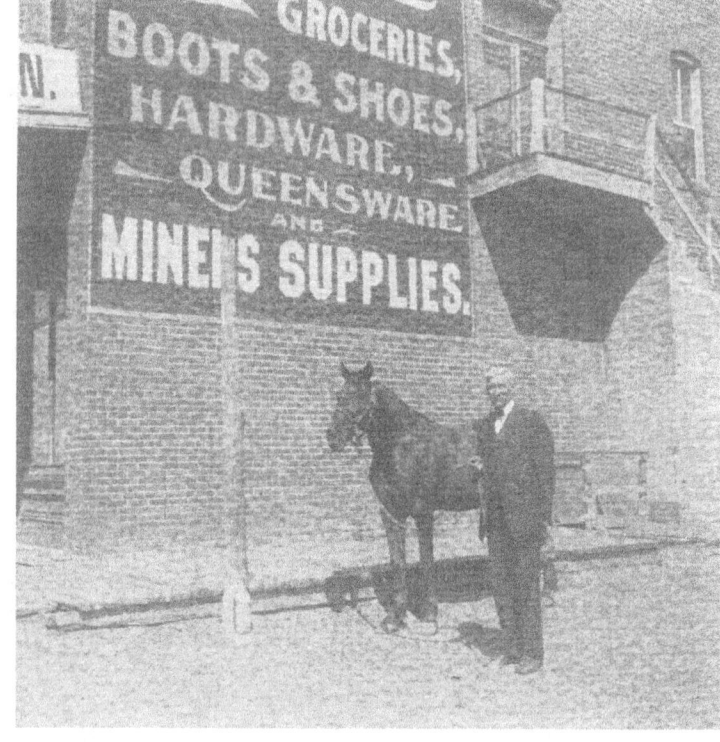

NOTES

1. Ina Dow, granddaughter of Charles Mayer, to Haldane, August 4, 2000, and April 24, 2001.

2. Mary Jo Jamison, granddaughter of Fred Mayer, to Haldane, August 9, 1995.

3. Edgar Turrentine, grandson of Frederick Mayer, Turrentine's genealogy research of the Mayer families, in author's collection.

4. U.S. Bureau of the Census, *Special Schedule, United States Census of 1890*. Enumeration Dist. No. 36, "Surviving Soldiers, Sailors, and Marines, and Widows," W. F. Blanchard, Enumerator.

5. Gertrude Mayer, daughter of Fred Mayer, to Edgar Turrentine, undated, copy in author's collection.

6. Ibid.

7. Paul Hile to Haldane, March 16, 2004.

8. Edgar Turrentine to Mary Jo Jamison, August 31, 1987, copy in author's collection.

9. Mayer to Turrentine.

10. *White Oaks Golden Era*, December 20, 1883.

11. Mayer to Turrentine.

12. Turrentine to Jamison, August 31, 1987.

13. *Lincoln County Times*, January 15, 1892.

14. Mayer to Turrentine.

15. *White Oaks Eagle*, September 25, 1902.

16. Ina Dow to Mary Jo Jamison, August 10, 1994, copy in author's collection.

17. *Willard Record*, February 17, 1910.

18. Haley, *1913 Year Book*.

19. *White Oaks Golden Era*, February 28, 1884, and March 18, 1884.

20. John Hall to Haldane, July 22, 2001.

21. Fred Tully, interview by Haldane, June 26, 2000, Belen, New Mexico; Jamison to Haldane, June 20, 2000.

22. *White Oaks Eagle*, January 25, 1900.

23. *White Oaks Eagle*, April 1, 1900.

24. *White Oaks Eagle*, July 3, 1902.

25. Ellen Key, Records of Green and Gilmore Families, n.d., in author's collection.

26. Haley, *1913 Year Book*.

27. Turrentine to Jamison, August 16, 1988, in author's collection.

28. Hall to Haldane, July 22, 2001.

29. *Carrizozo Outlook*, June 8, 1945.

30. Charles D. Mayer, interview by Edith Crawford, July 19, 1938. In this Works Project Administration interview, Charles recounts his early life in White Oaks and his careers as blacksmith, deputy sheriff, and mercantile owner.

31. Ina W. Mayer, interview by Edith Crawford, 1938.

32. Dow to Haldane, April 24, 2001.

33. Ibid.

34. *Carrizozo Outlook*, February 3, 1944, and September 12, 1958.

WILLIAM CALHOUN McDONALD
First Elected Governor of New Mexico

When he first arrived in White Oaks, William C. McDonald was a twenty-two-year-old former schoolteacher and newly minted lawyer with the sandy hair and blue eyes of his Scottish ancestors.[1] Another trait, not immediately apparent, was his ambition. Once McDonald had settled in White Oaks, he never again taught school or formally practiced law. Instead he set his sights on bigger things.

McDonald lived large the western American Dream before his life ended just short of his sixtieth birthday, boasting enormous land and cattle holdings, wealth, the esteem of his fellow citizens, a happy marriage, and—to crown his achievements—election as first governor of New Mexico after it became a state in 1912.

Beginnings

William was born to John and Lydia Marshall Biggs McDonald on July 25, 1858, on a farm near Jordanville, New York. He was educated in the public schools of Herkimer County, New York, and at Cazenovia Seminary in Cazenovia, New York. In his native state, he taught school for three years while studying law.

He was not content with that life, however, and, like the other restless and ambitious young men of his era, he began to look toward the West. For a while, he stopped off in Fort Scott, Kansas, where he had friends from New York and was admitted to the bar in 1880. He had been in Fort Scott only a short time before hearing of the mining excitement at White Oaks.

Building a Life in White Oaks

In the spring of 1880, McDonald arrived in White Oaks completely broke, like so many of the other young lawyers arriving at the gold camp. He was still a bachelor and bunked with two other young men who were to have bright futures: Emerson Hough, later a famous author, and George Ulrick, who would become a successful

William C. McDonald. Photographer unknown. From *Illustrated History of New Mexico* (Lewis Publishing: Chicago, 1895). Courtesy Chris and Jack Harkey.

lawyer with an office in the town's Exchange Bank.

McDonald found Lincoln County to be a largely undeveloped and primitive region. Mac, as he was known to his friends, immediately recognized the area's potential. He was not content to remain just another starving lawyer; instead, he opted for the steady-paying government job of U.S. Deputy Mineral Surveyor for New Mexico.[2] He also knew civil and mining engineering, skills that served him well in his job as surveyor. And, like everyone else, McDonald speculated in mining. Most importantly, he was already building a network of friends and useful contacts. There were troubles along the way, and McDonald made a few enemies. On

February 16, 1883, McDonald appeared before the local justice of the peace to claim that Will Hudgens, owner of a local saloon, intended to kill him.[3] A warrant was issued for Hudgens, who was hauled into court to answer the charge. Nothing further is heard of the threat after a peace bond of $500 was issued.

All was not serious business, however. McDonald was also elected captain of the popular White Oaks baseball team.[4] People from all over Lincoln County—Roswell, Fort Stanton, Lincoln—traveled miles to swell the crowds and cheer on their home teams.

Politics in Earnest

On his steady upward climb, McDonald was honored with many positions of trust, being the type of man who could command the attention and confidence of business and professional associates. McDonald had always found politics fascinating and was a dedicated Democrat. As early as October 1884, when he ran for county assessor, the *Golden Era* observed,

> No more deserving and capable man can be found in Lincoln county than our candidate for assessor, W. C. McDonald. Mac came to White Oaks nearly five years ago where he has lived ever since. He has been employed as a deputy United States mineral surveyor and is known as a man who thoroughly understands his business, and has given general satisfaction. He represented Lincoln county at the Denver Exposition in 1882 giving the best satisfaction. His public and private life have been above reproach, which should count for a good deal when it comes to choosing a man for so important an office as that to which the people will elect him next Tuesday.[5]

McDonald won the race and served as assessor from 1885 to 1886. In the fall of 1886, the entire Democratic county ticket was elected—with the single exception of the candidate for assessor for a second term, William C. McDonald.[6] Throughout his life, both as businessman and public official, McDonald always acted on his principles. During his first year as county assessor, he had made enemies of some influential cattlemen by trying to place reasonable valuations on county property. He also went after the many tax dodgers then living in the county. He paid the price when cattlemen whose assessments McDonald had raised led a movement to defeat him for reelection.

Knowing that a Republican could not be elected in Lincoln County, the Republicans put up an independent candidate named B. J. Baca, son of Capt. Saturnino Baca and brother-in-law of Jim Brent. Baca had a large following among Spanish Americans. This coalition of supporters succeeded in electing Baca. It was the only time McDonald was ever defeated for public office.

In September 1888, at Lincoln County's Democratic convention held in Lincoln to nominate candidates for county offices, McDonald was unanimously endorsed as Lincoln County's candidate for the territorial House of Representatives. He declined the nomination on grounds that he would not run on any ticket headed by then sheriff Jim Brent.[7]

Decoration Day Speech, May 1890

White Oaks and all of Lincoln County began to notice the young politician in their midst on Decoration Day (the former name of Memorial Day), May 1890. The Grand Army of the Republic, together with a large number of White Oaks citizens, assembled at the Congregational Church at three o'clock. From there, they proceeded to the cemetery to decorate the graves of dead soldiers and friends with flowers and evergreens.

That evening, Dr. Alexander Lane called to order a large assembly, and Judge Hewitt made a strong address to the crowd, dwelling especially on the pension question for Civil War veterans. Then, in the surprise of the evening, Mac McDonald delivered a masterly oration to the assembled audience. The speech was peppered with allusions to ancient history and to the events that had brought about the Civil War, still fresh in the minds of many in the audience. McDonald first evoked the history of past deeds of heroism:

> It is well for us to recall . . . deeds and events which, while they carried sorrow and sadness to loving hearts . . . at the same time swept from our land the dangerous and pernicious doctrines of secession and nullification and the nationally degrading institution of slavery, making these United States one country of freemen all. . . . Alexander sought new worlds for conquest for himself. Caesar beat back the northern hordes from the gates of Rome to maintain more than Imperial grandeur; Napoleon made France a Republic that he might grasp at the Empire of the world. . . . Our soldiers fought to maintain and advance the rights of humanity and individual sovereignty. . . . A

quarter of a century has united the broken links of State and healed the wounds of individual misfortune. . . . There is no longer the North and South of even a few years ago. . . . There are some here who have stood in the ranks of war. . . . We honor the memory of the dead and the presence of the living, who were and are heroic in thought and action. Men who fought with high purpose were made great within themselves, by testing and developing their best qualities.

Then he turned his attention to the present and the future, revealing a forward-looking, optimistic mind:

A new generation has risen since the war and upon the wisdom of them and those who come after will depend the future harvests of the seed sown by their predecessors. . . .We say truly that this is an age of progress. . . . Life is what we make it in our strength or weakness, each one for him or herself. . . .We may read, study, observe, think—develop and enlarge the mind with the principles of right and justice and liberty, and the nobleness of man's station in the universe. The knowledge of the world is at our command to draw from. We have the past as a guide for our footsteps which reach out into the future.[8]

Manager of the Bar W

Sometime in 1890, a group of English businessmen asked McDonald to manage the Carrizozo Cattle Ranche Company (commonly known as the Bar W), a large ranch eleven miles southwest of White Oaks and just north of the then-nonexistent Carrizozo.[9] The ranch was a corporation held by stockholders in England. McDonald had been doing an excellent job as bookkeeper for these English owners since about 1882, knew the business backward and forward, and seemed a logical choice to step into the manager's shoes. The ranch, originally owned by a Spanish American family, was bought by Maj. Lawrence G. Murphy (later of the House of Murphy in the Lincoln County War). At that time, it was called the Fairview Ranch and covered about thirty thousand acres.[10] It later became the Carrizo (also Corizozo, Carasosa, Corisoso, and Carrizozo) Ranch, and still later the more familiar Bar W. The brand of the old Bar W is probably the oldest two-characteristic brand in continual use in the territory and state of New Mexico.[11] The running "W" is like that used by the King Ranch in Texas, with the addition of a bar across the top of the letter.

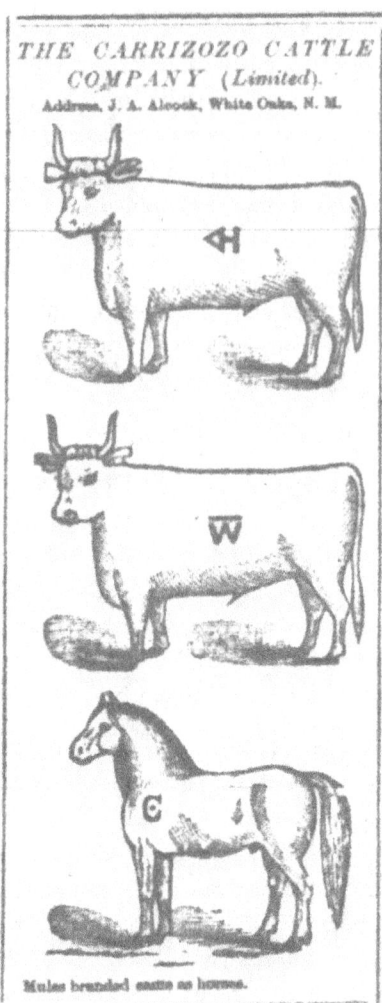

Brands of the Carrizozo Cattle Company in the *Lincoln Independent*, 1889. The present-day Bar W cattle brand is a running W; it is unknown whether the brand evolved or whether the newspaper incorrectly printed the brand. Author's collection.

Murphy's attorney and friend was Thomas Benton Catron of the notorious Santa Fe Ring. Murphy was heavily in debt to the First National Bank of Santa Fe and had given the bank a trust deed to the ranch to secure his loans. Catron, a director and major stockholder in the bank, had lent his personal endorsement to Murphy's notes. After Murphy died almost penniless in Santa Fe on October 20, 1878, the ranch eventually came into Catron's possession. Catron sold it in 1882 for $225,000 to James A. Alcock, representative of an English syndicate headed by a Captain Scott.[12] An Irishman himself, Alcock was part owner of the Carrizozo ranch and its first manager. A "fine chap," he was sadly out of place running a cattle ranch and almost drove the company on the rocks.[13] Alcock's foreman was Jim E. Nabours of Milam County, Texas. At that time the Carrizozo ranch (Bar W) could lay claim to between ten thousand and fifteen thousand head of cattle bearing its brand, all running on unfenced land. Without fences, the cattle roamed as far as down south to Tularosa Peak and north to Corona, to the Pecos on the east and the Río Grande on the west. (The free range privilege ended around 1907.) Later, the Bar W range ran to above Red Lake on the north, below Oscuro on the south, across the malpais on the west, and to Vera Cruz on the east—about forty-five miles by fifteen miles.[14]

McDonald bought out the British syndicate's interest and took complete charge. As a businessman, McDonald managed and supervised the ranch as an investment; he was not a cowboy. For day-to-day operations, he hired a ranch foreman to deal directly with the men and the cattle. He appointed as foreman Pete Johnson, formerly from the lower Pecos in Texas and one of the best cowmen ever to come out of that state.

In an era before wells and windmills, ranches in the upper Tularosa Basin either made it or not owing to water. A cattleman could have a ranch with thousands of acres of land, but it would be worthless if he did not control the watering holes. From the ranch headquarters just north of present-day Carrizozo, the Bar W owned most of the watering holes as far north as Gran Quivira and Montezuma not far from Corona. To the west, it owned the waterholes of Duck Lake, Indian Tank, and on the south they owned Mound Springs, Salt Creek, and Candelario Wells. This was an enormous range, but the Bar W owned outright only small blocks of land covering the waterholes.

It was enough. At its headquarters, the Bar W had Carrizo Spring, a water source that continues today to supply water to the headquarters, corrals, and surrounding pastures. Water was so critical to controlling his ranch that McDonald had as many as sixteen riders with weapons patrolling just one spring—Jake's Spring south of present-day Carrizozo.

The El Capitán Livestock Company

After his initial success at managing the Bar W, McDonnell gained control of El Capitán Livestock Company (the Block ranch), one of the largest cattle corporations in New Mexico. The Block covered some 225–250 square miles, or sections. The Block, originally known as the Richardson place, combined with the Bar W in 1908 when the stockholders of the Bar W bought out the Block.[15] Together these ranches covered much of Lincoln County and constituted two of the largest cattle companies in New Mexico in the early twentieth century. McDonald became manager for both places. The English syndicate ran its cattle on the Block from 1909 to 1913. Starting July 31, 1918, it was all called El Capitán Livestock Company.

At first McDonald managed the Block for other owners. These owners gathered enough cattle off the Block in two years to pay $250,000 for their shares. At the end of those two years, they notified McDonald that they would sell out their part of the Block to him. So he bought out their stock, and in another two years he had gathered enough cattle off the Block to pay for his part. In those years (ca. 1909–13), about half a million dollars worth of cattle were shipped to Kansas City. The cattle were moved back and forth over the land and (originally) to the rails at Las Vegas, New Mexico, for shipment to the Kansas City stockyards. The cattle, some of the finest that ever came out of the Southwest, were welcomed there because they had pure Hereford bloodlines. At one time, El Capitán ran twelve chuck wagons and thirteen supply wagons. Each wagon outfit also included a remuda, branding crew, and wagon boss. McDonald and other like-minded cowmen were leaders in organizing the New Mexico Cattle Growers' Association, and continued that leadership throughout their lives.[16]

Brand of the El Capitán Land and Cattle Company. Courtesy Nathalie Taylor, Lubbock, Texas.

Stock certificate, El Capitán Land and Cattle Company, July 1, 1885. Courtesy Chris and Jack Harkey.

In the big drought of the 1920s when all the cattlemen took their herds and went to Mexico, the Block and the Bar W cattlemen (now under the same management) were the only ones who survived and returned to New Mexico with all of their cattle.[17]

Frances Tarbell McCourt McDonald, the *First* First Lady of New Mexico

Frances Tarbell was born March 4, 1852, at Steven Point, Wisconsin, to George D. and Matilda Jane Wilcox Tarbell.[18] Her father, like so many in the early timber towns, was in the lumber business. Frances attended local schools and may have attended Appleton College. On September 25, 1870, in New London, Wisconsin, she married Thomas B. McCourt (born February 21, 1848, to the elder Thomas McCourt, a merchant in Oshkosh, Wisconsin).[19] Thomas learned the trade of machinist. (See sidebar for more on this family, especially Baby Doe.)

According to sworn testimony by Frances McCourt at her April 10, 1891, divorce hearing before the court of the Fifth Judicial District of New Mexico Territory, the couple's marriage had always been troubled. Thomas had been an alcoholic from the start. He failed to support

Thomas Benjamin McCourt. Date and photographer unknown. Courtesy Bill Mathews.

Frances Jane Tarbell McCourt McDonald. Date and photographer unknown. Courtesy Bill Mathews.

THE McDONALD CONNECTION TO BABY DOE TABOR, SILVER QUEEN OF COLORADO

Frances McDonald's first husband was Thomas "Tom" McCourt, Jr. Tom's father and his brother, Paul McCourt, were merchants in Oshkosh, Wisconsin. Paul was a well-to-do partner in a clothing and custom-tailoring store, McCourt and Cameron. He and wife Elizabeth Nellis McCourt were the parents of eleven children.

Their fourth-born child (born May 5, 1854) was Elizabeth Nellis McCourt, or "Lizzie." She grew up very close to brother Peter, who was two years younger. Lizzie was a striking beauty, with gold ringlets and china-blue eyes. In 1877 at the age of twenty-two, she married W. Harvey Doe, son of an important Oshkosh family. The newlyweds moved to Colorado to seek their fortune in Central City's gold camp.

Before long, Harvey's father had to support the couple, though "Baby Doe" (as she quickly came to be called in the gold camp) put on miner's clothes and worked hard to help her husband. After Harvey failed at his mining venture and began drinking heavily, the couple moved to Denver. Things went from bad to worse, and Baby Doe divorced Harvey in 1880.

Baby Doe then moved to Leadville, Colorado, a silver-mining boomtown, where one evening over oyster soup in the Saddle Rock Cafe she met Horace A. W. Tabor, Colorado's wealthy "Silver King" and state senator. Tabor had noticed the flamboyant young beauty, and invited her to his table, and before the evening was over, he wrote a check to her for $5,000. Late in 1880 Tabor moved Baby Doe to the Windsor Hotel in Denver, where he became a frequent visitor.

Tabor and Baby Doe carried on a torrid affair amid much tongue-wagging by Denver's social elite until Tabor's wife, Augusta, finally divorced him in January 1883. Tabor bought a fine house on Capitol Hill in Denver, where he installed Baby Doe after making her his wife in a magnificent wedding held in Washington, D.C., on March 1, 1883. In 1884, the couple became parents of daughter Lillie, and in 1889 another daughter, Rose Mary Echo Silver Dollar, was born.

Baby Doe's brother, Peter McCourt, surfaced in White Oaks along with his cousin, Thomas, and Thomas's wife, Frances, probably in 1888. Peter, who was born in 1856, would have been in his early thirties when he arrived in White Oaks. He must have been involved in his cousin Tom's hardware and tinware business, because the *New Mexico Interpreter* ran an ad in his name identical to that placed by Frances McCourt for "tin roofing, sheet steel roofing, iron roofing, and trough and conductor pipe put up."[1]

Colorado held more appeal for Peter, though, and he left for the gold camps of that state in mid-1892: "Peter McCourt, who not many months ago was running a tin shop in White Oaks, lately made a rich strike at Cripple Creek, Colo. and sold out for $900,000."[2]

Before long Baby Doe persuaded Tabor to hire Peter as his manager in Denver. Peter was fond of entertaining his friends for poker evenings at the Tabor house. A bad fight ensued once with Baby Doe when she stormed downstairs to berate him for allowing important men in Denver society to eat her food and make themselves at ease in her house, while their wives were snubbing her socially. Peter and Baby Doe patched up that quarrel.

In 1893, the silver panic caused by President Cleveland's demonetizing of silver made mines worthless. Tabor's millions disappeared. Baby Doe appealed in desperation to Peter to save the opera house that Tabor had built in Denver, but Peter refused. Tabor was reduced to menial jobs to support his family until he died of appendicitis in April 1899, but not before telling Baby Doe to stay with the Matchless Mine and someday it would recover and be valuable again.

From the day Peter refused to rescue the opera house until his death in 1929, Baby Doe refused to forgive him. When his will was read, he left a quarter of a million dollars but nothing to her except some worthless carriage stock.

However, after Baby Doe moved with her two girls to the Matchless Mine in Leadville without a cent to her name, Peter told a Leadville grocer he would pay the family's bills as long as the grocer kept his name secret. He also furnished train fare for Lillie Tabor to travel to Chicago to live with the McCourt relatives there after Lillie tired of the hard life at the mine. Peter staked niece Silver with enough money to go to Denver and to get a job on a newspaper. She failed as a newspaperwoman, and when she later died as a prostitute in Chicago at the age of thirty-five, it was Uncle Peter who wired $200 for her burial expenses.

Baby Doe was buried at Mt. Olivet Cemetery in Denver after she was found frozen to death in her unheated wooden shack at the Matchless Mine on March 7, 1935.

1. *New Mexico Interpreter*, January 17, 1890.
2. *Lincoln Republican*, September 2, 1892.

F. J. McCourt's ad for hardware, stoves, and tinware in *New Mexico Interpreter*, January 17, 1890. Author's collection.

his family, which grew to include John (born 1873), then Genevieve, Paul, and Margaret (or "Margie," born 1882). Frances's father set Thomas up in business and furnished money over the years. Out of the couple's business came the means to build a home in Medford, Wisconsin, which was deeded to Frances.

In an effort to improve her life, Frances mortgaged that home and moved the family to Kansas, a prohibition state. In spite of the state's laws, Thomas managed to find liquor even in dry Kansas. The McCourts stayed there only a year, returning to Wisconsin after Thomas promised once again to stop drinking. When the alcoholism continued back home in Wisconsin, the McCourts looked west to White Oaks, New Mexico, for a new start. They arrived in White Oaks on March 30, 1888 (one account says 1887).

Thomas set up a business that included tin and sheet steel roofing, iron roofing, and trough and conductor pipe work. Frances testified that because of his drinking, Thomas did not attend to business. By January 17, 1890, it appears from an ad in the *New Mexico Interpreter* that Frances had abandoned pretenses and was openly and competently running the business herself.[20]

Frances stated in her court testimony that Thomas had been drunk most of the time since he had been in New Mexico. The McCourts also became part owners of a gold-producing mine in White Oaks. In 1890, Thomas was elected Lincoln County School Superintendent.

McDonald Marries Frances McCourt

On April 10, 1891, Frances McCourt sued for divorce from her husband on the grounds of "cruel and inhuman treatment and failure to support" and was granted a divorce the following day.[21]

The *Las Vegas Daily Optic* of September 5, 1891, carried the announcement of Frances's marriage, attended by daughters Genevieve and Margie, to McDonald in East Las Vegas: "Wm. C. McDonald and Mrs. Frances J. McCourt, both of White Oaks, N.M., were united in marriage this morning, at 9 o'clock, at the home of Rev. A. A. Layton, pastor of the First Baptist church who pronounced the ceremony.... Mr. McDonald is one of New Mexico's leading business men, while Mrs. McCourt was one of the Territory's handsomest and most accomplished young widows. Mr. McDonald and bride left for Clayton and the East on the morning train."[22]

McDonald was thirty-three at the time of his marriage; Frances was six years older. The following year Thomas McCourt died at the age of forty-four on Sep-

Frances McDonald's living room in White Oaks. On the table, note an ornate lamp featuring a naked cherub lifting aloft a metal cutwork globe topped by a gauzy shade. Date and photographer unknown. Courtesy Bill Mathews.

tember 3, 1892, in White Oaks: "On Friday night of last week, T. B. McCourt, County School Superintendent, surrendered his hold on life at Brothers' Hotel, in this town.... On Saturday his funeral took place from the Congregational Church and was largely attended, and the remains of 'Tom McCourt' were followed to the grave by a large number of laden carriages."[23]

For a time, the McDonalds continued to live in White Oaks in a showplace on Willow and Lincoln Streets.

A daughter, also named Frances, was born in 1892. Two other babies were stillborn, one on February 6, 1895, and the other November 4, 1898.[24]

Life at the McDonalds' Ranch Home on the Bar W

Since McDonald was managing the Bar W, the couple decided to build their ranch home on the site of Carrizo Spring eleven miles from White Oaks just north of present-day Carrizozo. This ranch house would be the third at Carrizo Spring since the first built by L. G. Murphy in the late 1870s. (Jimmy Alcock's house followed the one built by L. G. Murphy.)[25] All three houses suffered the same fate: they burned. The McDonald ranch house, built in 1891 or 1892, burned in 1946, destroying many photos and items of historical interest.

A bubbly, fun-loving woman, Frances loved to cook and was a gifted hostess. She participated fully in the area's many community activities. Frances made the Bar W ranch house an important social center. Besides entertaining her husband's business and political friends, she organized frequent dinners, barbecues, parties, picnics, and even weddings there. In the days before the railroad came to Carrizozo in 1899, visitors thought nothing of riding many miles on horseback or in their buggies to the McDonald home to stay days at a time,

McDonald ranch house with view of pond. Date and photographer unknown. Courtesy Bill Mathews.

in true western tradition.[26] It was also a family haven. The family came to include George Tarbell (Frances's father, who moved west after the death of his wife), as well as the four McCourt teenagers and young Frances McDonald.[27]

In these years, Frances's daughter Genevieve married Morris Parker in 1893, and daughter Margie married Art Rolland. Son John married Vena Gumm and eventually moved to El Centro, California. Paul enlisted in the Spanish-American War under the name "Frank Tinney" in July 1900 and became a corporal.[28] After his discharge, he moved to Kansas and married wife, Louise, in 1908. Later on, they moved to the mining town of Butte, Montana, and lived there the rest of their lives.

George Tarbell would once a year walk two miles to Carrizozo's local saloon, where he would buy liquor and walk back. He lived to almost one hundred, dying in 1923. He too was fun-loving; he would climb the apple trees on the ranch and bombard the children with juicy red apples.[29]

Some time in his thirties or forties, McDonald began suffering the effects of what was then called Bright's disease (now known as polycystic kidney disease, or PKD).

Continuing to Climb the Ladder of Political Success

Now a leading and influential citizen, McDonald won election to the Twenty-Ninth Territorial General Assembly as the representative from Lincoln County in 1891. He was elected chairman of the Lincoln County Board of County Commissioners from 1895 to 1897, further honing his political skills.[30] Later he was appointed to the Cattle Sanitary Board (1905–11) by then territorial governor George Curry.[31]

Curry appointed McDonald to the Cattle Sanitary

Governor McDonald's ranch house, Carrizo Springs. Date and photographer unknown. Courtesy Bill Mathews.

Board despite Mac's defection back in May 1892 in which he and fellow White Oaks politician John Hewitt left the Lincoln County Democratic Convention to support Gus Schinzing of White Oaks, Curry's Republican opponent for sheriff.[32] Schinzing was a former foreman of the Old Abe Mine, and it may have been that White Oaks loyalty won over party loyalty. McDonald continued his upward climb until he was appointed chairman of the New Mexico Democratic Territorial Central Committee in 1910.[33]

Election as First Governor of the State of New Mexico

The president of the United States notified the last territorial governor of New Mexico, William J. Mills, of the passage of the Smith-Flood resolution admitting New Mexico and Arizona as the forty-seventh and forty-eighth states in the Union, respectively.[34] Governor Mills promptly issued a proclamation calling for an election of state and county officers, members of the legislature, and two congressional representatives. Election Day was fixed as November 7, 1911.

Immediately, the Republican and Democratic parties issued calls for delegate conventions. The Democratic state convention was held in Santa Fe on October 2, 1911. A group calling themselves the "Independent Republicans" and headed by Herbert Hagerman and Richard Hanna joined to form a "Fusion Ticket" with the state's Democrats, an offer eagerly accepted by the Democrats.[35] William McDonald was nominated for governor of New Mexico on this strong fusion ticket that included Harvey Fergusson for Congress, E. C. de Baca for lieutenant governor, and other well-known Republicans and Democrats.

With what may well be one of the longest single sentences ever delivered to a convention in Santa Fe, John A. Haley, a close friend from White Oaks, had the honor of nominating McDonald for Governor of New Mexico:

I have known this man ever since I have been in the Territory of New Mexico, and many of you have known him equally as long. No one can question his integrity; there is no doubt of his ability, and his firmness of character and his dealing with his fellowman warrant us in saying that [if you] elect him Governor of New Mexico there will be no more scandals connected with the administration of the affairs of this state, peace and prosperity will reign, and when you do that, gentlemen, as long as you live you will be proud of the fact that you were instruments in placing before the people of New Mexico the opportunity for them to vote and elect a man in whom the Democratic party has every confidence and a man whom they will be honored for conferring this honor upon Lincoln County, and upon Honorable William C. McDonald, who has so long been a resident of New Mexico.[36]

According to Ralph E. Twitchell, New Mexico historian,

The Democratic convention met at Santa Fe . . . and, after two days' deliberation, placed in nomination a ticket, which . . . its best friends confessed was weak. Great political sagacity, however, was displayed in the nomination of William C. McDonald for governor. A business man of sound conservative judgment, his candidacy at once appealed to the business interests of the new state. The chief criticism urged against his opponent [Holm Bursum] was a laxity of business methods in the conduct of public affairs when in charge of the territorial penitentiary as superintendent. There were many capable business men . . . who had always regarded the question of statehood as one of doubtful benefit to the whole people. This class of citizens, anxious to witness the success of state government, believed that the affairs of the state were safer in the hands of Mr. McDonald than in the control of Mr. Bursum and cast their votes for the former.[37]

Though it was a bitter campaign against Holm Bursum for governor, the election result of November 7, 1911, was clearly a Democratic victory. McDonald received more than 31,000 votes to Bursum's 28,000.[38] After McDonald's victory, the December 1911 issue of the *Earth* quoted the man who would soon become the state of New Mexico's first governor: "New Mexico is standing upon the threshold of a great future. For many years we have waited for admission into the union. During those years we have been progressing slowly but steadily; we have been learning the extent and character of our natural resources; eliminating frontier conditions; improving social and industrial conditions; building up our school system; prospecting our mineral areas. The period of probation has been long and trying; but it has been healthful."[39]

McDonald entraining at Carrizozo for Santa Fe, 1912. The new governor is second from right. At his right, daughter Frances; at his left, Georgia Lesnet. Mrs. McDonald is the first woman seated on the left, front row. Former Territorial Governor George Curry is fourth man standing from the left. Photographer unknown. Courtesy Chris and Jack Harkey.

On January 6, 1912, President Taft signed the proclamation admitting New Mexico into the Union. Among the New Mexicans present at the signing were congressmen George Curry and Harvey B. Fergusson.

The Inauguration

The one detail the planning committee for the inauguration could not agree upon was the date. Time after time it was changed, and as late as December 29, 1911, had not been set. The date was finally set for January 15, 1912.[40]

Some of the ladies who attended had their dresses made by designers in such distant cities as Paris. The new First Lady chose a gold satin gown veiled in black marquisette and gold lace, with a satin train. Daughter Frances would wear yellow satin with an overdress of gold trimmed with yellow satin roses. William C. McDonald, formally attired in a full dress suit, was inducted into office as the first elected governor of the State of New Mexico just after noon in the presence of a wildly enthusiastic crowd of seven to eight thousand people. The ceremonies were conducted at the main entrance to the state capitol.

It was a typically gorgeous New Mexico day, and the first governor of the Sunshine State (later known as the Land of Enchantment), together with William Mills, the last territorial governor, entered an open carriage at the Palace Hotel. Accompanied by the First Battalion of the First New Mexico Infantry, the party started for the capitol in a parade route lined by citizens. The Santa Fe City Police and the New Mexico Mounted Police led off the procession, followed by the First New Mexico Regiment band and the Santa Fe battalion of the National Guard. Governors McDonald and Mills followed, escorted by Adjutant General Brookes and his aides.

When the head of the inaugural parade approached the main entrance to the capitol, the waiting thousands broke into loud cheers. The guardsmen formed into a double line along the sidewalks, with the regimental band stationed beneath a platform raised on the main stairway. Mrs. McDonald, daughter Frances, Mrs. Mills, Judge N. B. Laughlin, and Attorney General F. W. Clancy were seated on this platform. In the rear sat New Mexico's clergy, including Archbishop Pitaval and Vicar-General Fourchegu. Chief Justice Clarence J. Roberts administered the oath of office. The responses from Governor McDonald were "ringing and intense."

In his inaugural address, Governor McDonald said,

> Laws and rules can help direct, but cannot make good people—happy and prosperous—but right-thinking, honest citizens can, under our form, make good government. You are entitled to be served by a

Governor William McDonald and three Taos men. The man on McDonald's left is governor of Taos Pueblo. Date and photographer unknown. Governor William C. McDonald Photograph Collection, image no. 38493. Courtesy of the New Mexico State Records Center and Archives.

mind unbiased by inordinate party zeal, which may be unjust to those who differ, but are equally sincere and honest; by will unhampered by careless or questionable promises that might compromise the best efforts for a free government of the expressed will of the people; and by a heart free from malice or hatred toward any, and which beats in sympathy with the great cause of humanity. . . . Awed by the solemnity, and yet sustained by the enthusiasm of the occasion, fully impressed with the importance of the responsibilities that are placed upon me, I have taken the oath that binds me to your service. And now, trusting the Power that controls the destinies of men and nations, and the encouragement and inspiration that comes through the confidence of a generous people, I shall take up the work that with the blessing of the Almighty, I trust may redound to the benefit of our new state and to the good of the whole people.[41]

That evening two thousand people attended a general reception from eight to ten held in the Palace of the Governors. Present as well were all of the living territorial governors and their wives. The regimental band furnished music for the reception. Everyone hobnobbed joyously without concern for political party or class division. Governor McDonald had a cheery greeting for everyone, and Mrs. McDonald was graciousness itself. However, in keeping with the firm wishes of the new governor's wife, no intoxicating liquor was served at any of the events.

After the reception, the party traveled under a canopy from the Palace of the Governors to the National Guard Armory, transformed for the occasion into a ballroom that looked like dreamland. According to the *Santa Fe New Mexican*: "The dates '1848–1912' were spelled out in great letters of evergreen on walls of green paneling, a background for masses of poinsettias, trailing vines, baskets of ferns and silver candelabra. There were garlands of roses and smilax. The Ramírez Orchestra played in an exotically decorated dining room that had been created to accommodate fourteen tables holding chicken salad, rolls, cakes and coffee. Morrison's Orchestra furnished the music in the ballroom."[42] The inaugural ball closed the festivities of one of the greatest days in New Mexico's history.

Governor of New Mexico

McDonald would serve a five-year term, the only elected governor to do so. One of his first acts as governor was to convene the legislature into session on March 11, 1912. During his administration, McDonald maintained his ranch home on the Bar W just north of Carrizozo. To get to Santa Fe, he rode in a private railroad car to Lamy Junction, seventeen miles southeast of Santa Fe. There, he would meet the buggy that would take him on into Santa Fe.[43]

Life around First Lady Frances was fun. One of Frances's great-granddaughters was told that in Santa Fe, McDonald hopped onto a sled with the children in the nearby mountains. Hair and dress flying, she careened down the slopes in deep snow, screaming and laughing all the way to the bottom.⁴⁴

McDonald's tenure was honest and capable. Former territorial governor Curry said that "He had given New Mexico a sound business administration. His appointments . . . had been good. His economical administration had been popular with the people, but unpopular with some influential Democratic politicians. . . . I have no doubt he would have been reelected."⁴⁵

Historian William Keleher called McDonald the "nearest approach to an ideal governor during my time of personal acquaintance with governors, extending over a period of more than fifty years."⁴⁶ McDonald's honesty and strong character were a rare combination.

McDonald's Thanksgiving Day proclamation of 1912 took pride in declaring to the people of New Mexico:

> The American holiday, Thanksgiving Day, is near at hand. New Mexicans have much to be thankful for this year. They have enjoyed many blessings and experienced few calamities. We have had no disastrous storms and have been free from dangerous conditions of any sort. Crime has decreased and order . . . maintained under the law. . . . During the past twelve months New Mexico has entered the Union as a sovereign commonwealth; she has assumed the functions of state government and her financial credit is exceptionally good. . . . Crops and produce have been plentiful and the public health is in good condition.⁴⁷

One well-known visitor to the mansion in 1916 was Elizabeth Garrett. The blind daughter of famous former sheriff Pat Garrett, she was asked by the governor to sing at the opening session of the 1916 legislature. She sang her own composition, "O Fair New Mexico," accompanying herself on the piano. Not long after, an enthusiastic legislature proclaimed it the official state song.⁴⁸

But there were ongoing problems, caused mainly by the state of the governor's health. Son-in-law Truman Spencer remembers that "he was a sick man, sick all the time he was governor, damn near died two or three times. I don't know as you ever saw it, cystic kidneys? A.N.'s got it now [A.N. was Truman's son]. Frances [McDonald Spencer], my wife, died of it. . . . My oldest daughter . . . had it."⁴⁹

In another interview, Truman elaborates:

> I'd go up there [to Santa Fe]. I did a lot of stuff for him. . . . The laws on the books right now I signed because he was so damned sick. . . . They'd say, the Governor ain't able to get out of bed, to come over here. I says, "Give me those bills that have got to go in today and I'll take 'em over and get 'em signed." The secretary of state gave 'em to me and I would take 'em over . . . and the Governor'd look at 'em and say, "Where do I sign?" I says, "Here's where you want to sign . . . right here." He says, "You sign 'em. You can write my name." So I signed 'em and took 'em back to the secretary of state. He touched a pen, and he says, "It's legal if you touch the pen." So the Governor touched the pen.⁵⁰

A Case of Sex Discrimination

In August 1912, the first official case of sex discrimination in New Mexico was filed in district court in Santa Fe.⁵¹ When he arrived in Santa Fe as a Democratic governor, McDonald, as was customary when administrations changed, began replacing Republican incumbents. He submitted the nomination of Mary Victory for state librarian to the senate for approval.

The incumbent, Lola Chávez de Armijo (a member of an old, aristocratic, politically powerful New Mexican family), continued to serve in that position after the Republican chairman of the Committee on Executive Communications rejected the governor's nomination. The governor had formerly professed that women *were* qualified to hold important appointive positions. But the committee chairman's rejection had raised McDonald's ire, and he sought new ways to retaliate by removing Mrs. Armijo.

McDonald began legal action to force Mrs. Armijo to either prove her right to the position by undertaking a legal battle or resign. He stated in his brief that Mrs. Armijo was "a woman, and not qualified under the constitution and laws of the State of New Mexico to hold office," and he nominated a man (William Thornton) for the post.

Mrs. Armijo chose to fight and took her case to court. In October, less than three months later, Judge E. C. Abbot (a Republican) dismissed the governor's claim and ruled flatly that "Mrs. Armijo is not ineligible to hold office on account of her sex." Public interest was now aroused.

Governor McDonald would not let the case go and carried his fight to the state supreme court. A year and a half later, by a two-to-one decision, the justices found that Mrs. Armijo had a legal right to keep her job. Unfortunately, part of their decision was based on the claim that the office (of librarian) was purely ministerial, calling for neither judgment nor discretion, and for that reason, the duties of the office were not incompatible with the ability of a woman to perform. Mrs. Armijo held her job until 1917.

There were long-range consequences from this case. On March 15, 1913, the Republican-dominated state legislature approved House Bill 150, which said in no uncertain terms that "women may hold any appointive office." In light of the political and social beliefs of the time, this was a significant statement by the first New Mexico state legislature on behalf of women's rights.

Son-in-Law Truman A. Spencer

Truman Spencer came from Kansas City in 1910, where he had worked in marketing and brokering livestock houses.[52] He answered an ad in a paper for manager of the Bar W, and at the age of twenty began working on the ranch. McDonald quickly assessed Spencer as a very shrewd and ambitious young man, much like himself when he had first come to White Oaks.

At that time, shorthand was a new invention popular with aspiring young businessmen. According to his son, Truman, Jr., "Big" Truman taught himself shorthand in twenty-four hours when he came out to New Mexico by train to be groomed as manager for the ranches. McDonald needed a trusted aide, someone smart who could chauffeur, take shorthand and type, and eventually run the ranches. So before Big Truman climbed on the train to New Mexico, he bought himself a Gregg shorthand book, and by the time he got off in twenty-four hours he could take shorthand.

According to Truman Spencer, Jr., the business at that time had twenty thousand cattle and an equal number of sheep. McDonald groomed Spencer to take complete control of the ranch, giving him sage advice along the way, such as "if you can't stand to watch 'em die, son, don't get in the cattle business."[53]

After McDonald went to Santa Fe in 1912 as governor, he turned over all the ranch management to Spencer, then only twenty-two. He also turned over the keys to his safe, saying, "You know how to use the safe? I'm going to Santa Fe and I haven't time to fool with the ranch."[54] Truman Spencer cemented his standing with the governor by marrying McDonald's only child, Frances. They married soon after the governor went to Santa Fe to serve his five-year term. The Spencers made their home on the Bar W but spent much of their time in Santa Fe. Their first two children were born in the governor's mansion. Truman and Frances would have four children: William, who died in high school; Jane, who married Doctor Turner of Carrizozo; Truman, Jr.; and A. N. Spencer, who would become a doctor.

The End of an Astonishing Life

In 1917, when the Democratic state convention met, Governor McDonald considered both his illness and the lack of any strong Democratic Party backing for a second run at the governorship and chose not to run again. It was time to return to the Bar W. He did, however, continue his patriotic service during World War I as federal fuel inspector for the state of New Mexico. Early the week of April 8, 1918, the governor became severely ill and was driven to Hotel Dieu in El Paso for treatment. His wife and son-in-law accompanied him.

Nothing further could be done. He died of advanced polycystic kidney disease April 11.[55] Funeral services were held at the McDonald ranch house on April 14. Afterward, the governor was buried in Cedarvale Cemetery at White Oaks, where he had arrived thirty-eight years before as an ambitious young Eastern tenderfoot.

Governor McDonald's headstone in Cedarvale Cemetery. Photograph by Bob Workhoven. Author's collection.

Frances in Later Years

After the death of her husband, the widow Frances continued a busy life in her community.[56] She stayed active in the Episcopal Church and the Carrizozo Women's Club, serving a term as president. She also remodeled the ranch house, a white, two-story building with a red tile roof and a total of nineteen rooms. The project cost $15,000. Generous of spirit, she gave money, food, and time to people "across the tracks" in Carrizozo and paid the bills of many who could not afford to pay on their own. Frances died November 28, 1936, of a cerebral hemorrhage and is buried in Cedarvale Cemetery in the McDonald family plot.

The Bar W, managed at the time of this publication by McDonald's great-grandson Stirling Spencer, remains a successful family ranch.

NOTES

1. Details of McDonald's early life come from "William C. McDonald," 591–92; and Haley, *1913 Year Book*.

2. One of McDonald's advertisements as U.S. Deputy Mineral Surveyor appeared in the *New Mexico Interpreter*, January 17, 1890.

3. Justice of the Peace Court Records for White Oaks, 1883, Criminal Action No. 2, *The Territory of New Mexico, W. C. McDonald, Plaintiff, vs. W. H. Hudgens, Defendant*.

4. Curry, *George Curry*, 68.

5. *White Oaks Golden Era*, October 30, 1884.

6. Curry, *George Curry*, 58–59.

7. Ibid., 63.

8. *New Mexico Interpreter*, June 6, 1890. The full text of this Decoration Day address of May 30, 1890, is contained in the files of the Lincoln County Historical Society, Lincoln.

9. Biographical note, William Calhoun McDonald, Lincoln County, New Mexico, Collection, 1885–1976, copy in author's collection.

10. Thorp, *Along the Rio Grande*, 47–48.

11. Steven Spencer, "History of the Bar W Ranch," transcript of speech presented to the Lincoln County Historical Society, April 13, 1996. The full transcript of this taped speech is contained in the files of the Lincoln County Historical Society.

12. Nolan, *West of Billy the Kid*, 227.

13. Thorp, *Along the Rio Grande*, 47.

14. Steven Spencer, "History of the Bar W Ranch," 7. Spencer describes the range of the Bar W, its water holes, its foremen, and its management in detail.

15. Truman Spencer, Jr., "The Block and Bar W Ranches," transcript of speech presented to the Lincoln County Historical Society, March 13, 1974; Articles of Incorporation No. 5722, Office of the Secretary, Territory of New Mexico, Book B, 452–54, Cháves County, El Capitán Live Stock [sic] Company, filed December 28, 1908.

16. Curry, *George Curry*, 237. Territorial governor Curry appointed McDonald as chairman of the Cattle Sanitary Board ca. 1909. In those days, most of the prominent cattlemen were Democrats, while most of the sheep growers were Republicans.

17. Truman Spencer, Jr., "The Block and Bar W Ranches."

18. Kalloch and Hall, *First Ladies of New Mexico*, 15.

19. Sworn testimony given by Frances J. McCourt before the District Court of the Fifth Judicial District of the Territory of New Mexico, Case No. 819, April 10, 1891. This three-page document contains the details of the divorce bill of complaint filed by Frances against her husband, Thomas.

20. *New Mexico Interpreter*, January 17, 1890.

21. Final Divorce Decree No. 819, Frances J. McCourt from Thomas B. McCourt, District Court, Lincoln County, April 11, 1891.

22. *Las Vegas Daily Optic*, September 5, 1891.

23. *Lincoln County Leader*, September 10, 1892.

24. Record of Deaths at White Oaks, 1895 and 1898.

25. Steven Spencer, "History of the Bar W Ranch," 22–23.

26. Kalloch and Hall, *First Ladies of New Mexico*, 16.

27. Corn, "Mama Donnally."

28. McCourt, *Ghost Towns & Characters*.

29. Betty Corn to Haldane, February 17, 2006.

30. Haley, *1913 Year Book*.

31. Curry, *George Curry*, 237.

32. Ibid., 73.

33. Haley, *1913 Year Book*.

34. Ernestine Chesser Williams, "Writer Tells How New Mexico Became a State," *Roswell Daily Record*, January 23, 1994.

35. Curry, *George Curry*, 260–61; Miller, "Colorful Governors," 26.

36. John A. Haley's Nominating Speech.

37. Twitchell, *Leading Facts of New Mexican History*, 599.

38. Ibid., 602.

39. "First Governor Visualized NM's Great Potential," *Alamogordo Daily News*, January 5, 1987, quoting the December 1911 issue of *The Earth*.

40. Details of the inauguration are contained in Twitchell, *Leading Facts of New Mexico History*, 604–606; and Kalloch and Hall, *First Ladies of New Mexico*, 17–18.

41. Twitchell, *Leading Facts of New Mexico History*, 605–606.

42. *Santa Fe New Mexican*, January 16, 1912.

43. Steven Spencer, "History of the Bar W Ranch," 19.

44. Betty Corn to Haldane, February 17, 2006. Corn, great-granddaughter of Frances McDonald, learned about the sledding episode from Roy Harman, former resident of White Oaks living in Carrizozo, when she sat with him on his front porch many years ago and talked with him about his youth in White Oaks.

45. Curry, *George Curry*, 282.

46. Keleher, *Memoirs*, 138.

47. William C. McDonald, Thanksgiving Day Proclamation, November 28, 1912.

48. Kalloch and Hall, *First Ladies of New Mexico*, 18.

49. Truman A. Spencer, Sr., interview by Barbara Jeanne Reily-Branum, October 11, 1980, La Luz, New Mexico, 10, in author's collection.

50. Truman Spencer, Sr.; Truman Spencer, Jr.; and Marion Spencer interview by Barbara Jeanne Reily-Branum, July 24, 1981, Bar W, Carrizozo, 6, in author's collection.

51. Adler and Adler, "How a Plucky Woman Foiled a Governor."

52. Steven Spencer, "History of the Bar W Ranch," 12–16; Truman Spencer, Jr., "The Block and Bar W Ranches," 8–9, 22; Truman Spencer, Sr., Truman Spencer, Jr., and Marion Spencer, interview, 6–7; Truman Spencer, Sr. and Jr., interview by Mary Krattiger, August 31, 1980, 3, in author's collection; Truman Spencer, Sr., interview, 10–11.

53. Steven Spencer, "History of the Bar W Ranch," 27.

54. Ibid.

55. *Carrizozo Outlook*, April 12, 1918.

56. Kalloch and Hall, *First Ladies of New Mexico*, 19.

CHARLES MADISON MERRILL
A Prospector's Prospector

Charlie Merrill was born a preacher's son in Texas in 1867.[1] More full of himself than of the Lord, his father would leave church and severely beat his children before they could reach home. Once he even yanked a large plug of blonde hair out of his daughter's head.[2] It was no surprise, then, when Charlie left home in 1880 at age thirteen. Twelve years later, in 1892, he married fellow Texan Martha Ann Amos.

By 1900, Charlie was a miner in White Oaks. The June 1900 census lists Charles and Martha Merrill and four children—Rosa, age seven; John, age five; Joseph "Brownie," age four; and baby Charles Edwin "Ed," two months. A fifth child, Nellie, was born in 1902.

Martha's mother, Martha Jane Amos, also lived with the family. Daughter Rosa remembers the Yaqui Indians from Mexico often coming into White Oaks in those days, and the children would talk to them. The Yaquis would trek north from El Paso and other places in the Chihuahua and Sonora Deserts.[3]

Charlie's wife, Martha, yearned to return to Texas. One day in 1904, while Charlie was at work, his mother-in-law pulled a wagon and team up to the front of the Merrill house. She and daughter Martha loaded it with the household goods, gathered the children, and left. When Charlie came home later that day, the house was empty and everyone gone.[4]

Not long after his family left White Oaks, Charlie traveled to Utah and acquired a little mine that he later traded for a butcher shop. Deciding that he was not cut out to be a butcher, he moved to southeast Oregon, where he owned a house and part of a mine in Jordan Valley near the Idaho state line. Finally, he headed for Nome, Alaska,[5] a small town on the southwest corner of Seward Peninsula on Norton Sound. In 1926, while still prospecting and mining there, he hitched up his dogsled and drove into Nome to get supplies.[6]

Charlie Merrill at Gila campsite after his return from Nome, Alaska, ca. 1926. Photographer unknown. Courtesy Loris Norris.

In town, Charlie joined other onlookers as the train came into the station and watched passengers alight. A woman with small children got off the train, looking for her husband who was to meet them, but he was nowhere to be found. Her dilemma, and the sight of the children, triggered something in Charlie, and he gave the woman his dogsled, loaded with six months' supply of staples and groceries, so she could look for her husband. At that moment, Charlie made a snap decision to buy a ticket, board the train, and return to Texas to search for his own wife and five children. After twenty-two years, his children were grown and he would be a stranger to them.

In mid-July 1926, Charlie arrived in Goldthwaite, a small town near Brownwood, Texas. Upon arrival,

Charlie Merrill in front of 1928 Dodge. Date and photographer unknown. Courtesy Loris Norris.

Charlie had asked about town if anybody knew John, Joe, or Ed Merrill, his three sons. He was directed to Center City, where word had already reached son John of his father's arrival. Now thirty-two years old, John did not recognize his father, but the family decided Charlie was authentic when he named all the horses that had pulled his wife's wagon back to Texas. Soon after, Charlie's wife, Martha, arrived on the scene and began walking toward her long-absent husband with open arms, saying, "This is Charlie."[7]

Son John's wife, Margret, took Charlie under her wing. A wonderful cook, she made sure he had fresh milk and eggs and shared whatever she cooked with him. A couple of years later, Charlie bought a 1928 Dodge for his wife.

But the old itch to roam resurfaced, and it seemed that Charlie could never stay in one place for long. Somehow Merrill ended up in the tiny hamlet of Weed, New Mexico, in the Sacramento Mountains in 1929 or 1930. His wife, Martha, visited him there.[8] Weed was also a temporary stop. Soon Charlie was prospecting in southwest New Mexico about thirty miles out of Gila, near the Gila River.

Merrill's grandson, Loyd Haven, drove Martha there in her 1928 Dodge. Finally getting directions to his camp, they walked five miles to his tent and found Charlie standing naked as a jaybird outside the tent cooking breakfast. He ducked into the tent and came out with his trousers on and buttoning his shirt, telling grandson Loyd that he did not recognize him. Loyd stayed to prospect with his grandfather in the Gila area for a short time. He said Charlie made his own syrup with coffee and sugar and the best sourdough bread he had ever eaten. He also noticed that Charlie believed in a daily bath in cold water no matter the temperature outside.

By 1931 Charlie was living in McGirk, Texas, in the hill country west of Waco. His accommodations were a tent down the hill from son John and behind granddaughter Ruby's grocery store.[9] The tent had a wooden floor and was warm in winter. He still owned a gun, a .22 Savage, clip-type, bolt action that held six shells and one in the barrel. It is said that he could strike a match with that gun at twenty feet. After satisfying himself that grandson Glen had learned how to take that gun apart and clean all of it well, Charlie gave him the Savage.

One of Charlie's idiosyncrasies was that he always slept with his head to the west. He believed that since the Lord was going to come again from the east, he wanted to be sure and see Him coming.

In 1939, after his wife died, the family moved Charlie to Cross Plains, a short drive southeast of Abilene, where he lived in a little three-room house across from daughter-in-law Margret's house. John and Margret were the parents of five lively girls—Ruby, Laura, Velma, Ina, and Loris—and Charlie enjoyed their antics. His big fenced yard was the playground for all the neighborhood children following their dash home from school.

He taught all the grandchildren and their friends—about twenty-two of them—to sign the alphabet on their hands using his version of sign language.[10] On entering the gate to Charlie's yard, no one was allowed to talk with their mouth, only with their hands. Charlie said, "I'm deaf. I don't care how many of you come to play, but you will talk on your hands so I know what's going on." There was jump rope, hopscotch, or just sitting on the porch signing with their hands as a group while Charlie smoked his pipe, rocked, and watched or sometimes played checkers with them. Granddaughter Ina had breakfast—flapjacks and molasses—with Charlie nearly every Saturday.

In 1942, daughter-in-law Margret and granddaughter Loris (then five years old) found Charlie propped up between wooden boxes on the floor in the middle room of his little house naked, a blanket folded over his lap. A small stream of blood from a self-inflicted gunshot wound ran down his right temple.[11] Loris related that she had never heard anyone say anything bad about Charlie Madison Merrill. Everyone loved him, and he returned that love in abundance to all his family after he returned home from his long Alaskan adventure.

NOTES

1. All Merrill family information is from a Merrill Family History package compiled by Loris Norris, of Abilene, Texas, July 2002 (referred to hereafter as MFH). Loris is a granddaughter of Charles Merrill and daughter of John Merrill. Other members of the Merrill family contributed their recollections of Charles.

2. Ruby and Glen Merrill, MFH.

3. Loyd Haven, telephone conversation with author, recorded January 30, 2002, and placed in the MFH. Loyd is the eldest son of Rosa Merrill.

4. Ibid.

5. Ruby Merrill, MFH.

6. Loris Norris, MFH.

7. Loris Norris, "White Oaks, New Mexico—2002," MFH.

8. Loyd Haven, MFH.

9. Glen Merrill, MFH.

10. Ina Merrill, MFH.

11. Loris Norris, MFH.

URBAIN OZANNE
A Frenchman with a Nose for Business

Urbain Ozanne was born May 8, 1835, in the small French town of Chateaubriand in what was then the province of Brittany and is now the province of Loire-Atlantique. He received a common-school education in his native town.[1] His father, a government teacher of considerable merit, received three medals of honor from King Louis Phillippe. His father's love and respect for education were ingrained in son Urbain, who later saw to it that his own sons were well educated.

Ozanne's father migrated to America in 1850. His father, mother, and the five children in the family left Le Havre, France, in a sailing vessel November 1, 1850, arriving in New Orleans on Christmas Day, 1850.[2] Urbain was fifteen years old at the time. The Ozannes lived in New Orleans for several months before moving to Paducah, Kentucky, where Ozanne's father entered the mercantile business. While in New Orleans to buy more stock for his store in 1853, Ozanne's father died of yellow fever. The ship that brought the Ozannes to America carried other French émigrés, in particular the Bouvards. Six years later, on June 25, 1857, Urbain married Frances Bouvard, whom he no doubt first noticed on that ship. Five children were born to Frances and Urbain between 1858 and 1871. The first child, Adeline, was followed by four boys (birth years are estimated): Henry, born 1859; Emile, born 1863; Alfred, born 1865; and Paul, born 1871. Adeline died in January 1864.

The Urbain Ozanne family moved to Kentucky, and by the time the Civil War erupted in 1861 they were living in Nashville in the border state of Tennessee, where Urbain supplied beef to the Union army.

Reconstruction

In June 1865, just after the close of the Civil War, the Ozannes visited France briefly before returning to the United States. Urbain's youngest son, Paul, was born in

Urbain Ozanne and his four sons. Date and photographer unknown. Courtesy Johnson Stearns.

Paris. Urbain then moved to the far northwest corner of Mississippi about fifty miles south of Memphis and acquired a plantation as part of the post–Civil War Reconstruction. He was a cotton planter in both Mississippi and Alabama, despised as a "carpetbagger" by his Southern neighbors.

In a letter to Republican senator T. Stevens on July 9, 1868, Urbain complained that white Republicans and black Mississippians were "on the same footing in the eyes of the Rebels."[3] He went on to report that men like himself were being persecuted because of their support of the Union cause, especially in the reconstruction of state civil governments. Urbain persevered, however, and became a power in Mississippi politics. He was elected sheriff of Panola County, Mississippi, in 1869 and reelected in 1871 and 1873. He was also elected a member of the Mississippi Constitutional Convention in 1869. That same year he was appointed probate judge by the governor of Mississippi. Family tradition holds that Urbain, an astute businessman, sold his agricultural interests before the Panic of 1873.[4]

After his wife died in 1873, Urbain revisited France in 1874 with the thought of returning to live permanently. In a short while, he "found that he was too thoroughly an American to ever be content in any other country save the one of his adoption."[5] Leaving Panola, Mississippi, Urbain moved to Memphis, Tennessee, making it the family home for a few years and investing in the wholesale feed business.[6] During these years, he made enough money to finance the education of all four of his sons at Saint Louis University. The Ozanne educational legacy followed down the generations: son Emile even ensured that his girls were college graduates in a time when few women were highly educated.[7]

Trailing the Railroad West

Sometime after 1875, Urbain trailed the grain and cattle business on to Wichita, Kansas. Wichita was situated on the railroad line, and Urbain was never far behind the railroad as it pushed into new territory.[8] He concentrated on his new feed business here while eldest son Henry started a dry goods store. Urbain entered the territory of New Mexico in early 1880, a year or so after the railroad arrived at Las Vegas, and lived there for two years. He began acquiring mining interests near White Oaks.[9]

Sometime in 1882, Urbain moved to White Oaks and opened a continental restaurant and bakery in partnership with son Alfred. They called their business "Ozanne & Son."[10] Urbain's sister-in-law (his brother's widow) became the cook for his restaurant. Back in Wichita, son Henry even supplied champagne and oysters for his father's White Oaks restaurant, which was later housed in the Hotel Ozanne that became famous throughout southeast New Mexico.[11]

Urbain soon became involved in a dizzying variety of business enterprises. For instance, the *Lincoln County Republican* observed a thriving lumber business at the close of 1882: "U. Ozanne and Charlie Anderson were in from the McPherson sawmill at the Capitáns for several days during the week. The mill has been doing a good business down there, supplying lumber for the Pecos country and intermediate points. They have also filled one order of eighty thousand feet from the Government to be delivered at Fort Stanton."[12]

Ozanne also dove into community affairs. He served on the academy board of the first high school in White Oaks. He held several public offices, including that of road superintendent of the Eighth Precinct (White Oaks). He placed the following notice in the *Lincoln County Leader* in the summer of 1890: "Every ablebodied male inhabitant of Precinct No. 8, between the age of 21 and 50, and subject to roadwork, is hereby notified to meet at Ozanne's office on Monday morning, July 14 at 7 o'clock, for the purpose of working the public roads.... Those who prefer to pay... rather than work the roads must settle... on or before July 17, '90, with the road supervisor."[13] All the while, he cast about for lucrative business opportunities.

Hotel Ozanne. Date and photographer unknown. Courtesy Donald M. Queen.

John M. Decort with Ozanne stage hack, ca. 1888–1889. Decort stands by his team of four mules near the Red Canyon stagecoach stop close to the old iron mines northwest of the malpais. Photographer unknown. Courtesy Larue Lane Wetzel.

Predecessors of Ozanne's Passenger & Stage Line

The White Oaks stage line changed hands several times before Urbain began running his line. When gold was found in White Oaks in 1879, the first mail and stage lines originated in Las Vegas. In 1881, an ad appeared in the *Las Vegas Gazette* for a hack line to White Oaks: "Strausner's hack line running weekly to White Oaks. Passengers carried 165 miles for $15. Leave orders at Sumner House Las Vegas or Burk's Hotel, White Oaks. Will make the trip in three or four days according to weather."[14]

By October 1880, what was needed was a Lincoln County wagon road from the new Río Grande valley rail terminal at Socorro to the east. Strausner continued to run the Las Vegas–to–White Oaks line for a short time, carrying mail and passengers.

Strausner's line was succeeded by the National Mail Company, managed by a man named W. H. H. Lewellyn. No one thought much of this hack line. The National Mail Company ran for about two years, until late 1882, amid many complaints. Finally, the Southwestern Stage Company, managed by E. W. Parker, took over; it was the same company running a line from the mining camp of Hillsboro through the Black Range. With a route from San Antonio located thirteen miles south of Socorro to White Oaks, Nogal, and Fort Stanton, the Southwestern Stage Company fared no better than the National Mail Service.[15]

Businessmen began going to court to recoup bills owed by the company. The justice docket of Lincoln County's Eighth Precinct recorded no fewer than ten lawsuits from November 1883 to August 1885.[16] The disputes were over money for merchandise sold and delivered, orders from merchants, promissory notes, or even (as in a suit brought by Urbain himself) for labor done. The *Golden Era* described an ailing business: "We have received mail in White Oaks two days in succession, but this morning the same old cry—no mail—was heard again. How long will this state of affairs exist? We give it up. Ask the postmaster."[17]

Ozanne's Passenger & Stage Line: Flagship of the Family Businesses

Urbain quickly translated this state of affairs into a business opportunity and began a competing independent hack line from Socorro to White Oaks. In November 1885, the *Golden Era* recorded his arrival in White Oaks from Socorro driving a four-horse team and with a new hack. The same newspaper noted at the end of December, "U. Ozanne came in Saturday night from the railroad. If any man in the county deserves success, it is Mr. Ozanne. Persons wishing quick transportation to the railroad will find Mr. Ozanne a very accommodating gentleman."[18]

Urbain Wins the Government Contract to Carry the Mail

In 1886, Urbain wrested the stage business from Southwestern. He secured the government contract to carry the mail from Carthage south of Socorro to White Oaks, Nogal, Fort Stanton, and Lincoln. Most of the money a stage line would make came from its government mail contract. Of the $10,000 or $11,000 a year earned by Ozanne's Passenger & Stage Line, $6,500 came from its mail contract.[19] The stage line also offered passenger and express service daily between San Antonio and Lincoln.

Ozanne's first contract ran from July 1, 1886, through June 30, 1890. Later Urbain won a second contract from July 1, 1890 through June 30, 1894. Passenger fare from San Antonio to Lincoln was $19.50 one-way or $34.50 round-trip. Passenger fare from San Antonio to White Oaks was $14.50 or $25 round-trip. San Antonio, New Mexico, was at this time a railway station on the Atchison, Topeka, and Santa Fe Railroad (AT&SF).

The AT&SF had entered New Mexico north of Ratón in December 1878, pushing steadily southward through Las Vegas and Lamy and on through Socorro and San Antonio to Rincón. At Rincón the line branched to Deming by March 1881 and to El Paso by June 1881.[20] A short branch was built to the coal-mining town of Carthage twenty-seven miles southeast of Socorro. Over two decades, the White Oaks mail route's western terminus changed from Socorro to Carthage to San Antonio. These changes reflected AT&SF rail construction along the Río Grande and the ups and downs of the spur lines to the coal mines at Carthage.

A First-Class Stage Line

Urbain strove to run a first-class stage line. He advertised "good teams, sound rigs, careful and sober drivers."[21] (One might infer that other drivers were neither careful nor sober.) Customers were much happier with his line. Stops were about every ten miles, qualifying this line as first class. In contrast, other lines, such as the Butterfield Overland Stage, stopped every eighteen to twenty miles.

During the first contract period, Urbain ran his stage through the malpais, crossing at Lower Crossing, for a total of about eighty-nine miles. In his second contract period, to avoid the rough trip over the malpais, Urbain

Ozanne's ad for passenger and express line in the *White Oaks Eagle*, January 9, 1886. Author's collection.

OZANNE'S PASSENGER & STAGE LINE ROUTE
during the First Mail Contract Period
(1886–1894)

In the first government mail contract period, the stage line crossed the malpais at Lower Crossing, or to the south, for a total of about eighty-nine miles:

SAN ANTONIO TO KINNEY WELL—
  eleven miles; water stop
KINNEY WELL TO CARTHAGE—
  two miles; horse change
CARTHAGE TO MONTOYA'S WELL—
  nine miles; water stop
MONTOYA'S WELL TO HANSONBURG—seven miles; horse change
HANSONBURG TO MOUNTAIN STATION RANCH—twelve miles; midway overnight stop plus water stop and horse change
MOUNTAIN STATION RANCH TO RED CANYON—ten miles; water stop and horse change
RED CANYON TO MALPAIS STATION—fifteen miles; water stop and horse change
MALPAIS STATION TO ANCHOR SPRING—eight miles; water stop
ANCHOR SPRING TO WHITE OAKS—fifteen miles to the home station and overnight stop

---

OZANNE'S
Tri-Weekly Passenger and Express Line
—FROM—
SAN ANTONIO to LINCOLN
—VIA—
White Oaks, Nogal and Fort Stanton!

Elegant new coaches have been put on this line, which will leave San Antonio every MONDAY, WEDNESDAY and FRIDAY, immediately after the arrival of the train, for which it will wait, however late the train may be; and will reach San Antonio from White Oaks every TUESDAY, THURSDAY and SATURDAY and connect with the eastbound train. No more night travel. Passengers will stop over night at the Mountain Station ranch, and reach White Oaks in time for dinner next day. None but careful sober men are employed to drive, and no expense will be spared to make passengers safe and comfortably. Coaches will leave White Oaks every Monday, Wednesday and Friday for the railroad. In all my eight years experience in carrying the U. S. mail I have never had a single accident resulting in injury to any one. Passengers who regard their comfort and safety will do well to patronize the Ozanne Stage Line, and when they reach White Oaks to

Stop at the Hotel Ozanne

Where they will be taken care of as well as if at their own homes. We strive to serve the public.

U. OZANNE, Prop.

Stagecoach station at malpais Lower Crossing. This crossing was about six miles south of Carrizozo on the McDonald Ranch. Date and photographer unknown. Courtesy Johnson Stearns.

A TYPICAL JOURNEY ABOARD OZANNE'S PASSENGER & STAGE LINE

Today's traveler, who speeds over our country's highways with ease, likely has no idea what a stage coach trip in the West in the late nineteenth century was like. Fortunately, an occasional passenger left a firsthand account of such an experience. In 1887, an unidentified correspondent identified only by the initials "H.S." wrote for the *El Paso Daily Times*:

At Carthage the stage line of Ozanne & Co. starts for White Oaks, Nogal, Ft. Stanton and Lincoln, the county seat of Lincoln county, the distance being:

White Oaks 85 miles
Nogal 103 miles
Fort Stanton 120 miles
Lincoln 130 miles

The stages usually leave the train at 10:30 A.M. and arrive at White Oaks the following morning at 6 o'clock.

They are pleasant hacks and make fast time. Although the route is ... long ... it is by no means tiresome. The route is over the Jornada, or "death-stretch" plains, of twenty miles.

East of the Jornada the Oscura mountains are passed, through a canyon in which the residence of H. Goujon is located.

The next is Taylor's ranch of 800 head of cattle, on which is a spring ample for watering his stock.

At twenty-eight miles is the stage stand where Judge Ozanne has 320 acres of land and his son A. J. 260 acres with a magnificent spring of cold mountain water sufficient for one thousand head of cattle and horses. ... This is the supper place on the route, and a most romantic and beautiful spot.

Crossing the Oscuras ... another prairie called the "Oscura plains" extends east to the Mal País, 27 miles, in full view of the White mountains.

The Mal Pais, or "Bad Lands," is the strangest formation I have ever beheld anywhere. As the stage crosses the vale passengers are politely requested to alight and exercise their pedal extremities in order to allow the coach to roll over the "rough and rugged" way. It is a relief.

Crossing the Mal Pais, the route is near the White Mountain range, with snow visible for miles. It is called the Carrizozo prairie and extends 24 miles to the canyon leading into White Oaks between the Baxter and Carrizo mountains.

East of this is the Carrizozo cattle ranch ... with the popular and efficient young manager, J. A. Alcock, whose residence is palatial and the surroundings more like Kentucky than New Mexico. [The ranch] comprises a stretch of country 45 miles by 15 and is watered by thirteen mountain springs.

Leaving that point where water was obtained for our horses, and the hospitalities of the Carrizozo were so handsomely extended to our little party composed of Captain C. H. Gallagher, of Wilmington, Del., Judge U. Ozanne, a Frenchman ... and your correspondent, we passed along 12 miles into White Oaks ... through the canyon entrance to White Oaks, our present objective point.[1]

1. *El Paso Daily Times*, 1887.

ran a northern route around the head of the malpais and what is now the Gallacher (J + H) Ranch to White Oaks, for a total of about ninety miles. The passenger hacks, called "celerity" wagons or hacks, were pulled by two to four Spanish mules and held four to six passengers.[22] In 1888, Urbain signed a note to M. Satterwhite for nine mules, three hacks, and four sets of double harness to be used on the line between San Antonio and Lincoln.[23]

The *Lincoln County Republican* described a typical stage ride on an Ozanne hack: "A stage ride of one hundred and twenty-five miles is a ... formidable undertaking, and this is the distance traversed in going from Lincoln to Carthage and except to one who sees the country for the first time, the way is long and wearisome, tho' all is done that can be to make the trip less fatiguing. The accommodations at Ozanne's [Mountain Station] Ranch in the Oscuras ... are all that can be desired.... The drivers are gentlemanly and obliging and in that respect are a marked contrast to some."[24] The route through the Oscura Mountains and the malpais had to have been rough, requiring a great deal of strength and energy from the drivers as well as from mules and horses.

Until 1894, Ozanne's open-air stagecoach was the primary means of transportation for hire between San Antonio and Carthage to Lincoln. Passengers could overnight at Mountain Station Ranch, White Oaks, or Lincoln (disembarking at the Aragón store by the Wortley Hotel).

Mountain Station Ranch

Ozanne's Passenger & Stage Line was a partnership between Urbain and his son Alfred. About halfway between Carthage and White Oaks, Alfred lived at and operated Mountain Station Ranch as a hotel and way station. Maintenance facilities and provision of food at Mountain Station Ranch constituted what we might today call "full service." Other swing stations along the way offered only water and changes of mules or horses. Stage horses were kept there, and the drivers had a room where they slept when not on the road. The drivers made the trip from the railroad at Carthage to Lincoln (the County seat) 140 miles away every day, driving day and night and stopping at way stations for fresh horses.

In December 1887, Alfred married pretty Olive Rencher of Mississippi and brought her to Mountain Station, where the couple lived and ran the stage stop. They had in residence a cook to prepare continental meals for tired and hungry travelers. A nearby well, lined with stone, produced good water.[25] Alfred filed a homestead claim at Mountain Station Ranch, but he discovered in 1890 that he had filed on section sixteen, the section reserved by law for erecting schools. Alfred's homestead claim was denied even though he had lived on it for five years.

Connections to Roswell

According to historians Elvis Fleming and Ernestine Williams, connections could be made at Lincoln with an Ozanne subsidiary stage called the "Buck Board Line" for Roswell, Seven Rivers, and Pecos country.[26] Little is known about the extent of these operations, except for one account.

In July 1890, when the first bank was about to be established in Roswell, a man named E. A. Cahoon (Calhoun) wrote that he had arranged to meet Urbain

Ozanne Mountain Station, Socorro County, ca. 1890. Photographer unknown. Archives and Special Collections, RG95-175, New Mexico State University Library.

at Carthage to freight his party and their $36,000 in bullion and paper money east to Roswell in a four-day trip.[27] The first overnight was to be Mountain Station Ranch; the second, Ozanne Hotel in White Oaks; and the third, Lincoln. This was most likely a special one-time contract between Ozanne and Cahoon.[28] Cahoon implied that his party was "well-heeled," meaning heavily armed. They and their cargo reached Roswell safely.

Hotel Ozanne and the Carrizo Annex

The two-story Hotel Ozanne was built in White Oaks of brick as a major end point for passengers on the stage in 1888. By 1891, Urbain had remarried, and it was his new wife, Ella, who actually operated the hotel. Rates were $2.50 and $3.00 per day. The *Lincoln County Republican* in 1892 called the hotel "a handsome brick building presided over and admirably managed by Mrs. Ozanne."[29] Hotel Ozanne stationery boasted, "The only first class fireproof hotel in White Oaks. Is a brick building recently renovated and refurnished in excellent style. Well screened against flies. No bedbugs."[30]

In March of 1892, the *Lincoln County Leader* reported that Urbain had bought the Carrizo Hotel (minus furniture) for $4,500.[31] He expanded his stage line business to include a better restaurant and hotel overnight stop. The Carrizo was to be used as an annex to Hotel Ozanne, with wife, Ella, running both establishments. Hotel Ozanne was located on the southeast corner of Carrizo and Livingston Streets, the annex a block away on Carrizo and Harrison.

End of the Line

The Sherman Silver Purchase Act, the basis of the West's prosperity, was repealed in 1893, causing a nationwide depression in 1893 and 1894. The largest smelter in the Southwest at Socorro closed; mining towns were deserted; and White Oaks itself went downhill.

Urbain was done in by the Panic of 1893 when carrying the mail began to lose money.[32] He also realized that the railroad would reach Lincoln County before long and knew the mail route contract, his major source of income, would vanish as well. In July 1894, he let the mail contract go to William Lane. In June 1894, he resolved to stop subsidizing the U.S. Postal Service and instead began a triweekly daytime express and passenger stage service between San Antonio and White Oaks.[33] He also eliminated the dangerous and bone-jarring night trips over the mountains.

The *Old Abe Eagle* sadly bid its favorite White Oaks mail contractor farewell: "On Saturday next Mr. Ozanne, who for the past eight years has been performing mail service between San Antonio and Lincoln, will close his contract with the government. . . . The patrons of the post office in this section of the country have been favored with a most excellent mail service under Mr. Ozanne, regardless of the elements . . . and notwithstanding that the mail service has been performed for the last two years at a heavy loss to the contractor."[34]

Ozanne bought new, more comfortable coaches from Carriage Works of Fort Scott, Kansas. None were Concords; so far as is known, Urbain never ran the elite Concord stagecoaches (the "Cadillac" of stagecoaches).[35] He also switched from mules to horses.

Though passengers believed that hacks were better than walking or riding burros, the hacks were still far less comfortable than Pullman railcars or Concord coaches. Nonetheless, Urbain tried hard to make his line as fast and comfortable as possible. Struggling to hold on, he operated at a financial loss through 1894 and 1895. It was all for naught; he and son Albert were finally forced to sell Mountain Station Ranch in 1895, followed by the Hotel Ozanne furniture and the Ozanne hacks and horses in 1896. When he sold Taylor's Well station on the stage route, he also sold seventy horses, sixteen mules, three four-horse hacks, five two-horse hacks, and two road carts.

Finally, in 1896, Ozanne sold the stage line business itself to William Lane, who replaced the more primitive stage hacks with new Concord models.[36] Lane's line, hit twice by robbers, survived in business only to early 1899.[37] Paul Mayer succeeded Lane, and a third contractor, A. H. Hilton, followed Mayer.

Marital Disasters

After Urbain's first wife died in Mississippi, he finished rearing his sons alone. A couple of years after arriving in White Oaks, Urbain began courting the local women: "Mr. U. Ozanne is quite popular with the girls, judging by the number he takes riding each Sunday."[38]

In approximately 1892, in his fifties and still quite well off financially, he married a gold digger named Ella. As time went on and business went downhill faster, Ella asserted more control over the family property. Urbain's marriage collapsed with his businesses and turned into a full-scale disaster.

STAGE ROBBERIES

In a dangerous time for stage lines, robberies occurred both before Ozanne started his line and afterwards. The only robbery known to occur while Urbain owned the line was in December 1892. O. N. Parker was charged in that month with holding up the Ozanne stage between Nogal and White Oaks. Deputy Sheriff Langston of White Oaks traveled to Mesilla Park to make the arrest.

Langston led the prisoner from jail to a room at the Commercial Hotel in Mesilla Park, handcuffed him, and chained him to the bed. The deputy and his posse, tired from their long ride, slept too soundly. Parker slipped the key from the deputy, unlocked his handcuffs and shackles, and escaped. No trace of the fugitive was found. Embarrassed, Langston claimed that this was the first time any prisoner had ever escaped from him.

Notable holdups after Urbain sold the stage to William Lane of White Oaks occurred in 1896. The notorious High Five Gang held up Lane's stage a total of four times, all within two weeks in October. Four members of the gang—the two Christian brothers, George Musgrave, and Bob Hayes (alias John West)—stopped Lane's stage on October 7, 1896, en route east between Mountain Station Ranch and Wash Hale's stop. Rifling through four mail sacks, they found a package shipped by the First National Bank of Las Vegas, New Mexico, to the Exchange Bank at White Oaks. The package contained $500, which the four promptly pocketed.

Six hours after the first holdup, the gang stopped the westbound coach two miles east of Taylor's Well Station and, after robbing it, rode off toward the Capitán Mountains. On October 19, at the Circle Diamond roundup camp on the Río Feliz, Musgrave shot and killed former Texas Ranger George Parker.

Musgrave and Hayes robbed the White Oaks stage twice again within the same day—October 21, 1896. They first hit the westbound stage from White Oaks to San Antonio a little past 8:00 P.M. near Cavanaugh's Lake in the Oscura Mountains for a haul of $152 in gold dust in one package and a letter holding $5 in cash. Two hours later they robbed the oncoming eastbound stage.

Pursuing posses lost the outlaws' trail and abandoned the chase.

In August 1895, Urbain placed the following notice in the *White Oaks Eagle*:

TO WHOM IT MAY CONCERN

The public is hereby notified not to trust my wife on my account, as I will not be responsible for any debts she may make or any contract she may enter into either as my wife or as a member of the firm of Ozanne & Co., without my consent. Under the terms of our partnership contract, which is recorded in Lincoln county, I am constituted the sole agent and manager of said firm, and any debt, contract, collection of debts or sale of any personal property, will not be recognized without my endorsement.

U. OZANNE[39]

Things went from bad to worse. In August of 1895, Ella charged that Urbain had assaulted her by throwing her on the floor and against the wall. However, on January 25, 1896, clerk George Curry recorded that "We the jury find the defendant not guilty as charged in the indictment. Lee H. Rudiselle, Foreman."[40]

Sometime before 1897, Ella and Urbain divorced and he married for a third time. Helen, fifty-six years old, hailed from Minnesota. Now Urbain's fortunes were definitely on the decline. He became embroiled in running financial battles with Helen, who continued to spend in spite of Urbain's being broke. In 1899, the various Ozanne properties were sold for back taxes. Even Hotel Ozanne had to be sold, and more horses along with the hotel. Helen sprang into action, suing everyone connected with the Hotel Ozanne—and won.[41]

Family tradition holds that Helen fed her husband ground glass to hasten his eventual demise. In any case, sick and worn out, Urbain died at the age of sixty-eight in White Oaks on August 11, 1903. The undertaker recorded his death as caused by Bright's disease. Col. George Prichard, a friend and fellow Knights of Pythias member, paid for the $62.15 coffin.

After Urbain's death, Helen even received the tax deed for Hotel Ozanne in 1904, although it had already been sold once in July 1901 to F. M. Lund.[42] In 1905, she took whatever money was left and skipped town one step ahead of the sheriff, who was searching for property to attach. There was nothing left of the Ozanne business interests. The location of Urbain Ozanne's grave in Cedarvale Cemetery in White Oaks has been lost with time, and there is no headstone for this truly remarkable French businessman.

First Son Henry's Businesses in New Mexico

In 1884, eldest son Henry partnered in a wholesale and retail drug business in Las Vegas and advertised in the *White Oaks Golden Era*: "Goodall & Ozanne, Wholesale and Retail Druggists, 510 Railroad Ave., Near Depot, Las Vegas, N.M. Orders by mail receive prompt and careful attention. Low prices guaranteed. Your patronage solicited."[43]

Henry must have lived in Las Vegas and shared the actual operation of the drugstore business there because the family reported his later departure from Las Vegas for Denver, Colorado. Henry may also have held onto his drug business back in Wichita. In any case, he eventually returned to Wichita and operated a drugstore there. On one occasion, when Urbain was very ill in White Oaks, the *Las Vegas Daily Optic* mentioned a son arriving from Wichita, "where he is engaged in the drug business" to attend his father's bedside. (Soon after, the *Eagle* reported a much-improved Urbain.)[44] In 1888, the *New Mexico Interpreter* ran an ad for Henry, "Dealer in all kinds of Perishable Goods":

> Having made arrangements with the Express Co. [meaning his father's stage business] for Quick Transportation, I am prepared to take orders for and furnish all kinds of Perishable Goods, such as BUTTER, EGGS, POULTRY, VEGETABLES, FRUITS, OYSTERS and FISH, at lowest market prices. Consignments of all the above class of Goods are received regularly from the Best firms East, and will be sold both at Wholesale and Retail.
>
> Fresh OYSTERS will be received at least twice a week, and parties having Standing Orders will have same delivered at their homes. No effort spared to have only Fresh Goods.
>
> Leave Orders at EXPRESS OFFICE. White Oaks, N.M.[45]

Second Son Emile

Sometime between 1880 and 1882, Emile Ozanne made a business deal with Dave Coleman, overseer for the Atchison, Topeka, and Santa Fe Railroad, to gather a group of men for railroad work at Cerrillos.[46] He rounded up two partners, Smith and McCohie, and thirty-five other men and arrived in Cerrillos with a letter of introduction from Dave Coleman to Coleman's brother, Dick, asking Dick to assign Emile to the engineers' corps. When an argument erupted after Dick refused to honor this request, Emile and his partners returned to Las Vegas. On arriving, the men went from Las Vegas's New Town over to Old Town, where they came upon the sobering sight of three cowboys hanged in the plaza some time before and still dangling in the air.

Emile then traveled north to Trinidad, Colorado, to work as a surveyor and drafter for seventy-five dollars a month, until he heard of the Rico strike in the San Juan Mountains northwest of Durango. He bought a team of mules and started out for Rico, but he did not like the place and returned to Trinidad and on to Alamosa. Broke by this time, he rode the blind baggage like a hobo until the conductor discovered him, and, taking pity on the young man, put him in the coach. On arriving back in Trinidad, he borrowed fifty dollars and worked there for two or three months.

Then Emile worked in a lawyer's office for a while. Urbain and Emile must have been together at this time, because Emile said that he and Urbain traveled to Las Vegas with three wagons and six horses, where they rented a large corral and worked around for a time. Urbain spotted an ad in the Las Vegas newspaper asking for someone to drive John Wethers and his wife and daughter and Henry Wilson, an assayer, to the gold-mining town of White Oaks. He drove the party to White Oaks and there met a man who needed Urbain and his son Emile to go to Cerrillos and work. In Cerrillos, Urbain oversaw the camp while Emile ran the commissary. Then a herder quit and Emile was asked to go out in the hinterland and herd. The Apache warrior Geronimo was terrorizing the countryside at that time, prompting Emile to herd only briefly before quitting and heading again to Trinidad.

Emile next traveled to Springerville in eastern Arizona to visit a friend on a nearby ranch. After two months, he went on to White Oaks, arriving April 6, 1882. In

Emile Ozanne at Jicarilla store and post office, ca. 1898. Photographer unknown. Courtesy Johnson Stearns.

the Oaks, Emile took up sign painting and wallpaper making in partnership with William Chase. An ad in a White Oaks paper announced the dissolution of the firm of Ozanne & Chase, to be succeeded by: "William Chase, Sign Painter. Anyone having painting, papering, glazing, or any other work in my line. Give me a call."[47]

All was not work; Emile played baseball on the White Oaks team during the 1880s and 1890s as the pitcher. In July 1889, Emile married Belle Wiener, a Roswell schoolteacher and daughter of White Oaks businessman Sol Wiener. The year 1889 was also the same year Cháves County was carved out of Lincoln County. George Curry, a rising politician living in Lincoln, asked Emile to accompany the Cháves County portion of the county record books to their destination.

Although he was a college-trained civil and mining engineer, by 1891 Emile had started up a dry goods store in the west end of the Brown Store. His sign read: "E. L. Ozanne, Clothier, Notions, Apparel, and Dry Goods Store."[48]

In 1895, Emile moved his family north a short way to Jicarilla. He became postmaster here (1894–98) and also worked the Dark Cloud mining claim, which he and brother Henry jointly owned. When the postmaster appointment in Jicarilla ended, Emile traveled south to Silver City, near the Santa Rita mines. Emile finally became caretaker at Orogrande Iron and Steel at Orogrande. When that company closed in the 1920s, Emile returned to Jicarilla to spend his last days. He died in 1941.

Third Son Alfred

Alfred Ozanne married Olive Rencher, a young, pretty Mississippi schoolmarm who came to White Oaks drawn by the appeal of a glowing account of the town in a *Home and Farm Magazine* that she had read. In a little story called "A Romance," Olive described her trip from the South as a girl of eighteen years of age to teach in New Mexico:

> On arriving at S, a small railroad station, she found ninety miles of "staging" between that point and her destination. The night was passed at the little hotel kept by a garrulous old Scotchman and his wife. . . . [This "old Scotchman" was the father of Conrad Hilton, who later became an internationally famous hotelman.] During the conversation, and a short while before the stage was to leave, a young man [Alfred Ozanne] entered the little parlor and was introduced as the proprietor of the stage line and informed by the landlord that the gentleman would be her traveling companion throughout the journey.
>
> It is hardly necessary to state that a more considerate companion or a more pleasant stage ride would be hard to imagine. . . . But for the sake of brevity we will only add that less than seven weeks after that meeting in the hotel parlor at S. the proprietor and the passenger were pronounced "man and wife" and the village of N. [Nogal] was minus a teacher.[49]

The newlyweds were to live at Mountain Station Ranch. In December 1890, a little over three years later, the couple left Mountain Station Ranch in care of a driver and moved to White Oaks for the birth of their son, but the baby was stillborn. The couple began making plans to build a permanent home in White Oaks. Olive enjoyed living in her new brick house, but Alfred still had to spend a lot of time at Mountain Station Ranch. A girl named Frances was born to them two years after the death of their son.

In the early 1890s, after Alfred discovered that Mountain Station Ranch and the land on which it was built was not legally his, more trouble loomed as Urbain remarried and the new Mrs. Ozanne schemed to take all the family property. Ozanne Passenger & Stage Line had run at a loss for two years, and the Ozannes did not apply for another four-year contract. Alfred decided instead to leave White Oaks for California. The couple sold all their cattle and what real estate they could.

Three years later, they followed Olive's sister and her doctor husband to Birmingham, Alabama. There Alfred built the West End Real Estate Company, and another baby girl, Ruth, was born. Alfred sold his real estate business after some years and looked after his private interests in land and rentals. Times became hard, and the couple finally lost their houses for the mortgages. Olive was forced to take in boarders as Alfred's health began to fail. He died of cancer of the spleen and bowel.

Fourth Son Paul

Fourth son Paul went to Portland, Oregon, married Etta Bradley, and settled down in Butte, Montana.[50] As county clerk and recorder of Silver Bow County, Paul was in charge of moving the voting machines of the state of Montana from Deer Lodge to Helena when the state capital moved to Helena. Paul also had sixteen mining claims in Gold Creek, Montana, which he had to keep up by doing the annual assessment work. It took a day by wagon to get to Gold Creek from Butte. He would load up his boys and make them help him do the annual assessment; they hated it. As was the case with Urbain Ozanne, Bright's disease killed Paul.

NOTES

1. Haines, "Biographical Sketches: Urbane [sic] Ozane [sic]," 431.

2. "Entered Into Rest: Death of Urbain Ozanne at White Oaks, N.M.," from a partial copy of an unidentified newspaper obituary clipping in the Garry Owen collection. Garry Owen was a grandson of Emile Ozanne, second son of Urbain. The Garry Owen collection includes several items pasted into a scrapbook. Copy furnished courtesy of Robert Hart, former Curator of History at the New Mexico Farm and Ranch Heritage Museum in Las Cruces. Most of the facts in the first two pages of this chapter are taken from the obituary.

3. Eidenbach and Hart, *Number of Things*. Hart did extensive research on the Ozanne Stage Line while in the employ of Human Systems Research, headquartered in Las Cruces. The chapter about the Ozanne family in this book owes much to his research and to his kindness in furnishing this writer material about the Ozanne family. Hart now lives in Eugene, Oregon.

4. Hart, Robert, "Ozanne Family Summary," a three-page article dated March 8, 1994, written by Hart while in the employ of Human Systems Research, Tularosa, New Mexico.

5. "Entered Into Rest."

6. "Ozanne Family Summary."

7. Hart, "Ozanne Stage Line and a Brief History," speech presented to the Carrizozo Women's Club, September 23, 1995, transcript in author's collection. A transcript of this speech is on file with the Lincoln County Historical Society, Lincoln.

8. "Entered Into Rest."

9. "Ozanne Family Summary."

10. *White Oaks Golden Era*, February 11, 1882.

11. *New Mexico Interpreter*, March 23, 1888; Hart, " Ozanne Stage Line and a Brief History."

12. *Lincoln County Republican*, December 30, 1882.

13. *Lincoln County Leader*, July 12, 1890.

14. *Las Vegas Gazette*, day and month unidentified, 1881.

15. Hart, " Ozanne Stage Line and a Brief History"; *White Oaks Golden Era*, January 3, 1884.

16. Justice Docket, Lincoln County, New Mexico, Precinct No. 8, November 1883 to August 1885. Dorothy Leslie of Carrizozo donated this ledger to the Lincoln County Clerk.

17. *White Oaks Golden Era*, June 12, 1884.

18. *White Oaks Golden Era*, November 5, 1885, and December 31, 1885.

19. According to Olive Ozanne, in Eidenbach and Hart, *Number of Things*.

20. Bryan, Howard, "The Coming of the Railroad," 3.

21. Ozanne Stage Line schedule, ca. 1888, E. V. Long Collection, State Records Center and Archives, Santa Fe.

22. Hart, "Ozanne Stage Line"; Robert Leslie to Haldane, March 25, 1999.

23. Mortgage, June 26, 1888, signed by Ozanne & Co. Mortgagor: M. Satterwhite, copy in author's collection.

24. *Lincoln County Republican*, August 5, 1892.

25. Hart, "Ozanne Stage Line."

26. Fleming and Williams, *Treasures of History*, 27–28.

27. *Lincoln County Leader*, July 26, 1890.

28. Hart, "Ozanne Stage Line and a Brief History."

29. Clipping from *Lincoln County Republican*, undated.

30. Letterhead from a letter written by U. Ozanne to George Curry, Socorro, New Mexico, July 1, 1895, Río Grande Historical Collection, Ms223, Box 3/4, New Mexico State University Library.

31. *Lincoln County Leader*, March 19, 1892.

32. Eidenbach and Hart, *Number of Things*, 135.

33. *Santa Fe New Mexican*, June 26, 1894.

34. *Old Abe Eagle*, June 28, 1894.

35. Donald Queen to Haldane, November 29, 2000.

36. Ibid.

37. Tanner and Tanner, "San Antonio–White Oaks Stage Robberies," 20–23.

38. *White Oaks Golden Era*, July 3, 1884.

39. *White Oaks Eagle*, August 5, 1895.

40. Warrant for arrest of Ozanne on charges of assault against wife Ella, August 21, 1895; and Case No. 1488, handwritten note from Foreman Lee H. Rudiselle, both in Criminal Court Cases for Lincoln County, New Mexico State Archives, Santa Fe.

41. Hart, "Ozanne Stage Line," 22.

42. Eidenbach and Hart, *Number of Things*, 133; *White Oaks Eagle*, July 18, 1901.

43. *White Oaks Golden Era*, October 16, 1884.

44. Undated *Las Vegas Daily Optic* newspaper clipping; and undated *Old Abe Eagle* newspaper clipping, both in Garry Owen collection.

45. *New Mexico Interpreter*, March 23, 1888.

46. Traylor and Runnels, *Saga of the Sierra Blanca*, 48–51.

47. *New Mexico Interpreter*, March 23, 1888.

48. Hart, "Ozanne Stage Line," 19.

49. Olive R. Ozanne, "A Romance," 102–14.

50. Paul Ozanne, telephone conversation with Haldane, November 6, 2004.

DR. MELVIN G. PADEN
Legendary Country Doctor

Melvin G. Paden was an old-time country doctor who served Lincoln County for more than half a century during his long life and left a historic treasure in the form of the Paden's Drug Store in Carrizozo, New Mexico. This drugstore is one of the oldest still operating in the state.

Paden was born February 4, 1851, on a farm near Padensville in Wetzel County, West Virginia, and was educated at local public schools. Adventurous and freedom-loving, he ran away from home at the age of seventeen in 1868 and went to sea.[1] He sailed around Cape Horn, the southernmost point of South America, twice in sailing ships carrying loads of grain. These ships would sometimes not put ashore for six months.

Paden returned to West Virginia after his seagoing days but longed to be off to California, the new "Golden Land." To get there, he went first to New York and then traveled by boat to the Isthmus of Panama. He crossed the Isthmus by rail and made his way to California and later to British Columbia. After some months on the Pacific slope, he sailed for Liverpool, England, by Cape Horn. He stayed in Europe for three months and then returned to America and his studies, finishing his literary education at the age of twenty-one. Soon he again crossed the continent to California to stay a year and a half with an older brother who was already a doctor in Napa. It was here that he began his study of medicine at the age of twenty-four. Napa Valley could not contain him long, and he was off for an eight-month stay in Gainesville, Texas.[2]

A Pioneer in White Oaks

Before arriving in White Oaks in late 1880 when he was twenty-nine, Doctor Paden met a young cowboy at a campsite near Fort Sumner.[3] A fellow camper introduced Paden to a William H. Bonney. Paden later said this was the closest he ever came to Billy the Kid.

No sooner had he arrived in White Oaks than he faced

A young Dr. Melvin G. Paden. Date and photographer unknown. Herman B. Weisner Papers, Archives and Special Collections, Ms249, New Mexico State University Library.

a raging smallpox epidemic, which he managed to bring under control.[4] Then, saving his money and helped by friends, he attended medical school at Louisville, Kentucky, for advanced training. The *Golden Era* reported in 1883, "M. G. Paden, who is attending the Louisville Medical College, in a letter to a friend, says that he and W. N. Moore... are getting quite portly, the latter weighing 204 pounds and the former 164. We are pleased to learn that these young gentlemen are improving so rapidly in weight, and trust that in the same proportion, if not more, they will advance in their studies."[5]

After completing this course of medicine, he came back to White Oaks for a time and then returned to Louisville to continue his studies. In January of 1884,

Belle Williams Paden. Date and photographer unknown. Herman B. Weisner Papers, Archives and Special Collections, Ms249, New Mexico State University Library.

another report by the *Golden Era* observed, "Dr. Melvin Paden, now at Louisville, Ky., has been appointed Surgeon of the 1st Reg. N.M.V.M. We know that Dr. Mel. could doctor the above organization all right, if he could find it, but its whereabouts just now are rather mystical."[6]

And in May of 1884, the *Golden Era* noted the return of the good doctor to White Oaks: "Dr. Melvin Paden, who has been attending the Louisville Medical College the past winter, returned via Las Vegas Monday morning. He left the wagon Sunday night the other side of Red Cloud and made the rest of the journey afoot, some forty miles. Doctor Paden has a host of warm friends who are glad to welcome his safe return."[7] He graduated in 1886, earning a gold medal in surgery. Some time in these years he also found time for medical work at Hudson Street Hospital in New York City and special studies at New York Postgraduate College.

A Solid Citizen of White Oaks

After completing his courses at Louisville in 1886, Doctor Paden returned to Padensville in his native West Virginia to marry the charming and popular Belle Williams.[8] She would later make a large contribution to the social and civil life of White Oaks. Doctor Paden brought his bride back to White Oaks, where he practiced general medicine and soon built a drugstore to serve the growing community.

One of his first medical-related duties was to officiate at the execution in Lincoln, New Mexico Territory, on November 19, 1886, of one Dewitt C. Johnson at what was probably the last legal hanging (by Sheriff Brent) in Lincoln County. The eighteen-year-old Johnson had killed storekeeper Alfred Howe in Tularosa Canyon while trying to rob him. As Johnson stood on the gallows, he delivered a long address of warning to wild young men and "died game."[9] Paden received twenty-five dollars for his services from Lincoln County official

Payment to Dr. Paden, after Lincoln County's last legal hanging, December 20, 1886. Signed by Jones Taliaferro, Lincoln County Clerk. Author's collection.

Dr. Paden with sons Melvin and Morgan. Date and photographer unknown. Courtesy Robert N. Mullin Collection, Ms249, Haley Memorial Library and History Center, Midland, Texas.

Jasper Baird. William C. McDonald, the future first governor of the state of New Mexico, notarized the payment document.[10]

Doctor Paden also partnered with lawyer George W. Prichard to work to improve the town. The two men tried to bring in a spring of mountain water for domestic use, and later they drilled a well north of town in anticipation of the needs of a future railroad.[11]

At his own home, Paden was noted for his fine vegetables and fruit, including grapes.[12] He also had an apple orchard that was routinely raided by local boys. Like most other businessmen in White Oaks, Paden continued his interest in mining. He became president and director of Rico Gold Placers Company of Lincoln County and was associated through business with Edward L. Fernsten, trustee of Lincoln Gold Placers of Jicarilla.[13]

Doctor Paden and Belle had three children, all born in White Oaks. The oldest, Morgan (always called "Brent"), was born in 1887. Melvin, Jr., "Mel," was born in 1889. Baby Virginia followed in January 1895 but died after only two months. As his boys grew up, their need for advanced education prompted Paden to send Brent to the New Mexico Military Institute in Roswell. Brent was a cadet there in 1900 and 1901.[14] Mel went away to medical school.

In 1890, Paden built and opened his new red-brick drugstore on White Oaks Avenue. The *Lincoln County Leader* observed, "Dr. Paden . . . is daily making additions to his already large stock. This is a much needed addition to the mercantile establishments of this town and will doubtless receive the patronage it so richly deserves."[15]

Doctor Paden, Avid Hunter and Horseman

It is not surprising that Doctor Paden was appointed game warden given his strong interest in hunting. A local newspaper reported in 1900, "Dr. M. G. Paden went to Nogal Monday, where as Game Warden, he went to prosecute A. C. Storm for a violation of the fish law. The charge against Storm was destruction of fish in Eagle Creek by allowing sawdust from his mill on that stream to deposit in the river. . . . The defendant

Gus Wingfield and Dr. Paden with deer trophies, ca. 1900. Photographer unknown. Courtesy Sara Jackson, My House of Old Things, Ancho, New Mexico.

Interior of Dr. Paden's drugstore in White Oaks, ca. 1890. *Left to right*: Dr. A. A. Bearup, dentist; Dr. Paden; Emile Ozanne. Photographer unknown. Courtesy Lincoln County Historical Society.

was fined $25.00 and costs of the suit."[16] Paden had two good friends who were also great hunters. One was Gus Wingfield and the other was White Oaks' only dentist for a few years, Doctor Bearup. With one or the other, Paden would tramp over the mountains of Lincoln County, bring back trophies of the hunt, and do his own taxidermy. People who traded at Paden's Drug Store in White Oaks, and later at his drugstore in Carrizozo, remembered all the stuffed heads of wildlife on the walls.

Besides the deer heads (and the horns of one magnificent longhorn) adorning the walls of his drugstore, trophies of lynx, quail, wildcats, wolves, and wild turkeys were arranged on stands. Later on, these trophies made the journey from the drugstore in White Oaks to the one in Carrizozo. Eventually, though, Doctor Paden became concerned about the growing carelessness of hunters and his own safety and gave up hunting.

In addition to hunting, one of Doctor Paden's special hobbies was training horses, particularly in his younger days. Horses were his pride and joy, and he drove one of the finest teams in Lincoln County. He would hitch his beautiful, well-groomed animals to his sturdy buggy and drive over rough washboard roads far and near in rain or shine to help anyone who needed help—from the owner of the largest ranch to the most humble sheepherder.[17]

The Move to Carrizozo

In 1906, as the mines began to fail and the population declined in White Oaks, Doctor Paden moved to Carrizozo to fulfill his appointment as division surgeon for the El Paso and Southwestern Railroad. He also became county health officer for Lincoln County from 1910 to 1913, and a member of the Surgeons' Association of the Southwest.[18]

For two years, the Padens lived in a little wooden

Portrait of Dr. Melvin G. Paden. This portrait hangs in Roy Dow's drugstore, formerly Dr. Paden's drugstore. Date and artist unknown. Courtesy Roy Dow and Mr. and Mrs. A. M. Carl.

Marble-topped back bar, Paden's drugstore (in the current Dow drugstore). Note the stained glass insets and the old green mixers. Photograph by Bob Workhoven. Author's collection.

building next to the present-day Paden Drugstore at the corner of U.S. Highway 54 and Twelfth Street in Carrizozo.[19] In 1908, Paden built the first floor of Paden's Drug Store out of red bricks from Ancho to the exact measurements required for his original cases and fixtures from White Oaks, including the antique soda fountain.[20]

He moved all these fixtures to the new drugstore when it was finished. On one side of the drugstore, there are still sixty drawers with white porcelain knobs. Each knob bears the name of the original contents: corks, syringes, bandages, shellac, twine, cabinet glue, insect powder, pilula, gelatin, zinc metal, suspensor. Besides local medicinal herbs, like *mansanea*, the store stocked household and farm potions. These were available along with school books and supplies, personal items, and penny candy.

In 1917, a second floor that was to be used as a hospital was added. The *Carrizozo News* detailed, "The work on the Paden hospital is proceeding rapidly—in fact, the upper story is going up as fast as masons can lay the brick. All material is on hand, the window and door frames are in and all cement lintels ready to be placed in position. The lower story is being remodeled to give more room for the drug store proper and the upper story will have six wards, besides the following rooms: a laboratory, an operating room, a kitchen, and linen closets, and three large office rooms. . . . In every way it will be a modern,

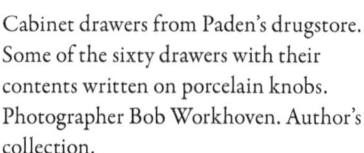

Cabinet drawers from Paden's drugstore. Some of the sixty drawers with their contents written on porcelain knobs. Photographer Bob Workhoven. Author's collection.

up-to-date affair, equipped with all latest devices used by leading hospitals."[21]

There were six small patient rooms with individual doors and push-button wall switches. A nursing station had an alarm box with bells connected to the patients' beds, and there was an operating room for the patients who needed their tonsils or appendixes removed. Laura Sullivan Scharf was the registered nurse for the hospital. Though many times Doctor Paden still went by horse and buggy to make house calls in Carrizozo, he also hired a pharmacist, Ross, who assisted when Paden had to travel by freight train as railroad doctor to help hurt or sick railroaders. During these years, the doctor was said to have also taken postgraduate courses at Rush University in Chicago and Johns Hopkins in Baltimore to stay current in his field.[22]

Disintegration of a Family

In 1910, Belle Paden fell ill with pneumonia—her husband, the doctor who had saved so many others, could not save her. She died at the age of forty-seven, and the funeral was held in her beloved adopted hometown of White Oaks. A long procession followed Belle's casket to Cedarvale Cemetery, and mounds of beautiful flowers were placed atop the casket as it was lowered into the ground. One of the most touching incidents at the cemetery was when an aged Hispanic man, who had known the Paden family for a quarter of a century, deposited a wreath of evergreens in which was entwined one tiny wildflower (this at a season when wildflowers were almost unknown).[23] He must have searched for many hours to find it.

Son Mel went to medical school like his father and was house surgeon at Nassau Hospital, Nassau, Long Island, in 1913.[24] Mel returned to Carrizozo to assist his father in the practice of medicine. Shortly after his return, he became seriously ill and died in 1914 at the age of twenty-four, destroying his father's long-held dreams of a father-son medical partnership.[25] Brent too did not live a long life. He worked for a while as bookkeeper at the Carrizozo Trading Company. He died in 1930 in his early forties and is buried with the rest of his family in Cedarvale Cemetery.

Last Years

Lonely, almost totally blind, and the sole survivor of a once-thriving family, Doctor Paden sold his Carrizozo drugstore around 1934 to Elmer S. "Red" Eaker and returned to White Oaks to spend his last days in the place that held so many memories for him. He lived for a while with family friends Ed and May Queen. Only when he became too feeble to care for himself did he move to Carrizozo, where he died in 1943 at the age of ninety-two. He is buried with the rest of his family in Cedarvale Cemetery.[26]

After his death, old friend Dave Jackson presented Amanda Aguayo Treat Mayer with a silver watch worn smooth by use, its crystal scratched and dulled, on one of his last visits to El Paso. The watch no longer told time. Before his death, Doctor Paden had entrusted the watch to Dave, asking him to deliver it in person to Amanda. He thought Amanda should have it "because her children were born by that watch."[27]

The Paden Drugstore in Carrizozo

After he bought Paden's Drug Store in 1934, Red Eaker replaced the original soda fountain with the present one in 1935.[28] He operated the drugstore and soda fountain for approximately thirty years. Alamogordo pharmacist E. W. "Curly" Burns bought the historic building from Red and ran it for several years. In 1978, Roy Dow bought the place and moved his Gift Gallery from the Roselle Building on Avenue E to the Paden Drug Store.[29]

Some of the historical treasures remaining in the drugstore are the 1935 soda dispensers made by the Liquid Carbonic Company, a 1911 cash register, and several Depression-era green Hamilton Beach mixers. The old pharmacy desk where Doctor Paden stirred up his compounds is there as well.

At Paden's Drug Store, customers can still perch on a stool across from the beautiful antique backbar and indulge in one of many cold treats—ice cream cones, sodas, limeades, banana splits, sundaes, milkshakes, and floats—or solve the world's problems around a table sitting on an old wire-backed chair. The runaway bestseller is still the old-fashioned chocolate malt.

NOTES

1. Haley, *1913 Year Book*; "Melvin G. Paden, M.D.," 572–73.

2. "Melvin G. Paden, M.D."

3. As told to Albert Roberts by Dr. Paden, *Lincoln County News*, April 26, 1979.

4. David Townsend, *Tularosa Basin News*, October 16, 1996.

5. *White Oaks Golden Era*, December 13, 1883.

6. *White Oaks Golden Era*, January 10, 1884.

7. *White Oaks Golden Era*, May 8, 1884.

8. Stearns, *Carrizozo Story*, 87.

9. Rasch, *Violence in Lincoln County*, 8.

10. Payment by Territory of New Mexico, County of Lincoln, to Dr. M. G. Paden, dated December 20, 1886. Filed by Jones Taliaferro, Clerk, January 7, 1887, copy in author's collection.

11. Chris Harkey, undated note to Haldane.

12. "Melvin G. Paden, M.D."

13. Biographical note, Melvin G. Paden, Lincoln County, New Mexico, Collection, 1885–1976, copy in author's collection.

14. *White Oaks Eagle*, September 6, 1900.

15. *Lincoln County Leader*, October 4, 1890.

16. *White Oaks Eagle*, January 4, 1900, and April 18, 1901.

17. Amanda Treat Mayer, "One Special Doctor Served Lincoln County," *Lincoln County News*, March 27, 1969.

18. Haley, *1913 Year Book*.

19. Polly Chávez, "The Quiet Splendor of a Past Age—Paden's Drug Store," *Lincoln County News*, April 26, 1979.

20. "Melvin G. Paden, M.D."

21. *Carrizozo News*, April 13, 1917.

22. Chávez, "Quiet Splendor."

23. Undated newspaper obituary from Amanda Treat Mayer's papers, 1910.

24. Chávez, "Quiet Splendor."

25. Mayer, "One special doctor."

26. Dow, "This Is the Story"; Chávez, "Quiet Splendor."

27. Mayer, "One special doctor."

28. Dow, "This Is the Story."

29. Marty Racine, "A Soda Fountain That Serves Up Nostalgia," *Ruidoso News*, May 21, 1999.

E. W. AND EMMELINE PARKER
Civilizers of a Raucous Mining Camp and Parents of a Mining Dynasty

Erastus Wells Parker (always called "E. W.") was born in Dansville, New York, September 28, 1843, to James and Cherrille Wells Parker and named for an uncle, Erastus Wells.[1] His mother's side boasted many prominent and wealthy New York people including her grandfather, once governor of Connecticut.

In 1852, when E. W. was a boy of nine, his parents started for California. On the way, his mother became ill, forcing them to stop at Saint Louis. There his father set up a sawmill and lumber business on the Mississippi, floated logs on big rafts down the river, and accumulated a fortune during the Civil War filling the timber needs of the U.S. government. Steamboats, war vessels, stores, and warehouses needed to be built, as did new bridges and even residences. Unfortunately, James Parker lost most of his money in the Panics of 1873 and 1877. E. W. was educated in the public schools of Saint Louis and later attended Center College in Kentucky.

In physical appearance as an adult, E. W. was six feet tall and slender and sported a giant walrus mustache. He always wore a gentleman's Stetson hat that had to be medium soft, creased, with a high and semiwide brim. His clothes were of good quality, and he liked to be neat. To complete his favorite look, he wore low-topped half boots of soft leather calfskin with medium thin soles. Physically active and adept at socializing, Parker was a highly moral man of good will, basically honest and decent. His biggest fault was a weakness in managing financial affairs. Late in life he commented, "My greatest handicap in life has been rich relatives. How often . . . have I heard friends say, 'If the deal is good, why doesn't Erastus Wells [E. W.'s wealthy uncle] go into it?'"[2]

Perhaps owing to the influence of the gateway town of Saint Louis, throughout his life, E. W. was a westerner in spirit and work. He had a lifelong love for fast horses. As a youth, he delighted in showing off high-steppers on the streets of Saint Louis and held his own in the town's frequent horse races. Parker drove a single-seat, light, two-wheeled carriage around the city.

In his youth, Parker traveled to many places—Boston, New York, Washington, D.C., Louisville, Chicago, and elsewhere. From 1860 until September 1862, E. W. was a midshipman in the U.S. Navy. During the Civil War, Newport in Rhode Island was home to the U.S. Naval Academy and became the site of the first naval training school. There E. W. was trained on the USS *Constitution* and the USS *Marion*. No records exist as to whether he saw combat service, but one ocean trip took him as far as the Mediterranean.

In the latter part of 1862, E. W. joined the Parker's Express business, a shipping line running from Saint Louis to New Orleans on the Mississippi and its tributaries. The Luwa Nawda Express line, another branch of the family business, transported goods by ocean from New York City to New Orleans. Parker was agent of both companies until 1865, when the lines were sold to the Adams Express Company.

In April 1865, E. W. traveled with his father, a lifelong Republican, to attend the funeral and burial services of President Lincoln in Illinois. Lincoln was assassinated just six days after General Lee's surrender of the Confederate forces. E. W. never forgot the crowd of people who came to pay their last respects to the Great Emancipator.

Marriage to Emmeline Brown

In December 1866, E. W. married Emmeline Brown, pampered daughter of Judge Morris Brown. Judge Brown, a prominent landowner, attorney, and politician, had served as one of the attorneys in the notorious Tweed Ring case. The Browns lived at scenic Penn Yan on the north end of Lake Keuka, one of New York's Finger Lakes. Emmeline (born May 14, 1848) grew up with at least one maid in attendance and enjoyed a life of ease and comfort.

E. W. first spotted Emmeline as she was alighting from a carriage. E. W.'s son Morris later reported, "He took note of the prettiest foot and leg he had ever seen and proceeded to fall in love with them. 'That's my wife if I can get her,' he said. Evidence of keen eyesight and perfect judgment!"[3] Emmeline was a talented singer and pianist who could read and play music, but she was at her best improvising by ear, simply making up music as she went along. She could actually play both "Yankee Doodle" and "Dixie" at the same time, each with a separate hand. More importantly for her future home of White Oaks, Emmeline also grew into a woman of strong character and religious principles.

For a wedding present, Emmeline's father gave the young couple a well-cultivated vineyard of Catawba grapes on the western shore of Lake Keuka five miles south of Penn Yan. The place came to be known as Parker's Landing. By 1872, E. W. had tired of growing and marketing Catawba grapes—he hated being a farmer, missed his former friends, and found country life slow and dull. Frequent trips back to Saint Louis helped maintain contact with his western friends. He lived a double life, cultivating grapes in New York while stealing as much time away as he could in Saint Louis.

During the first seven years of marriage, three sons and one daughter were born to the Parkers: James Henry, August 17, 1868; Frank Wells, September 19, 1870; Morris Brown, August 20, 1871; and a little girl named Lina who was born August 17, 1878, and died at six months.

In New York, Emmeline Parker developed "lung trouble," and her health deteriorated alarmingly.

The National Mail Company

In 1873, E. W. found a way to leave the Catawba grapes in Penn Yan for good. Friends in Atchison, Kansas—top managers of the National Mail Company—offered him a job as manager of their southern division. His region would extend from Fort Worth, Texas, to San Diego, California. This southern division also included the territories. The adventurous E. W. accepted without hesitation. The National Mail Company was one of the large stagecoach companies carrying mail for the U.S. government, along with express mail and goods, baggage, and passengers. (The railroads had not yet reached the routes covered by this company.)

The headquarters of the new southern division of the National Mail Company would be Saint Louis. Accordingly, E. W. moved his family west to Missouri. The mid- and late 1870s saw E. W. away from home most of the time and traveling constantly—especially west of the Mississippi River—as he supervised the mail company, checked existing routes, and found new ones. He came to know the country very well and became one of the best-known and most widely traveled westerners of his time.

Parker made the journey several times from Fort Worth, Texas, to San Diego, California, in the old-time Concord coaches. The Concords traveled day and night, and the passengers never took off their clothes for the entire trip of some fifteen hundred miles (twelve days and nights of continuous travel). Bandits held up the stage somewhere in Texas on one of these trips from Fort Worth to San Diego for the mail company. E. W. was sitting on the high front seat with the driver. As one of the bandits rode horseback alongside the coach waving a .45 Colt revolver, E. W. coolly gauged the precise moment when the revolver was pointing toward the passengers inside the coach. He sprang from his seat, grabbed the weapon as he jumped, and carried both bandit and gun to the ground. In the ensuing melee, some of the passengers whipped out their own guns and chased off the other bandits. The robber E. W. had unhorsed was tied up and taken to the next town to be turned over to the town marshal.

Because his influential uncle, Erastus Wells, lived in Washington, D.C., where he served as a member of Congress during President Grant's two presidential terms, E. W. occasionally made trips to the nation's capital to further the interest of his company.

The First Government Star Route Mail Line in Texas

A certain Dr. Henry F. Hoyt used to practice medicine in Old Tascosa in the Texas Panhandle.[4] Hoyt first went there in 1877 when Tascosa consisted only of two stores, a blacksmith shop, and some adobe houses on a plaza. Doctor Hoyt had an interesting practice, treating frequent outbreaks of smallpox as well as gunshot wounds. The town's citizens were largely what the doctor called "Mexicans," and trouble often developed at their frequent *bailes*. Doctor Hoyt also worked on the LX cattle spread, treating the Texans there. His Tascosa clients kept him occupied while he waited for someone to come along and give him a ride to Las Vegas, New Mexico, where he planned to set up shop and practice medicine.

One day a man came in and introduced himself as E. W. Parker. E. W. had just secured the very first government star route mail line in Texas (routes at that time were designated by stars on post office records, hence the name). The line was to run from Fort Elliott to Las Vegas, New Mexico. E. W. needed directions to Fort Elliott. Word got out at Tascosa, erroneously, that E. W. was a detective, and consequently not one cowboy could be found to guide him to Fort Elliott. Hoyt came to the rescue, offering his services as a guide.

E. W. let the contract for carrying a different mail route—from Tascosa to Fort Bascom—to an ex-cowboy in the country named Roy Copeland. All this mail was saddle mail, carried weekly. In August or early September 1878, Copeland made a few trips on his contract before becoming sick. He summoned Doctor Hoyt to treat him and told the doctor that he was distressed about his mail contract.

Still waiting for his ride to New Mexico, Hoyt offered to make one or two trips in Copeland's place. Because the Indians were still a threat, he rode mostly at night. Along the way, E. W. had set up camps to rest and change horses. On one such trip, as Hoyt started back from Fort Bascom on his night ride that was due in Tascosa at sunrise, he found himself about ten miles from Tascosa when he met five men. They stopped, and the man in the lead threw up his hand in customary cowboy salutation. The doctor noticed the lead man was only a smooth-faced boy, while the others were older. All, however, had splendid horses and were well armed.

The men inquired pleasantly about the country and especially the big ranches, claiming to be New Mexico cattlemen who had heard there were some big ranches around Tascosa that needed horses (the leader had horses he wanted to sell). The good doctor told the party of five where the ranches were. He found out later that these men were, allegedly, Billy the Kid and his gang.

The Famous Star Route Civil Suits

Big trouble loomed for National Mail Company in 1882 because of the star routes. The "Star Route Frauds," as they were called, involved a series of alleged fraudulent contracts for the transporting of mail on largely uninhabited western stage routes.

In mid-1885, the *Dodge City Times* reported that the United States had brought three major civil suits against the Parkers:

> to recover money said to have been improperly secured on mail contracts from Vinita to Las Vegas [via Fort Elliott].... It is claimed that by false representations Post Offices were established and the routes increased in length, and that the Parkers secured from the Government, without any equivalent whatever, the sum of $113,000. The defendants by their attorney... denied all the allegations regarding the false representations of fraud, and claimed, on the other hand, that the Post Offices were petitioned for by large numbers of the inhabitants of the country traversed, and not through any conclusion of theirs.... There have been witnesses summoned, many from the Indian Territory and a number from Washington. *It is one of the most important cases ever tried in Kansas, and therefore a special term was provided.* United States Attorney Hallowell and Assistant Attorney Geo. L. Douglass are prosecuting.[5]

The government spent probably two to three years preparing for the star route trials. All this caused major anxiety for the Parkers before the cases finally came to trial. The *New York Times* reported post-trial:

> THE STAR ROUTE CIVIL SUITS
> VERDICT FOR THE DEFENDANTS IN THE
> PARKER CASES IN TOPEKA, KAN.
>
> *Correspondence of the Kansas City Journal.*
>
> The celebrated star route trials have just been completed, resulting in a unanimous verdict in favor of the defendants.
>
> The cases which have just been tried in the Federal court here were three civil actions, the first being the case of the United States of America against V. W. Parker and J. W. Parker, in which judgment was claimed in the sum of $179,000, with interest from August, 1882. This was concerning ... the Vinita route. The second case was the United States against E. W. Parker and J. W. Parker, in which judgment was claimed in the sum of $31,000, with interest from January, 1883, concerning services on ... the Arkansas route. The third was the case of the United States against G. H. Pease and J. W. Parker, in which the sum of $6,000 and interest from Aug. 8, 1882, concerning service on ... the Nevada route, was asked.
>
> The cases by order of the Judge were consolidated and tried as one. The entire week has been consumed in the evidence and argument of the counsel. The Government has raked the whole country from New

Mexico to Rhode Island for witnesses.... A number of Special Agents, Post Office Inspectors, and detectives have been employed for several months to work up these cases.

It is but just to say that . . . the verdict of the jury voices the sentiment of all who heard the trial. It was evident almost from the outset that the Government had weak grounds for action, which became more weak as the case proceeded.

"Justice prevails."[6]

A Developing Interest in Mining

The 1870s were an era of widespread mining discoveries of gold and silver in the West. As an official of the National Mail Company, E. W. made frequent trips over company routes.[7] While on a tour of inspection from Vinita in Indian Territory (present-day Oklahoma) to Las Vegas, New Mexico, he heard of the discovery of gold at White Oaks. In 1879, he took a team and drove toward the Oaks, stopping at Jicarilla. E. W. was satisfied with what he saw and subsequently located mineral claims in the Jicarilla Mountains.

At the time of the discoveries in White Oaks and nearby Jicarilla, both camps were a long way from anywhere. The main line of travel south from Las Vegas to El Paso followed the Río Grande. The U.S. mail stage route did so also, following El Camino Real, the old Spanish trail. The nearest post office was San Antonio, ninety miles west of White Oaks. San Antonio applied for a post office in 1879 after the discovery of gold, and the mail contract was awarded to E. W.'s National Mail Company. At White Oaks, he picked up a handful of ore that was two-thirds pure gold, worth $400,000 per ton. The gold bug bit instantly and hard.

Speeding back to Saint Louis, Parker resigned his job with the National Mail Company on the way and quickly persuaded a band of men to partner with him to buy and develop the South Homestake Mine. The partners decided to abandon the Jicarilla claims and concentrate on operating the South Homestake at White Oaks.

The Parkers Move to White Oaks

E. W.'s family first caught sight of New Mexico in June 1882. The Parker boys—James, Frank, and Morris—imagined that they were already westerners. After all, their father had once sent them a small Indian pony from Indian Territory complete with saddle, saddle blanket, bridle, and quirt.[8] He also sent buckskin breeches and jackets, moccasins, and a feathered bonnet.

The Parkers headed west to their new home, crossing the state of Kansas by train and traveling toward Las Vegas, New Mexico, in those days the New Mexico passenger terminal for railway travelers. The last town of any size that they encountered on this trip was Newton, Kansas, which had a Harvey House, a dozen frame buildings, and many tents. Beyond Newton, the Great Plains stretched clear to Ratón Pass on the border between Colorado and New Mexico. Dodge City was the only town of any importance in the five hundred miles between Newton and Las Vegas.

Along the way, the Parker boys stood on the rear platform and watched the flat plains go by until far to the west they finally spied their first real mountains: snow-capped Pikes Peak in Colorado's Front Range and the Spanish Peaks. When the Parkers arrived in Las Vegas, they noticed that the air was already clearer and fresher. Compared with what they had just seen, Las Vegas, with perhaps five hundred houses, looked like a city. They remained blissfully unaware that they had just arrived in one of the most lawless towns in the West.

A few hours after reaching Las Vegas, it was on to White Oaks to the south, accompanied by two twenty-gallon half-barrels of water. E. W. and Emmeline rode in a light buggy; the three boys traveled with their great-uncle Marshall in a light spring wagon. Two horses pulled each vehicle. At night they camped in the open, sleeping on the ground. Only two human dwellings appeared along the way. The first, a one-room rock house, served as night camp for a sheepherder on the Spence-Maxwell Ranch. They reached the second, Jerry Hockradle's ranch, on the afternoon of the fifth day.

E. W. carried with him a .45-caliber Colt six-shooter and a .44-caliber Winchester rifle. He chose his camps carefully because of the Apache Indian outbreaks in 1882. (More than one hundred white settlers had been killed in Arizona and New Mexico that year.) The Parkers arrived in White Oaks a few hours after their last stop at the Hockradle place.

Their first home was a two-room log cabin with a flat mud roof and a lean-to kitchen at one end. Floors were of rough one-inch boards with wide cracks between; pieces cut from tin cans and nailed to the floor covered holes where the knots had fallen out. Emmeline Parker must have been appalled, but she never complained. She sent back to Saint Louis for enough furniture for scaled-down cabin living.

Erastus W. Parker family, 1890. *Left to right*: James, E. W., Morris, Emmeline, Frank. Photographer unknown. Courtesy Bill Mathews.

After three years the family finally moved to a much larger eight-room adobe house in the southeast part of town.[9] The adobe bricks were eighteen inches thick, making the house cool in summer and warm in winter. There were wide covered porches in both front and back. The house was in the center of a forty-acre pasture in the middle of which grew a piñon tree. Only a short time before the family arrived in White Oaks, vigilantes had used the tree as a gallows to hang a black man named Richard Jackson, a reputed horse thief. Behind the house were a stable, cow shed, and chicken coop. Also in back of the house was a pond of about an acre filled with water from a pipeline from White Oaks Spring.[10] In cold winters, the ice would freeze four to six inches thick; then it was cut into blocks, stored in the brick icehouse near the main residence, and covered with sawdust. Emmeline kept her perishable food there.

E. W. had the rest of their furniture shipped by rail to Socorro from Saint Louis. Mules and oxen hauled the furniture in a lumber wagon 105 miles across a rough stage road to White Oaks. Emmeline's Chase piano and Wilcox & Gibbs sewing machine were the first such items in White Oaks.[11] Because she was still sick, Emmeline trained her three boys to help with the housework until she could hire servants, even teaching her boys to cook. After a time, the high altitude and dry climate of White Oaks helped restore her normal strength, though her reserves of energy were never great.

E. W.'s horses were the best in White Oaks. For everyday traveling he had a light, two-wheeled carriage and a big bay mare named Polly. Later on, a large black horse named Charlie took over Polly's duties. For longer trips, or for traveling with Emmeline and the boys, there was a span of two matched grays called Dick and Whitey. As young men, the three Parker sons vied for the privilege of taking out the local young ladies for rides.

The South Homestake

Hoping to raise more capital through the sale of stock in their company, the South Homestake partners went public in December 1883.[12] Original owners of the

South Homestake Mill. Date and photographer unknown. Courtesy John and Wesley Mottinger.

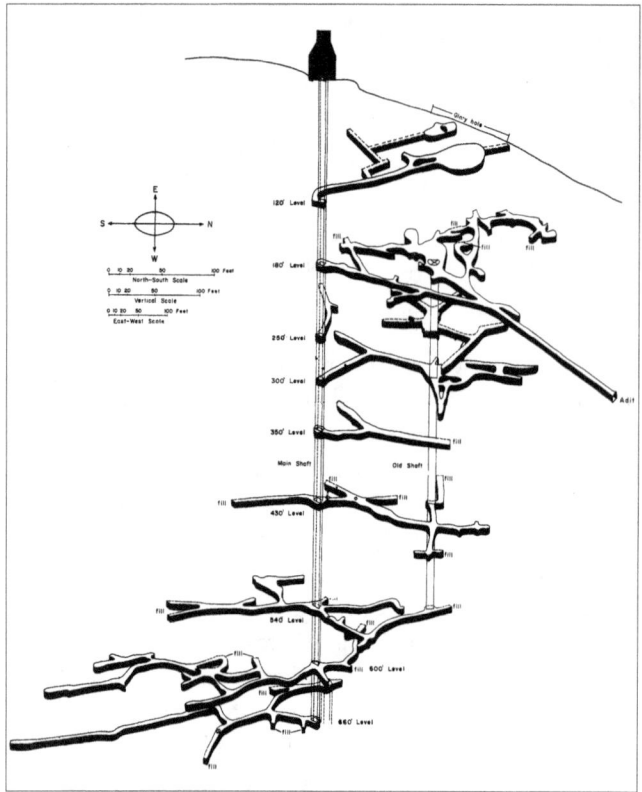

Block diagram of the South Homestake Mine. From George B. Griswold's *Mineral Deposits of Lincoln County, New Mexico*, Bulletin 67 (1959), plate 4. Courtesy New Mexico Bureau of Mines and Mineral Resources, a division of New Mexico Institute of Mining and Technology, Socorro, New Mexico.

South Homestake were E. W. Parker, J. W. Parker, Erastus Wells, E. S. Chester, T. W. Heman, William Watson, C. Ewing Patterson, and John A. Wilson. The partners began plans to develop the mine. They erected a stamp mill and equipped the mine with the latest improvements in machinery and facilities.

A novice when it came to practical mining, E. W. had no idea of the unreliability of high-grade ore specimens and no understanding as to how hard it was to mine gold. He believed that Eastern capital was the key to everything. He began by acquiring heavy equipment from Saint Louis, Denver, and Chicago:

Two portable sawmills

Thirty to forty span of oxen

Heavy logging wagons

A well-drilling outfit with cable and standard rigging (he brought to Lincoln County the first artesian-well outfit, intending to operate it in placer mining)

A pole-pipe boring outfit to make a wooden pipeline for water

The usual mine hoist, cable, sheave, buckets, cars, rails; a boiler, engine, iron pipe and fittings, a blacksmith shop and tools[13]

Once this equipment was en route by road to White Oaks, machinery and equipment for a complete ore mill were ordered. Most of it was purchased in Saint Louis and shipped by rail to western Kansas, then transferred to wagons and hauled by mule or ox teams to Las Vegas, New Mexico. From there, the equipment went south to San Antonio and finally to White Oaks.

Work continued in the meantime on shops, barns, and storehouses. Sawmills were built ten miles east of White Oaks near the foot of Patos Mountain, with a water supply brought from White Oaks Spring. Two and a half years went by before the ore mill was completed and production could start. By summer of 1882, a trial run was at hand, with high hopes of extracting bushels of gold. It did not happen; the mill was a dismal failure, and now E. W. found himself broke in the middle of nowhere in New Mexico Territory with no prospects in sight. Emmeline could have returned to her privileged life back East at her former home during this stressful time. She did not, choosing instead to face the future in White Oaks with her husband.

The owners reorganized, contracting to Fraser, Chalmers and Company of Chicago, well-known builders of mine and mill machinery. They planned a new twenty-stamp amalgamation mill with a capacity of forty to fifty tons in a day. E. W. surrendered most of his control to these experienced mine managers but stayed on as on-site manager in White Oaks.

After three hard years of undoing mistakes, a new mill was up and running well. Recovery of gold was 80 percent to 85 percent of the assay content of the ores. The builder of the new mill, J. B. Schronz, remained as superintendent of both mine and mill. This mill operated steadily for almost ten years (and then intermittently until finally destroyed by fire in 1925).

Always seeking business opportunities, E. W. reverted to the stage business as he was getting the South Homestake Mill under way. In September 1884, the *Golden Era* reported, "E. W. Parker is appointed stage superintendent and will have charge of the San Antonio and Fort Stanton line. If there is anything that E. W. understands—and there are a number—it is the stage business."[14]

Emmeline: A Civilizing Force in a Mining Camp

True to her upbringing and character, Emmeline set about organizing a Sunday school, using mail-order lessons from a Boston missionary society, after arriving in White Oaks while her husband busied himself with the South Homestake. In her Sunday school classes, Emmeline insisted that her charges behave as civilized young ladies and gentlemen.[15] Throughout life, these children carried with them a certain civility and distinctive demeanor that was readily apparent to others. In fact, many of these Sunday school children and their descendants held her largely responsible for the civilization of White Oaks. She personified the beneficial influence of pioneer women on the Old West.

The Plymouth Congregational Church and Community Affairs

Throughout the seventeen years of her stay in White Oaks, Emmeline led church and Sunday school activities. She was a prime force in organizing the building of the Congregational Church that served the entire community for many years.[16]

Organized by twenty charter members belonging to nine different denominations, the Congregational sys-

Christian Endeavor Society, Congregational Church. Top row, *left to right*: Johnnie Lane, Corrinne Hulbert, Ruth Tompkins, Rosa Capuano, Addie LaLone, Eddie Tompkins, Mimi Capuano, Richard Hamilton. Second row, *left to right*: Florence Wharton, Jodie Biggs, Nettie Lee, Blythe Biggs, Carrie Simms, Ethel Walsh, Carrie Timoney, Marshal Parker. Third row, *left to right*: Mr. Hollars, May Lee, Bessie Reed, Ida Hoyle, Edith Parker, Mabel Walsh, Mabel Stewart, Winnie Capuano, Mrs. Sidney Parker. Date and photographer unknown. Courtesy Chris and Jack Harkey.

Congregational Church of White Oaks. Date and photographer unknown. Courtesy Donald M. Queen.

tem of government and beliefs was chosen by the charter members because of its liberal doctrine. The members then began building their church. Material was both bought and donated, and a twenty-four-by-forty-eight-foot building was constructed. The Congregational Church had actually first met in the hall. The hall quickly became the center for all public gatherings—church, Sunday school, day school, political meetings, dances and theatricals, and Christmas celebrations. The first church services were held in this building in 1884. Emmeline sang in the church choir in a clear contralto (or, when needed, soprano) and supervised the children's special entertainments that required speaking, acting, and singing. She attended church morning and evening every Sunday no matter the weather and was always present for Wednesday evening prayer meetings.

Located at the corner of Livingston and Washington Streets, the church was the only one of this denomination in southeast New Mexico Territory. When the church was dedicated in the first week of May 1890, Reverend E. Lyman Hood, Superintendent of the American Home Missionary Society of New York for New Mexico and Arizona, officiated. The May 10 *Lincoln County Leader* observed, "The day was most beautiful. At 11 A.M. the church was well filled and the congregation joined in singing the beautiful Doxology. The house had cost . . . $2,300. . . . Mrs. E. W. Parker, President of the Society of Cheerful Workers . . . stated that $200 had been raised for the building fund from entertainments, suppers, etc. since the previous August."[17]

Meanwhile, E. W. became prominent in the affairs of White Oaks and Lincoln County. He served as a director of the White Oaks School District and a delegate to the Lincoln County Republican Convention. In 1896, he became Worshipful Master of White Oaks Masonic Lodge No. 20.

Front view of Erastus W. Parker home. E. W. Parker sits on the front porch; wife Emmeline is in the buggy. Date and photographer unknown. Courtesy Parker Kemp.

The Parker Boys Grow Up in White Oaks

At first, Emmeline homeschooled her boys with textbooks sent from Saint Louis. The boys studied during hours corresponding to those of a regular school season. Before long, the boys were all studying the same lessons, making Emmeline's job easier. Morris excelled at mathematics. He also studied geometry with William C. McDonald, the man who would later become New Mexico's first elected governor (and Morris Parker's stepfather-in-law). Emmeline sent for musical instruments—a violin, flute, guitar, and banjo. With her piano in the lead, musicals entertained the community as well as the family. She also instigated recitations, readings, talks, and even a literary club.

Along with helping their mother and doing many chores, the three boys found time to play. Frank and Morris tried to be cowboys by riding a young calf around their yard and the town. "Doc" Reid, a family

Morris B. Parker at age nineteen, ca. 1890. Photographer unknown. Courtesy Bill Mathews and Louise Kelt.

Actors in *Among the Breakers*. *Left to right*: James Parker, Wallace Gumm, Eugene Stewart. Date unknown. Photograph by Max H. Koch. C. L. Sonnichsen Papers, Special Collections Department, MS141, University of Texas at El Paso Library.

Among the Breakers playbill, White Oaks Dramatic Club. Author's collection.

friend, built a small, strong wagon for the boys that had iron axles, braces, and supports. The family owned a burro, calf, and goat, and the boys enjoyed trips around town on errands with one of these animals powering the wagon. One activity the boys greatly enjoyed as they grew older was their work with the White Oaks Dramatic Club.[18] Members would dress in costume to put on such plays as *Nevada, or The Lost Mine*, *Among the Breakers*, and *White Mountain Boy*. General admission was twenty-five cents, reserved seats were fifty cents, and the town band supplied all the music. The club lasted several years.

Three Young Mining Men in the Making

The Parker boys found an interesting diversion from their chores in working the placer deposits of Baxter Gulch less than a mile from home.[19] They would work their placers three or four hours a day, first sinking a hole to try to reach pay dirt, then shoveling the gravel into a cradle or rocker. All through the summer of 1883, the boys toiled, only to find that after they had spent several days digging a hole to pay dirt and were set to clean up the next day, their pay gravel would have disappeared overnight. Their entire summer's work produced one ounce, valued at eighteen dollars. E. W. took the gold with him to Saint Louis the next winter and had a heavy ring made from it, intended for the boys' mother, but it was too heavy and soon found a permanent home on E. W.'s little finger. He wore it proudly as long as he lived. The placer mining enterprise ended when the boys' wagon was wrecked by a runaway burro. Frank and Morris then commandeered another wagon, which they learned to drive with oxen. They got into the wood business, timbering in the hills around town. After a near disaster in felling a tree that almost killed the oxen as well as the boys, E. W. made them quit the wood-supply business.

Morris found a job as camp cook in a sawmill at Nogal fifteen miles away. He got up at 4:30 A.M. and was busy all day supplying the sawmill crew with hotcakes, bacon, eggs, steaks, roasts, bread, and pies. After six weeks, Emmeline sent for him, fearing bad results from contact with rough characters. Morris returned home with $37.50 in his pocket.

Once the South Homestake Mill was up and running, the boys worked hard there, arriving for work at 7:00 A.M. and quitting at 5:30 P.M., walking a mile and a half each way.

Discovery of a "Jewelry Shop" at the South Homestake

The South Homestake was operational by early 1885.[20] Then E. W. let his mill man go and, with Joe Grieshaber as mine foreman, assumed the management. For three years, the mine produced steadily, about 50,000 tons, because of a glory hole[21] known as Devil's Kitchen. The Devil's Kitchen ore body was 250 feet up the south hill slope of a gulch. The glory hole was excavated to 70 feet below the surface. Beneath the hole a 600-foot shaft was sunk. Work continued until, on July 1, 1891, an underground mine fire started in the hoist room at the collar of a two-compartment shaft sunk to 600 feet. Smoke from the fire, thought to be started by a lighted candle left on an oil-soaked shelf, killed the two men on night shift. The men had been eating lunch at midnight and were found reclining in death as though they had merely gone to sleep.

After the fire, the mine was reopened by drilling a new shaft. This shaft was sunk to a depth of 700 feet from a higher point on the hill above Devil's Kitchen. Then blasts were set off so that the miners in the South Homestake could move straight ahead toward the North Homestake. Another shaft 800 feet deep was sunk on the opposite hill near the ridge about 500 feet from the gulch and 300 feet from the dividing line with the North Homestake.

When the miners reached a point no more than 25 feet from the dividing line between the two Homestakes, two rounds of shots were fired off. Mine foreman Grieshaber and Morris waited until the smoke had cleared and the workmen on the day shift had left work for the day (there was no night shift on the dynamited shaft) before descending into the ore shoot. Morris described what they saw then as "a jewelry shop" of gold—leaf gold, wire gold, and coarse and fine gold. The two men filled a 50-pound sample sack and a powder box full of gold ore and carried them back to the shaft. Leaving the containers at the shaft level 300 feet down from the surface, they climbed ladders to the surface and lay down on the floor of the shaft house. They tried to calm themselves and catch whatever sleep was possible.

The next morning when the seven o'clock crew came to work, Grieshaber was lowered to the 300-foot level and sent up the ore he and Morris had collected the night before. Grieshaber stayed with the early crew to gather more material while Morris took the sample ore to the assay office and melted it. The night's take came to $18,000.

James "Jamie" Henry Parker

Jamie, the eldest Parker boy, stayed in White Oaks to be tutored by Professor Richmond while Morris was at school in Saint Louis. All three brothers then went to the Colorado School of Mines at Colorado Springs in the fall of 1890. After a work stint in White Oaks again, Jamie returned to the Colorado School of Mines and graduated in 1895. One year (1896–97) of postgraduate work followed at Columbia in New York. He returned to White Oaks for a time.

In 1903, Jamie married Olive Maria Shelden, daughter of a Manhattan, Kansas, jewelry merchant. After their wedding, the couple lived at El Paso for twenty years or so at 712 Prospect Avenue. Jamie traveled throughout the Southwest and Mexico doing exploration and reports on mining properties and also served as a consulting mining engineer. He made one trip to Bluefields in Nicaragua, sailing from New Orleans, to examine gold mines.

Another trip took Jamie to Germany in connection with gold property (owned by the German firm of Alemania) that happened to adjoin the famous Dos Estrella gold mine at El Oro, Mexico. However, he was forced to drop his mining deal when the Francisco Madero revolution got well under way in Mexico in 1911. Jamie operated iron mines at Orogrande forty-five miles north of El Paso for several years, selling iron to the Colorado Fuel and Iron Company at Pueblo, Colorado. He also did some work with titanium deposits near Marfa, Texas. He was considered a good consultant, geologist, and mineralogist.

Jamie died in Houston, Texas, in April 1948.

Frank Wells Parker

Second Parker son Frank carved out a career as a mechanical engineer. After two years (1886–88) at the Manual Training School at Washington University in Saint Louis, he returned to White Oaks for a time before joining his two brothers at the Colorado School of Mines in the fall of 1890.

On January 13, 1892, Frank married Bruce Lane, daughter of Dr. Alexander Lane of White Oaks. Later Frank went into mining, traveling to Orogrande, New Mexico; Richinbar, Arizona; Minas de San Pedro and Parral in Sonora, Mexico; and Dos Cabezas, La Fortuna, and North Tigre, also in Sonora. He was superintendent of the United States Smelting, Refining and Mining Company at the Rainbow Mine near Huntington, Oregon.

Frank died of pneumonia in El Paso on February 11, 1922. His family then moved to Oakland, California, to be near the Lane family.

Morris Parker, Youngest Son

In the fall of 1885, at fourteen, Morris was chosen as the first Parker boy to leave home for school. He was sent to Penn Yan in New York, his mother's hometown. After two years there, he went to Saint Louis, where his grandparents were living. His principal studies were languages (Greek, Latin, German, and French), mathematics, and chemistry. In 1888, he received his diploma in Saint Louis.[22] In 1890, he joined his brothers at the Colorado School of Mines at Colorado Springs. Morris took as much chemistry as possible, learned fire assaying, and worked out the correct fire assay of the telluride ores recently discovered at Cripple Creek. Then he attended Missouri School of Mines at Rolla in 1892 and 1893.

With their formal education completed, and back again at White Oaks, both Morris and Jamie resumed their old jobs in the mill where Morris advertised himself as a mining engineer and assayer. His predecessor in assay work, Henry Kearsing, disapproved of a green "kid" in charge. Soon, a stranger brought in for testing a paper sack containing a dark, powdery substance. When Morris put a hot flame to it on a prepared piece of charcoal, the sample flamed a bit and disappeared, but the fumes smelled like organic vegetable matter. There was no mineral matter in the sample. When the man later returned, Morris handed him a certificate that read "Vegetable matter. No Moisture." In one corner he also wrote, "Fee, $25." On the way home that day, Morris stopped in at Weed's store to tell him the story. Greatly amused, Weed laughed and said, "Make him pay!"[23] A week later, Weed handed Morris a check for twenty-five dollars from Kearsing. Kearsing left town, and the story came out. The stranger had submitted an old, dried-up potato for assay, thinking to have some fun with Morris. The incident boosted Morris's reputation as an assayer.

Morris married Olive Genevieve McCourt in the White Oaks Congregational Church on November 29, 1893. He and his wife lost their first child, an infant son, at birth on September 28, 1895. They had five more children: Lina, Frances, Genevieve, Margaret, and Morris, Jr.[24]

Home of Morris B. and Genevieve Parker, built 1894. Carrizo Mountain at an elevation of 10,500 feet can be seen in the background. Date and photographer unknown. Courtesy Bill Mathews.

The First Cyanide Plant in New Mexico

In the 1890s, Morris experimented with the cyanide process of ore concentration at the South Homestake.[25] In June 1892, he had stopped off at Denver on his way home from college at Rolla. He visited the McArthur plant there and saw the cyanide process used to treat gold tailings that was first used experimentally in New Zealand in 1889.

After reaching White Oaks, Morris sent a one-ton sample of ore from the South Homestake tailing pond to Denver by freight for testing, following to watch the tests. The sample yielded an average gold value of about $2 a ton. These results justified building the first cyanide plant in New Mexico—and one of the first in America—at the South Homestake in 1893. The plant consisted of six wooden tanks, each twenty-two feet in diameter, with a solution and sump tank for each tank, a precipitation box for zinc shavings, a small rotary pump, and a gasoline engine. Each leaching tank could hold thirty tons. Leaching took five days and one day to unload and fill again. It cost sixty cents per ton to operate, with recovery at $2.60, or a profit of $2 per ton.

The following year, 1894, a similar cyanide plant was completed for the North Homestake. (In February 1903, the *White Oaks Eagle* reported that 32,550 tons of tailings from the South Homestake had been "cyanided" up to that time, with $81,375 in gold saved by this method.)[26]

For nearly ten years the two Homestakes—E. W.'s South Homestake and the North Homestake owned by James Sigafus of Colorado—continued to produce and prosper. By early 1893, however, both Homestakes began to fail and the two mines were offered for sale to prospective buyers.

Morris Parker's Years as a Mining Engineer in Mexico

In 1895, the closing of the White Oaks mines forced Morris to leave and make a living elsewhere. It was that same year, at an optimistic twenty-four years of age, that he first became interested in northern Mexico's mining activities.[27] He entered Mexico through El Paso, a place that came to have much meaning for him and his family. He often traveled during these years in the Sierra Madre of Mexico as a consulting engineer for several mines, among them El Carmen, La Republica, La Fortuna, North Tigre, Rosario, and Dos Cabezas. These places were chiefly in the states of Sonora and Chihuahua where those northern Mexican states touch Arizona, New Mexico, and Texas.

Morris's most productive years as a mining man were those spent in Mexico. He spent most of his time below the border traveling along trails far from civilization in remote places among mountain people. His love affair with Mexico lasted until 1932 when Mexico's revolutionary chaos made it impossible to stay.

Three of Morris Parker's children were born in Mexico: Frances Jane and Genevieve in San Pedro; and Margaret Lovina in Pilares de Nacozari, Sonora. After their stay in Nacozari ended in 1903, the Parkers established a permanent home in El Paso, Texas, first at 1213 Myrtle, then at 806 East Rio Grande, 327 Upson Avenue, and, finally, 601 Upson Avenue in Sunset Heights.[28] This house also served as a home base where Genevieve could supervise her children's schooling while Morris continued to travel and work in Mexico.

The last residential listing for the Parkers in El Paso was the 1909 city directory, but from 1902 to 1915 the city directory lists offices for Morris and his brother, James, at the Masonic Temple and the City National Bank Building. The brothers had a loose partnership in a consulting engineering firm. In 1906, Morris served as president of the International Miners' Association, which had its home office at El Paso and held annual meetings in Chihuahua City.

Parker's time in Mexico was the golden age of mining there for Americans. President Porfirio Díaz had established policies that attracted American capital, as evidenced by an increase in American-owned mining companies from 13 in 1868 to 840 by 1907.[29] In 1880, Mexico produced less than 1 percent of the world's gold; by 1910, it peaked at 4.98 percent, with Americans owning three-fourths of Mexico's dividend-paying mines. Chihuahua ranked first among Mexico's states in minerals produced during this golden age. A great aid in the mining revolution was the completion of the Mexican Central Railroad between Juárez, Chihuahua, and Mexico City in 1884. Then in 1897, the Río Grande, Sierra Madre, and Pacific from Juárez to Casas Grandes was completed, and 1900 saw the completion of the Kansas City, Mexico, and Orient from Chihuahua to Minaca.

Parker's many adventures in Mexico included his kidnapping by Yaqui Indians and, later, a dramatic escape by canoe at night from a bandit raid on his mining camp.[30] He also once became very ill with jungle fever and had to be transported 150 miles on muleback to medical care. Over the years, Parker became personally well acquainted with many prominent political and business leaders in Mexico such as President Porfirio Díaz, Pascual Orozco, Francisco Madero, and Don Luís Terrazas.

Parker also came to know the revolutionary figure Francisco "Pancho" Villa very well through mining activities. Day after day the ornery Mexican pack mules had to be coaxed uphill and down over very rough stretches, and muleteers were rated by their ability to deliver the ore no matter the obstacle. Whereas Orozco was an ore cargo contractor, Pancho Villa was Morris's trail boss in charge of transporting heavy mining machinery by mule over the mountain trails. Morris rated both Villa and Orozco as hardworking, dependable overseers to hundreds of men and thousands of mules and their gear.[31] The job required a tough personality, rugged physique, courage, and brains—the perfect apprenticeship for any aspiring generalissimo. Morris remembered Villa as a hardworking go-getter, trustworthy and dependable. Eventually, the Mexican peasants revolted against Díaz and the foreign influence brought about by his policies.

A Haven in El Paso, Texas

Although the Parkers lived in El Paso for the sake of their children's schooling, Morris's profession took him far afield from Mexico to Arizona, Oregon, California, and even to Alaska.

During the Mexican Revolution, the Parkers were at home in El Paso when Orozco, Villa, and Juan Chávez, as heads of the insurgent forces in Mexico, captured Casas Grandes.[32] Then the rebels continued north to capture Juárez, sister city to El Paso, attacking from May 8 to May 10, 1911. In the evenings, the family and all of El Paso watched the fireworks across the border as the insurgents entered Juárez and took possession. Abraham González was appointed the new governor of Chihuahua. When the fighting was over, with a pass safely in hand from González (one of the first such passes issued), Morris and his wife crossed the border to view the carnage. Splashes of blood covered the sidewalks, and bodies were being carried to the cemetery for burial. Small detachments of armed Mexican rebel troops patrolled the streets. The Parkers met and talked briefly with both new President Madero and Governor González. They came away from their adventure with one memento: a small crucifix proffered as a gift from a boy soldier bedecked with cartridge belts and a rifle.

Later Pancho Villa crossed the border to El Paso and lived for several months in exile in a two-story brick house diagonally across the street from the Parkers. Well-dressed and respectable-looking, he and Morris often walked downtown together.

In 1916, Morris was consulting engineer for the King and Queen Copper Company at Steins, New Mexico. From 1919 to 1921, he consulted for California Rand Silver. Then he freelanced, traveling from the west coast of Mexico to the far north and the frozen tundra of Alaska.[33]

E. W. and Emmeline's Mid-Years

After the South Homestake played out, finances were tight for the elder Parkers. Finally, in the winter of 1896–97, son Morris arranged for the sale of the Helen Rae Mine at Nogal. E. W. realized $20,000 from this sale, assuring relief from financial worries. The money came at an opportune time; during the winter of 1896–97, E. W. became seriously ill with a severe case of pneumonia from which he never completely recovered.

He spent the winter of 1897–98 recuperating in the lower, warmer climate of El Paso. Back in White Oaks in 1898, he sold the South Homestake gold mine, mill, and coal mine, and also the Parker home and forty-acre pasture.

Retirement for E. W. and Emmeline

For several years after the sale of the South Homestake, E. W. and Emmeline visited with relatives in Denver, Saint Louis, and Penn Yan. Early in 1904, they relocated permanently to El Paso, where all three sons and their families were living. They bought a large brick home at 608 Stewart Place and became very active in the Presbyterian Church. E. W. continued to attend his Masonic Lodge meetings.

From Mexico, tucked inside his shirt Morris brought his parents a small Chihuahua dog called Spot. He was a smart little pet; he would lie perfectly still during family prayers at night, and the instant E. W. intoned "Amen," the dog would spring up and scamper into the bedroom, jump onto the bed, and cuddle down at the foot of the bed for the night.

Semiretired, E. W. began to read widely in biography and history. From his backyard, made lush with a garden, vines, shrubbery, and trees, he could look across the Río Grande to Juárez and Mexico where his sons often traveled and vicariously join them in their adventures. In 1907, after E. W.'s mother died in Saint Louis, the couple realized several thousand dollars from her estate. Together with what E. W. had saved from his Lincoln County mine and property sales, they lived comfortably. They bought a cottage in the cool woods of Cloudcroft, New Mexico, which they named "Penn Yan" after Emmeline's girlhood home. There they, like others of their El Paso friends, spent cool summers away from El Paso's blistering heat.

Erastus W. and Emmeline Parker in later life. E. W. was always "Rattie" to her; she, "Liny" to him. Date and photographer unknown. Courtesy Bill Mathews.

E. W. died in El Paso on February 2, 1920, at age seventy-six. Emmeline passed away at the Cloudcroft cottage on July 19, 1921. Both E. W. and Emmeline are buried in El Paso's Evergreen Cemetery.

Last Years of Morris Parker

In 1907, Morris and his wife began spending summers in Hermosa, California. Many years later, in the late 1930s, they made Hermosa their permanent residence.[34] Olive Genevieve Parker died at Hermosa in 1955; Morris followed on October 18, 1957. Morris Parker was the ideal chronicler of the mining frontier of his day. He left two books as a testament to his long and adventurous life as a mining engineer: *White Oaks* and *Mules, Mines and Me in Mexico*.

He wrote in the frontispiece of the latter book a reflection on his life: "The sixteen and three score years 1871–1947, the life span of the author to the present time, have been marvelous years in which to have lived: from horses and buggies, tallow candles and black powder, to pushbuttons and atomic bombs."[35]

NOTES

1. Details of E. W. Parker's early life are from "E. W. Parker," 590–91; and Parker, *White Oaks*, 2–3.

Morris Parker's *White Oaks* about his family's seventeen years in White Oaks affords many more stories and details about both home life and business at the mines in White Oaks than is permitted in this short chapter. This book is now out of print, but interested readers can consult the copy filed at the Lincoln County Historical Society in Carrizozo.

2. Parker, *White Oaks*, 81.

3. Parker, "Parker Family"; includes genealogy, dates, places, events, and business affairs. Parker's grandson, William Mathews of Redondo Beach, California, furnished a copy to Haldane. Information about the careers and lives of James and Frank Parker and about Morris Parker's years in Mexico is drawn from this work.

4. Information on Hoyt in this chapter comes from Henry F. Hoyt, interview by J. Evetts Haley, Long Beach, California, March 2, 1928, J. Evetts Haley Collection, Haley Memorial Library and History Center, Midland, Texas. Dr. Hoyt reminisced about his experiences in Old Tascosa in the Texas Panhandle. He encountered E. W. Parker in mid-1878 when Parker was just mapping out his Star Route mail lines from Fort Elliott to Las Vegas, New Mexico, and from Tascosa to Fort Bascom.

5. *Dodge City Times*, June 18, 1885, reproduced in *Panhandle-Plains Historical Review* 40:157–58. Emphasis added.

6. *New York Times*, June 17, 1885.

7. Parker, *White Oaks*, 6. Morris Parker describes the sequence of events that brought the Parker family to White Oaks.

8. Ibid., 9–15.

9. Ibid., 73–74.

10. Birdsong, "E. W. Parker House."

11. Parker, *White Oaks*, 30.

12. *White Oaks Golden Era*, December 6, 1883.

13. Parker, *White Oaks*, 7.

14. *White Oaks Golden Era*, September 18, 1884.

15. From Donald Queen's (descendant of the Lees and Queens of White Oaks) unpublished three-page essay about the Parker family written in correspondence to Haldane, 1998, copy in author's collection.

16. Parker, *White Oaks*, 84–85.

17. *Lincoln County Leader*, May 10, 1890.

18. Parker, *White Oaks*, 85–86.

19. Ibid., 51–57.

20. Ibid., 59–68.

21. A glory hole is a large open pit from which ore is extracted by dropping the ore through shafts or openings down into a haulageway beneath the ore body.

22. Parker, *Mules, Mines and Me*, xii–xiii.

23. Parker, *White Oaks*, 83.

24. Obituary for Morris Parker in the *Redondo Beach (Calif.) Daily Breeze*, October 21, 1957.

25. Parker, *White Oaks*, 102–104.

26. *White Oaks Eagle*, February 5, 1903.

27. The reader is referred to Morris Parker's book *Mules, Mines and Me in Mexico* for a complete account of his fascinating adventures during the thirty-seven years he spent as a mining engineer and consultant in Mexico.

28. *Mules, Mines and Me*, xiii.

29. Ibid., xv.

30. Obituary, *Redondo Beach Daily Breeze*.

31. Parker, *Mules, Mines and Me*, 104–105, 108–10.

32. Ibid., 105; Joan Parker Hughes to Haldane, December 1, 2000.

33. Parker, *Mules, Mines and Me*, xvi.

34. Obituary, *Redondo Beach Daily Breeze*.

35. Parker, *Mules, Mines and Me*, epigraph.

COLONEL GEORGE WORTH PRICHARD
New Mexico's Clarence Darrow

George W. Prichard had been appointed a lieutenant colonel in the Arkansas militia by Governor Clayton of Arkansas early in his career and consequently was always called "Colonel Prichard" as a term of respect.[1] Prichard's long and productive life spanned several careers: lawyer, mine developer, district attorney for the Fifth Federal District of New Mexico, U.S. Attorney for New Mexico, solicitor general for New Mexico, political reformer, ardent Republican, tireless worker for New Mexico statehood, and civic leader.

George W. Prichard was born in New Harmony, Indiana, August 5, 1849, twelve miles west of Pigeon Creek, where Abraham Lincoln lived as a youth.[2] Nine years later he moved to Pike County, Arkansas, where he attended country schools and received further instruction by tutors. After two years in Pike County, the family moved to Hot Spring County, Arkansas. He went on to law school at the University of Michigan at Ann Arbor and graduated in 1872. He began practicing law in Hot Springs, Arkansas—eventually branching out to include Little Rock—and was honored there by being named a presidential elector on the Republican ticket.

George married his first wife, Agnes Whitmor, in Hot Springs in 1876. She was college educated and musically talented.[3] A year after his marriage, his ill health forced a turn toward the West. He and Agnes went to Denver, and it was there that he first became interested in mining.

From Denver to White Oaks

In 1879, news of the gold strike at White Oaks reached Colorado. Prichard lost no time in making his way to the mining camp and obtaining gold specimens. By August 1879, he had assembled a group of potential investors back in Denver and then headed into Las Vegas, New Mexico, on the brand-new Atchison, Topeka, and Santa Fe Railroad to investigate mining property in Lincoln County. On this trip, Prichard brought along some of the richest gold specimens ever seen in Las Vegas—some as large as a man's hand and almost solid gold.[4] In Las Vegas, the party caught the stage on to White Oaks.

Then in its infancy, the gold-mining camp appealed to the young man so strongly that he relocated there after an intermediate stop of four years in Las Vegas. White Oaks would continue to be his home for twenty more years. He owned an interest in the Comstock Mine along with James M. Allen and H. C. Campbell and an interest in the Captain Kidd along with Colonel Watts and James M. Allen. More importantly, he quickly developed special legal expertise in mining laws and found a ready market for his services because of the litigation and many squabbles over mining claims in Lincoln County and beyond.[5]

Emerson Hough, another young White Oaks lawyer, used White Oaks as the locale in his novel *Heart's Desire* and populated the pages of the book with many of the people he knew, including Colonel Prichard, who was the model for a character in his novel.[6] Prichard, himself a writer and historian, published "The Report of San Miguel County" in 1882 in Las Vegas, New Mexico.[7]

In time, Prichard and fellow White Oaks townsman Dr. Melvin Paden sponsored a two-hundred-gallon-a-minute water well north of the Oaks that everyone hoped would service the anticipated railroad through that town.[8] But the railroad bypassed the town, and the partners finally put their well up for sale in May 1919.

A Distinguished Legal Career in the Territory of New Mexico

After building his reputation as a noted lawyer specializing in mining law, Prichard branched out into criminal law, becoming one of the best criminal lawyers in the region.[9] He easily swayed jurors on behalf of his clients. He was short and stocky, with bright, sparkling eyes, and

George W. Prichard. Date and photographer unknown. From *Illustrated History of New Mexico* (Lewis Publishing: Chicago, 1895). Courtesy Chris and Jack Harkey.

had white, silky hair and a distinguished goatee to match. When he was gesturing to a jury with his hands, his goatee followed his hand gestures and mesmerized jurors.

One of his best-known Lincoln County criminal cases was his defense of Will Titsworth, Hunt Hobbs, and Tom Simer in their trial for the murder of Robert Hurt of Capitán on January 24, 1923.[10] After a change of venue, the trial was held in Alamogordo; all three clients were acquitted.

Politics in the Territory of New Mexico

Early on, Colonel Prichard showed a strong interest in politics and got himself noticed. In 1882, President Arthur appointed him U.S. Attorney for New Mexico Territory. In 1886, he was elected to the twenty-seventh assembly of New Mexico Territory as councilman from San Miguel County.[11] He continued in politics following his wife's death in 1889, and in early January 1902, he was named U.S. District Attorney for Socorro, Cháves, Lincoln, and Eddy Counties (the Fifth District). This same year he campaigned for the Republican Party for a month throughout the territory of New Mexico. Territorial governor Miguel Otero named Prichard as solicitor general of New Mexico in 1904, prompting him to move to Santa Fe that year.[12] He was reappointed in 1906 by Gov. Wallace Reynolds. In 1909 and 1910, he again served as solicitor general. During his tenure, Prichard directed many investigations of the transfer of state lands to individuals. In 1884, the *Golden Era* relayed a report from the *Deming Tribune*:

> G. W. Prichard, United States district attorney for New Mexico, has completed his business . . . with the Interior Department, and has secured . . . documents that will aid him in prosecuting a long docket of land sharks. The policy inaugurated by Secretary Teller . . . of sending an army of special agents from the land office into the states and territories, has unearthed a multitude of cases where land entries were made through false affidavits, and these will revert to the government a large percentage of land covered by this system of steals. The practice . . . was carried on to an alarming extent in New Mexico . . . and he means to prosecute.[13]

Prichard also investigated the superintendent of prisons for abuse of convicts, resulting in the removal of this man from office.

The Otero Alliance

Dating from his early Las Vegas days, Prichard formed a strong friendship and alliance with Miguel Antonio Otero, who later served nine years (1897–1905) as governor of the New Mexico Territory.[14] Colonel Prichard joined with Otero in a fifteen-year fight for political control of New Mexico against the Santa Fe Ring, a group of powerful politicians who dominated the affairs of the territory of New Mexico for many years. The Santa Fe Ring was led by Thomas Catron; Catron's then ally against Otero was L. Bradford Prince.

Prince and his New Mexico Republican Reform League established headquarters in Washington, D.C. In 1901, the Prince faction lined up formidable allies to block Otero's reappointment by Congress as governor. Otero countered by persuading Albert B. Fall, one of the officers Otero had appointed as a captain during the Spanish-American War, to assist him in arguing his case before President Roosevelt. The meeting with the president was to defend Otero against charges brought by Prince and Catron that could sabotage his confirmation by the Senate as governor.

Prichard stayed by his friend's side during this critical time. Otero and his supporters, including Prichard, appeared several times before the Senate Committee on Territories to defend Otero against charges made by his enemies that he was uneducated and a supporter of Spain.

On January 16, 1902, the Senate committee's last hearing was held. Otero's friends were there in force. The sole representative of the Santa Fe Ring, William M. Berger, appeared with "enough documents to perfect a treaty with Great Britain" and was allowed to read all he had brought against Otero. Eventually, the committee chairman, Senator Beveridge, who did not ask any other question of Berger, asked if Berger was finished. When Berger assented and left the room, Senator Beveridge did not even put the Otero side on the stand but immediately convened his committee in executive session. Within two minutes, the door opened to smiling senators who shook hands with Otero, telling him that he had been unanimously recommended by the committee for confirmation. Prichard then joined Otero, and all his supporters drove to the Shoreham Hotel to celebrate. Otero was duly confirmed and received his new commission on January 24, 1902.

In 1903, the Catron-Otero feud entered another phase when President Roosevelt directed Otero to fire certain territorial officials who held office by Otero's appointment. George Prichard and four others were among those appointees. Though Governor Otero vigorously defended his appointments, only Colonel Prichard escaped elimination.[15] The removal of so many of his enemies naturally strengthened Catron's power.

Curiously, in years to come Prichard must have buried the hatchet and become friends with Catron, because it was Prichard who presented the eulogy at Senator Catron's funeral in May 1921. Among other tributes, Prichard said, "The impress of his great personality left on this State is ineffacable [sic]. The last half-century of history of the State would be incomplete with the work of his life left out.... Not only those of this generation, but those that will follow us will share knowledge of the life and fame of this great man."[16]

The Statehood Convention of 1902

The statehood convention held in Albuquerque in 1902 asked Governor Otero to appoint a delegation of an equal number of Republicans and Democrats for the purpose of visiting Washington and furthering the interests of New Mexico statehood. From the Republican side, Otero appointed his friend Colonel Prichard, who successfully worked to get the omnibus statehood bill passed by the House of Representatives on May 9. This bill provided for the admission of three territories: New Mexico, Arizona, and Oklahoma. But it was a victory that simply transferred the statehood fight to the Senate. The Senate committee sent L. G. Rothschild of Indianapolis to New Mexico to report on whether the territory was ready for admission to the union. Otero said "the Baron," as he dubbed Rothschild, would sneak around in the slum districts and talk to "impossible" people so that he could report adverse conditions.[17] He also visited the saloons and dance halls and photographed prostitutes and drunks. These exhibits convinced easterners that it would be a mistake to add New Mexico to the Union; once again statehood was delayed.

Later on, in 1910, Prichard, a tireless worker for New Mexico statehood, was again named a delegate to the New Mexico State Constitutional Convention. From the first, he also vigorously supported the extension of voting rights to women and drew up the plank in the progressive Republican platform of the territory proposing this reform.[18] He also advocated the popular election of U.S. senators and fought to put saloons out of business in New Mexico. Finally, as chairman of the Committee on Education for New Mexico, he authored most of the provisions on education in the state's first constitution.

The Final Social Events of the Otero Administration

In December 1905, Governor and Mrs. Otero gave a dinner in honor of Herbert Hagerman, who would succeed Otero in January as governor of the territory.[19] Attorney General and the then Mrs. George Prichard were among the distinguished guests present at this glittering affair. (Prichard must have been remarried by this time.)

On the evening of January 20, 1906, Attorney General and Mrs. Prichard in turn hosted a dinner in honor of Governor Otero at their residence on East Palace in Santa Fe.[20] It was the governor's final social affair of his administration. The dinner was one of the most delightful and exquisite ever to take place in Santa Fe. American beauty roses were everywhere in a color scheme of pink and green. The dining table was set for sixteen.

After the elegant dinner, a handsome and large solid-silver loving cup was presented to the guest of honor by Chief Justice Mills in a speech on behalf of the Supreme Court. Then, in a four-hour marathon, every member of the party delivered laudatory after-dinner speeches.

The old friends enjoyed themselves so much that it was the small hours of the morning before "Auld Lang Syne" was sung.

The Knights of Pythias

One of Colonel Prichard's major interests was the Knights of Pythias, which he introduced to the territory of New Mexico in 1881. As supreme chancellor and representative for New Mexico, he organized the order's first six lodges in the territory.

Eldorado Lodge No. 1, instituted at East Las Vegas in 1881, soon became one of the largest and most prosperous lodges in New Mexico.[21] Prichard and Miguel Otero organized a uniformed division that formed a resplendent sight as they marched on parade in black frock coats with red bandoleers, red belts, swords with silver trimmings, and helmets adorned with red and blue plumes. In midweek of the spring of 1881, a community affair required a parade made up of all the Las Vegas citizens, orders, and military organizations. It was the perfect opportunity to display the new uniforms of the Knights of Pythias. Colonel Prichard, front and center in command of the Knights, marched in the parade with a gorgeous yellow plume on his metal helmet. Everyone else wore red and blue plumes.

A Friend of Albert Jennings Fountain

Inevitably, Prichard made the acquaintance of other important political figures in the New Mexico Territory. He was a friend of Albert Jennings Fountain and was among the last persons on January 31, 1895, to see him alive.[22] Prichard had been in the village of Lincoln and had visited with Fountain, who was completing some legal business before driving himself and his nine-year-old son home to Las Cruces.

Colonel Prichard pointed out the danger of traveling alone with enemies circling round and suggested it would be better to follow the buckboard carrying the U.S. mail. But Fountain, gesturing to a shotgun he carried in the front of his buggy, said, "This will be my protection!" With that remark the Fountains, father and son, disappeared into White Sands, never to be seen again.

Later Years

Colonel Prichard retired to private practice in Santa Fe around 1912. In these years, he built his formidable reputation defending criminals and became known as the "Clarence Darrow of New Mexico."[23]

On November 6, 1917, Prichard married Maude H. Hancock at Tucumcari, New Mexico.[24] She was also college trained; for many years, she taught music in the public schools of Santa Fe and later at the university at Las Vegas, Mexico. At one time she was president of the Santa Fe Woman's Club.

Still practicing in his advanced years, Prichard died February 15, 1935, at the age of eighty-five in Santa Fe.[25] Justices of the New Mexico Supreme Court, district judges, and former governor Miguel Otero were among the pallbearers at the funeral. His wife, Maude, died in 1940 at Tucumcari. Their bodies were cremated and the ashes placed in a crypt at Fairview Park Cemetery in Albuquerque.[26] Colonel Prichard was a true adopted son of New Mexico—so far as can be determined, he never returned to Arkansas, not even for a short visit.

NOTES

1. Prichard, *Prichard Family*, 136.
2. "George Worth Prichard," 774.
3. Prichard, *Prichard Family*, 114.
4. Otero, *My Life on the Frontier*, 231.
5. Prichard, *Prichard Family*, 117, 137.
6. Hough, *Heart's Desire*, 46–47.
7. Bancroft, *History of Arizona and New Mexico*, 752.
8. Chris Harkey, undated note to Haldane.
9. *Lincoln County News*, March 10, 1961.
10. Frederick Nolan, e-mail to Karen Mills, August 2, 2004.
11. Bancroft, *History of Arizona and New Mexico*, 709.
12. "George Worth Prichard," 774.
13. *White Oaks Golden Era*, January 10, 1884, quoting from the *Deming Tribune*.
14. Otero, *My Nine Years as Governor*. Chapter thirteen of Governor Otero's book relates in detail his fight against the New Mexico Republican Reform League, and the constant support of his friend Colonel Prichard.
15. Ibid., 329–30.
16. Westphall, *Thomas Benton Catron*, 392–94.
17. Otero, *My Nine Years as Governor*, 212.
18. Marc Simmons, "Golden Ancestor Dug Up in Prichard Family's Past," *El Paso Times*, August 14, 1977.
19. Otero, *My Nine Years as Governor*, 261–62.
20. Ibid., 262.
21. Otero, *My Life on the Frontier*, 243.
22. Simmons, "Golden Ancestor."
23. *Lincoln County News*, March 10, 1961.
24. Prichard, *Prichard Family*, 140.
25. Hamilton, *El Paso Herald Post*, September 11, 1973.
26. Prichard, *Prichard Family*, 144.

THE QUEEN FAMILIES
Mining in Their Blood

The Queens were a multigenerational mining family.[1] The Queen men carried their passion for the mining profession down through three generations, from the Civil War until the early 1950s. In later life, the men would pay dearly as they developed silicosis of the lungs, the disease of hard-rock miners that gradually but inexorably causes shortness of breath and eventual death from suffocation.

At times the Queen women lived hard lives as they followed their husbands into the roughest camps and had to make do without modern conveniences. The Queen men always provided well as the women kept the home fires burning and raised their children to become decent, productive citizens. The couples remained deeply devoted to each other throughout life.

John Queen and Wife, Sophia "Sophie" Mary Turner

On May 25, 1863, John Queen, an eighteen-year-old native of Cincinnati, Ohio, enlisted in the Civil War as a private in the Union army. His unit had originally been mustered into service in September 1862 at Camp Portsmouth, Ohio, as the 117th Regiment, Ohio Volunteer Infantry, but was reorganized as the First Regiment Heavy Artillery, Ohio Volunteers on August 12, 1863. John was posted with a detachment of Company B to Virginia City, Nevada Territory, to prevent Southern sympathizers from taking over the mines there. Virginia City's Comstock Mine, then the largest producer of gold and silver in the West, furnished the bullion desperately needed to pay for the Union army's cost of

Sophie Turner Queen. Date unknown. Photograph by Max H. Koch. Courtesy Donald M. Queen.

George Queen, ca. 1900. Photograph by Max H. Koch. Courtesy Donald M. Queen.

the Civil War. In Virginia City, John met and married Sophia "Sophie" Turner, born October 1850. Sophie's roots were German, with both parents hailing from Bremen in Saxony.

After two years and two months of service, John mustered out with the rest of Company B on July 25, 1865, at Knoxville, Tennessee. Being posted to Nevada was lucky—most of the First Ohio Heavy Artillery died in the war. After the war, John and Sophie moved to Pueblo, Colorado. Two sons were born to the couple in Colorado: George Forrest at Gilpin in February 1880 and Edward "Ed" Lawrence at Pueblo on August 10, 1881.

John and his brother, Preston, both avid prospectors, located the Rawhide Placers near Central City, Colorado, mines that eventually produced $300 million in gold. In the early 1870s while prospecting on Russell Gulch near Idaho Springs, Colorado, the pair found the famous War Dance Mine, which produced around $3.5 million in gold. Family lore has it that John lost his interest in the War Dance in a poker game.

From Colorado, the John Queen family moved south to the territory of New Mexico—first to Las Vegas, then San Antonio—while John's brother, Preston, made his way west to San Francisco and became a millionaire.

In the New Mexico Territory

George and Edward attended public school in San Antonio south of Socorro, New Mexico, where the entire curriculum was in Spanish. They were the only Anglo boys in school and became proficient in Spanish; they also spoke German, the first language of their mother, Sophie. Their knowledge of Spanish would come in handy in the mining years ahead.

In 1888, John moved the family to White Oaks. The boys went to school first in the Methodist Church, where all the desks were homemade, next in the upstairs rooms of the Hewitt building, where the students sat in store-bought desks, and finally in the new red-brick schoolhouse built in 1895. The Queen boys were always into mischief, especially fun-loving Ed. When an unknown culprit sprinkled red pepper on the floor of the classroom, the teacher whipped the entire class of forty-two. Girls of that time wore their hair in long tresses, and some of them would flip their hair back onto the desks occupied by Ed and another boy. The girls flipped once too often; Ed and his friend cut off the beautiful curls.

John returned to Ohio around 1892 because of illness; he died some time before 1900 at the Soldiers and Sailors Home in Sandusky. The absence of their father meant that George and Ed had to go to work early to help the family survive. Without her husband, Sophie also led a hard life.

Ed Queen

Ed worked many jobs beginning at the age of eleven: a printer's devil and typesetter for the *Lincoln County Leader*, an apprentice in Charles Mayer's blacksmith shop, and a helper in a coal mine. After school and on Saturdays and vacations, he also sold papers and magazines such as the *Saturday Blade*, *Saturday Evening Post*, *Chicago Ledger*, and *Youth's Companion*.

Ed stood at a medium height of five feet and nine inches, weighed between 165 and 185 pounds as a grown man, and possessed striking blue eyes that often sparkled with merriment. According to his grandson Donald, backbreaking labor in the mines would cause Ed to grow like so many men of the Old West into a person of enormous strength and endurance; his body became as wiry and hard as the rock he mined. Those same mines took a huge toll on his lungs and heart; he developed silicosis from constantly inhaling silica dust.

Ed had a mischievous but not malicious sense of humor, and he loved animals. Though Ed was never religious in the conventional sense, he was a man who lived his beliefs. He was honest, tolerant, and generous to a fault, and his word was always his bond.

Ed Marries May Pearson Lee

On January 1, 1902, Ed married nineteen-year-old May Pearson Lee, the friendly, outgoing daughter of Captain and Mrs. John Lee, in White Oaks' Methodist Church. A brunette with soft brown eyes, May spoke fluent Spanish as well as English. May followed her husband from White Oaks to the mining life in California, Arizona, Nevada, and Zacatecas, Mexico. Three children were born to the couple: Ellyn on September 21, 1902; Lawrence Lee on March 15, 1905; and Donald "Don" John on December 6, 1909. Donald was born in the beautiful Gumm Mansion of White Oaks, which Ed had bought for his family and which was where they lived from 1908 to 1917.

In time, Ed's mining earnings in White Oaks allowed him to buy a coal mine, forty acres of timberland, a sawmill in Dark Canyon, two sheep ranches,

Ed and May Queen, 1927. Photographer unknown. Courtesy Donald M. Queen.

Don and Lawrence Queen with the family dog, ca. 1914. Photographer unknown. Courtesy Donald M. Queen.

twenty-five angora goats, and a forty-acre farm in San Simeon Valley, Arizona. Though Ed and May Queen lived many places in their lives, they always considered White Oaks home and returned often during the Great Depression and for the final years of their life after World War II.

Both Ed and May were the kind of people who, wherever they found themselves, improved conditions around them and helped the less fortunate. Both were staunch, lifelong Democrats and supporters of Franklin D. Roosevelt.

Life Was Not All Work and No Play

In 1912, when he was seven years old, Lawrence and friend Eddie Lane loaded up an old burro and rode five miles over to Texas Park to camp out. Little brother Don was along; he was five. On this trip, a bitter winter storm hit, and to protect themselves from the cold, the boys spread blankets on the ground and covered themselves with a canvas tarp. In the morning, they woke to four inches of snow and a frozen white world. The boys gathered their blankets and tramped to the power plant two

Don, Ellyn, and Lawrence Queen, ca. 1914. Photographer unknown. Courtesy Donald M. Queen.

miles away, where Ed and Wild Cat Leasing Company partners Lane and Jackson were preparing a search.

In that same year of 1912, New Mexico finally became a state. Ed and May packed up the family and headed south to El Paso, Texas, to celebrate the event along with May's Lee relatives and many others from the Oaks. Amid vendors hawking food, there was a parade that featured a steaming fire truck pulled by a white horse. One of the Lee boys bought a watermelon, and everyone in their party met at the plaza downtown to eat the melon and watch the alligators that lived as exotic captives in a pool there.

Ed and May bought their first car in 1914, a four-cylinder E-M-F (forerunner to the Studebaker), and one of the first cars in the Oaks. Cars were all underpowered because of the New Mexico altitude—6,400 feet at White Oaks and 8,000 to 9,000 feet at Bonito and Ruidoso. As a result, cars quickly steamed and had to be backed up the hills (reverse gear then had a lower gear ratio than now).

At the age of nine, Lawrence decided to take a joyride in the family's new car from their house atop a steep hill down into White Oaks. All went well until the boy descended to the valley, crossed the bridge near the schoolhouse, and changed to low gear. He stepped on the gas and started back up the hill. At the top, the engine threw a rod, which went through the cylinder head. As with most cars then, the brakes did not function well going backward and so Lawrence hung on for a roller-coaster ride back down the hill, finally hitting a bridge railing in order to stop.

The Wild Cats

From 1905 through 1917, Ed partnered in a very successful gold-mining business in White Oaks called the Wild Cat Leasing Company. The Wild Cats operated the North and South Homestake gold mines. In the beginning, Ed's partners were his brother George, Alfred James, and Allen Lane. The Wild Cats secured a lease on the North Homestake from James Sigafus of Colorado. Because Sigafus thought the North Homestake had been mined out, he agreed to lease it to the Wild Cats for only $35,000 with an option to buy.

Ed's first son, Lawrence, was born just as the mine's whistles began blowing for the men to go to work. This was the same day the Wild Cats were making their very first cleanup at the mill, which yielded more than two hundred tons of ore. That first cleanup represented about three months of hard work using hand steel and netted enough to pay all the bills and hire some help.

Around 1907, two years after the partnership was formed, both George Queen and Alfred James dropped out of the Wild Cats to run an operation at the Old Abe Mine. George and Alfred each made about $400 a week on that contract. That same year Dave "Happy Jack" Jackson joined the Wild Cats. Allen Lane stayed on as master mechanic, Ed oversaw mining operations, and Jack oversaw milling operations and also acted as secretary-treasurer.

It was an unusual partnership. Ed's father had fought for the Union in the Civil War; Lane's father supported the Confederacy; and Jackson was the son of former slaves. This new partnership was highly successful. Over the next fifteen years, the Wild Cats took out close to half a million dollars in both gold and tungsten—gold from the North Homestake, tungsten from the South Homestake.

In the same time frame, the Wild Cats managed to pay for the North Homestake, lease the South Homestake (buying it several years later), and buy a coal mine, forty acres of timberland, and a sawmill. They installed a power plant, a 500-kVA Westinghouse turbine generator set, and a 250-kVA Corliss valve engine with a generator at the coal mine. Originally the two mines and the mill were run by steam engines, but because the hard water clogged the boilers they converted to electricity.

In 1912, with proceeds from their mining operations, the partners disassembled the equipment of the old power plant on Vera Cruz Mountain and moved it to White Oaks. They built the very first power plant in Lincoln County at the coal mines on the slopes of Carrizo Mountain two miles east of White Oaks and brought electricity to the Oaks. To fuel the plant, the Wild Cats bought a coal mine on the same slope. Later they extended electricity to much of the rest of Lincoln County, including Parsons and Carrizozo in 1914.

In 1914, the Wild Cats sorted a fifty-ton carload of tungsten from the dump of the South Homestake and shipped it to Lakeview, Colorado, netting $50,000. The following year the Wild Cats shipped out ninety thousand pounds of tungsten ore from the South Homestake.

A disagreement over how to power the mines caused Ed Queen to leave the Wild Cats in 1917 and sell out for $20,000. Just prior to that, the Wild Cats had sold their power plant for $50,000 and the telephone company, subsequently made possible by the introduction of electricity, for $40,000 and so had no power to operate a mine and mill. Ed wanted to buy power from the company to whom the partners had sold the power plant; Jackson and Lane favored purchasing a diesel electric plant and a truck and furnishing their own power.

With the money from the sale of his interest in the Wild Cats, Ed planned to start a cattle ranch on land leased behind White Oaks from Lone Mountain as far as Red Lake. After someone else proved up on the land he had his eye on, Ed abandoned ranching. Next, he tried to drill a water well. At eight hundred feet down, the driller had still not hit water. Ed paid him anyway, sold the family home in 1917, and left, intending to move to California.

In Douglas, Arizona, and Zacatecas, Mexico

On the way to California, Ed and May stopped off in the mining town of Douglas, Arizona, to visit May's father, Capt. John Lee. Lee was quite ill and so the couple stayed. Ed bought a trucking company and operated it for four years before selling out to a wealthy Mexican. Then Ed left Douglas to take a job in Mexico as mining superintendent of the Cinco de Mayo copper mine at Zacatecas in the Sierra Madres above Hermicolla. The Bank of Douglas also hired him to work for them while he was in Mexico. He worked in Zacatecas for two years but was finally forced to leave because of a plague of bandits. During the Zacatecas years, son Don also became fluent in Spanish.

In 1922, Lawrence graduated from Douglas High School and went to work in a smelter. While in Douglas, Lawrence and his lifelong friend from White Oaks, Eddie Lane, accompanied Ed to look for a mine site in the Dragoon Mountains north of Douglas, where they found azurite and malachite. The boys hiked out from camp and stumbled on an aboveground Indian burial—an animal skin stretched on four poles holding the dead body of an Indian. After a year and a half, Lawrence had had enough of smelter work, and he and Don drove off together in a cut-down Ford Model T headed for San Bernardino, California, where uncle George Queen and his wife were living.

The Queen Family in San Bernadino and Los Angeles

The Queen boys' parents and sister Ellyn soon followed to San Bernardino. Then everyone moved to Los Angeles, where Ed and Lawrence found jobs at Baash-Ross Oil Well Tools in the machine shop. For several years,

Imperial Valley Irrigation Canal. The Queens built this canal near Indio and Palm Springs, California. Lawrence stands on the bridge. Date and photographer unknown. Courtesy Carol Queen Watt.

Lawrence Queen in an old Jennie airplane at Rogers Airport on Western Avenue, Denny, California, ca. 1924–1925. Lawrence is in the back seat. Photographer unknown. Courtesy Carol Queen Watt.

Vera Tracy Queen, 1926. Photographer unknown. Courtesy Carol Queen Watt.

Lawrence Queen, 1927. Photographer unknown. Courtesy Carol Queen Watt.

the brothers worked in various Los Angeles machine shops and prospered in the relative affluence of the early and mid-1920s in California. During these years, Ellyn met and married Earl Leonard Whitwell. Leonard fit in right away with the family and became a mining man too. During this time, Lawrence, always in love with machines, learned to fly airplanes in an old Jennie at Rogers Airport in San Diego. In 1927, Lawrence met Vera Lynn Tracy, a Los Angeles girl, at a church party in California. They were married in Santa Ana in 1928.

Don and Dorothy Queen

Don was a man who was six feet two inches tall, weighed between 175 and 190 pounds as an adult, and was dark-complexioned with brown eyes and hair. (In later years, his hair would turn a distinguished silver color.) At a Methodist church's Valentine's Day party in 1927, Don met Dorothy Bricker, a native of Los Angeles. The pair began a love affair that lasted for almost half a century. They were married in San Bernardino on January 1, 1928. Don had just turned nineteen; Dorothy was sixteen.

Don possessed a powerful, authoritarian personality that the more passive Dorothy complemented. A perfectionist endowed with a good mind and a cynical sense of humor, Don was always pacing, loved to talk, and lived to work. He was an exceptional mechanic and could fix anything. His love for cars was a lifelong passion; he would buy, repair, and sell them to augment his income. His special passion was Mercedes-Benz automobiles; he owned several during his life. A born haggler, he loved hanging out at used-car lots, pawnshops, and furniture stores, always looking for a bargain. Like his mother, May, he was musically talented and could play the guitar, banjo, piano, and organ. And—in common with all the Queens—he loved animals.

Don had an almost mystical affinity for gold, and any spare money went to buy gold coins. Buying at $35 an ounce, he lived to see gold rise to $200 (but not to see it later top out at much higher figures). He also amassed an excellent collection of mineral samples.

In 1935, Don and Dorothy's only child, Donald M., was born in Carrizozo. After their son was born, Don

Don and Dorothy Queen's wedding photo. *Left to right*: Don and Dorothy Queen, Vera Lynn and Lawrence Queen. Date and photographer unknown. Courtesy Donald M. Queen.

and Dorothy moved to White Oaks, first to the Taylor house and later to the Huffmyer place.

The Great Depression

In 1928, with the dawn of the Great Depression, Ed Queen as patriarch of the family decided it was time to leave city life and return to the family's first love—mining. Times were hard: Lawrence, then twenty-three, was employed only three or four days a week. After Ed found a job as superintendent of a placer mine in northern California, he sent for both sons and his daughter and her husband, Leonard Whitwell.

Ed and May's children by that time were all married, and the entire clan chose to join the parents. They outfitted themselves for an expedition to New River in northern California to do placer mining. New River was an unspoiled wilderness—a fisherman's paradise with wild berries growing everywhere and abundant game. The men had to pack in all their equipment and supplies.

May Queen and the other women rode eighteen miles on horseback in order to reach their placer mine in the Salmon Mountains of northern California near Denny. She rented horses from Burnt Ranch, and the women rode along New River, which flows into the Trinity River. Then they crossed over the Trinity, balancing on a swinging bridge. The surroundings in northern California were beautiful, but the mine was losing money, eventually forcing a shutdown.

Lawrence Queen at Big Bear, California, early 1928. Photographer unknown. Courtesy Carol Queen Watt.

Dam at New River, northern California, 1929. Photographer unknown. Courtesy Donald M. Queen.

Back and Forth between White Oaks and California

From 1929 through 1942, the families juggled work and life among White Oaks, Nevada, Arizona, and California. The young wives of Lawrence and Don were both city girls and had to learn a completely new way of life that included coping with freezing winter weather and no plumbing. The men worked the Smuggler Mine in White Oaks, leasing the mining rights from Judge Andrew Hudspeth, from 1933 through 1935. They broke a carload, or sixty tons, of ore, sent it to an El Paso smelter, and received a $1,100 check. Then Hudspeth, believing they had uncovered a major ore body, changed his mind about leasing and instead offered the Queens jobs sinking a shaft, installing equipment, and developing the mine further.

Unhappy with this turn of events, but short of funds, the Queens accepted Hudspeth's offer and produced more than $38,000 in gold over the next year. When the Roosevelt administration devalued U.S. currency in 1933 by raising the price of gold from $20 per ounce to $35, the operation became much more profitable. (This devaluation was one step that FDR took to mitigate the effects of the Great Depression—steps now known as part of the New Deal.)

The Queens became increasingly dissatisfied with working for Hudspeth and struck out again on their own at several mines in Arizona. Work at the Congress Mine in Congress Junction, Arizona, yielded enough

Leonard Whitwell at New River, ca. 1929. Leonard is washing down the side of the mountain preparatory to gathering the gold into a large wooden box. Photographer unknown. Courtesy Donald M. Queen.

money to pay off debts and buy new cars. The Congress Mine contained an especially high silica content that had an adverse long-term effect on their health. Don, in particular, developed a bad case of silicosis. Just as with Judge Hudspeth in White Oaks, the Congress Mine owners refused to give the Queens a lease and decided they would operate the mines themselves. This turned out to be good luck for Don and the others as it got them out of an unhealthy mine.

Lawrence's only child, Carol, was born in 1938 in

Arrival of supplies at New River Gold Mine, 1929. A Native American man and his mules packed in supplies for the Queens. Photographer unknown. Courtesy Donald M. Queen.

Alhambra, California, where Lawrence's wife, Vera, was staying with her parents until the baby arrived.

At the beginning of World War II, the Queens left White Oaks to work in the war plants in San Diego and were there when Pearl Harbor was bombed by the Japanese in December 1941. May Queen stayed at home caring for her two grandchildren while the adults were working. She and Ed also partnered to operate Hotel Ramona.

End of the Road for Ed and May Queen

In 1948 after the close of World War II in Ramona, California, May was stricken with cancer, and she and Ed had to give up running Hotel Ramona. During her illness, Ed never left his wife's side and attended to all her needs. The younger Queen men raised the money to finance a mining venture back in White Oaks, which meant that Ed and May could return to their beloved hometown for the last time.

Ed died of a heart attack on April 22, 1949, at Carrizozo during dinner at the home of May's sister Nettie and her husband, Ray Lemon; he was sixty-seven years old. His obituary lauded him as "a man of sterling integrity, a fond and loving husband and father."[2] The family buried him in the Cedarvale Cemetery at White Oaks. May followed her husband of forty-seven years in death on November 1, 1949, and was buried beside him in Cedarvale Cemetery.

Hotel Ramona in California. Date and photographer unknown. Courtesy Donald M. Queen.

The Last Years for Don and Lawrence

Don began a twenty-two-year career with the New Mexico Employment Security Commission in 1952 and was soon appointed manager. He and Dorothy lived all over southern New Mexico. Don maintained a high energy level until silicosis and a serious heart attack slowed him down. A cigarette habit of two packs a day caused him eventually to develop emphysema as well. The couple retired in Las Cruces in 1974 and lived there until Don's death on July 28, 1976, at the age of sixty-six.

Lawrence worked from 1950 to 1952 as Carrizozo's city manager before he, Vera, and daughter Carol returned to California in 1952. From 1952 through 1969, he worked at the Douglas Aircraft Company's Missile and Space Division in Lakewood, California, and then at McDonnell Douglas as an engineer scientist. He retired because of poor health in 1969, and in 1974 he and Vera moved to San Jacinto, a desert community separated from Palm Springs by the Santa Rosa mountain range.

Lawrence died November 15, 1988, at his home in San Jacinto at the age of eighty-three. After his funeral, his widow and daughter carried his ashes, his gun, the pets, and assorted treasures in the car to a new home, where they interred Lawrence's ashes in Sunset Memory Garden, Langley, South Carolina.

Ellyn Queen Whitwell Cady

Ellyn's husband, Leonard, died of silicosis October 9, 1951, at the age of fifty-one. She remarried a man named Cady, but this marriage ended in divorce. In 1959, Ellyn moved in with her Aunt Nettie and Uncle Ray Lemon in Carrizozo and helped care for them in their old age. After Nettie died in 1976, Ellyn continued to live in the house.

One winter night in February 1985, a howling snowstorm hit New Mexico, with the temperature falling to twenty degrees below zero. For some reason, the eighty-three-year-old Ellyn stepped outside during the night, slipped, and fell. She could not get up and was found the next day covered over with snow, frozen to death. She too is buried at the Cedarvale Cemetery in White Oaks.

NOTES

1. All information in this chapter on the Queen family comes from the following three letters: Carol Queen Watt to Haldane, October 14, 1998 (Mrs. Watt, daughter of Lawrence Queen [younger son of Ed Queen], also provided an addendum to her letter in the form of a two-page timeline of the movements of the Queen families and the significant events in their lives); Donald M. Queen, to Haldane, July 2001 (Donald M. Queen, son of Donald Queen [elder son of Ed Queen], wrote his version of the movements and significant events in the lives of the Queen families); and Lawrence Queen to John Kelt, August 18, 1971, copy in author's collection (Lawrence, younger son of Ed Queen, typed a fourteen-page document detailing the lives of the Queen family and his version of mining in White Oaks from 1905 to 1950).

2. *Ruidoso News*, April 29, 1949.

JAMES B. REDMAN/REDMOND, G. S. AND J. S. REDMAN/READMAN, S. J./ W. J. WOODLAND
Man of Many Names

Jim Redman, as he was most often called, lists himself under this alias in the June 1880 White Oaks census, which was taken when he was twenty-nine.[1] He was born in Pennsylvania of a German father and Irish mother and inherited his mother's Irish wit. His original name was a long German compound that he anglicized by literal translation to "Woodland." But his flaming red hair and ruddy complexion led contemporaries to dub him the "Red Man," a name that eventually transformed into "Redman" or "Redmond."[2] Jim used the name that suited him best at the moment, but in later years usually referred to himself as "Jim Woodland."

Before settling in Lincoln County in 1878, Redman traveled in various parts of the United States. After a time he settled in Texas, where he became a warm friend to outlaw Tom O'Folliard.[3] They traveled together through Lincoln, where O'Folliard signed on as a member of Billy the Kid's Regulators.

Redman called Lincoln home in these years, though he could as often be found on the Pecos, at White Oaks, or elsewhere in southwestern New Mexico. In Lincoln, he used the name "G. S. Readman"; this name appears on an accounts receivable ledger of the Tunstall store.[4]

Lincoln County in a State of Anarchy, 1878

Redman chose not to follow his friend O'Folliard's lead in actively joining the Kid's gang during the Lincoln troubles. The reason may have been the influence of another close friend, Bob M. Gilbert, a level-headed man trusted by such men as John Chisum and Lew Wallace.[5] Gilbert was not only a member of the coroner's jury empaneled to consider how the Englishman John Tunstall met his death on February 18, 1878, but was also a signatory to a eulogy written by Alexander McSween for Dick Brewer, killed April 4, 1878.[6]

The Murder of Huston Chapman

On February 18, 1879, the Kid wrote a letter to a member of the Murphy-Dolan faction in Lincoln "wanting to know whether they proposed peace or fight &c."[7] The men of the Dolan party sent word that they would meet the Kid on the plaza that evening. On the Regulators' side were the Kid, O'Folliard, and two or three others. Somehow Redman got caught up on the Dolan side, which included Jimmy Dolan, Jesse Evans, William Campbell, Billy Mathews, and George Van Sickle. Eventually, the two parties met in the middle of the street to shake hands and sign a peace treaty.

An immediate celebratory tour of the Lincoln bars was in order, resulting in most of the men being thoroughly drunk when they came across Huston Chapman, Susan McSween's lawyer, in the street. After Campbell jammed his six-shooter in Chapman's chest and demanded that Chapman dance, Dolan fired his gun (supposedly into the ground, as he later claimed), triggering a reflexive shot by Campbell into Chapman's chest that killed him. Two shots hit Chapman; Dolan may have fired the second. In the turmoil following Chapman's murder, the Kid slipped away with O'Folliard and rode out of town. The next day, Redman joined most of the law-abiding citizens in Lincoln in signing a petition to Col. Nathan Dudley at Fort Stanton asking for troops to be stationed immediately in Lincoln.[8] (Redman signed his name as "G. S. Redman" on this petition.) The colonel promptly dispatched Lt. Millard F. Goodwin and a detachment of soldiers to Lincoln.

Dolan, along with Campbell, was charged with Chapman's murder. His lawyer successfully pleaded for a change of trial venue from Lincoln, and Dolan was taken to La Mesilla for a habeas corpus examination.[9] At this hearing, Redman was among four witnesses who testified that they had been in "abject fear of

Campbell," afraid to express the least disapproval of his killing Chapman.[10]

Redman Picked to Escort Billy the Kid to Mesilla

In March 1879, both Redman and Gilbert were appointed as special deputy sheriffs, with one of their first tasks being to accompany the Fort Stanton soldiers dispatched to round up those on Governor Wallace's most wanted list of March 11, 1879.[11] Billy the Kid's name appeared on this list.

On the night of March 21, the Kid surrendered and agreed to turn state's evidence in return for a full pardon from Gov. Lew Wallace. A fake arrest was staged, and he was held in Juan Patrón's house at Lincoln awaiting events.[12] In June 1879, Sheriff George Kimbrell picked Deputy James S. Redman to escort the Kid to Mesilla, where the Kid was to testify against Dolan and Campbell about Chapman's murder.[13] The Kid had already testified in April at district court in Lincoln and again on May 28 before the Dudley court of inquiry. But, his trust in Governor Wallace's promises evaporating, and knowing it was impossible to get a speedy trial because it was too late to be tried during the current term of court, he no doubt saw the handwriting on the wall. He must have realized the futility of his request for a pardon from the governor and simply decided to return to Fort Sumner and resume his career as a rustler.

One version has it that his "jail" host, Patrón, said to Billy, "Kid, I'll give you a horse and a Winchester. Then let's see if they can capture you." Others believe that, after Sheriff Kimbrell picked James Redman as Billy's escort from Lincoln to Mesilla, the Kid thought the journey from Lincoln to Mesilla would give his enemy Redman too tempting a chance to kill him while "trying to escape." No matter, the Kid chose his own course of action on June 17, 1879. He told his guards, "Boys, I'm tired of this. Tell the General [Wallace] I'm tired."[14] Slipping off his handcuffs, he walked out of the building and across the street to where a horse was hitched and rode away in broad daylight toward Fort Sumner.

Redman did not appear in Socorro where Dolan was actually tried for the murder of Chapman in November 1879. In fact, no witnesses at all were present at this trial to testify. Dolan's attorneys secured a dismissal, and Dolan was a free man.

Over the Mountains to White Oaks

In the fall of 1879, rumors of the discovery of gold at Baxter Gulch near White Oaks were flying. For a time, Lincoln became almost a ghost town as most of its citizens rushed to stake claims in the mining camp. Redman was among the first to leave Lincoln for the Oaks. The town was laid out and officers elected. The always amiable James S. Redman was chosen president;[15] G. W. Gaides, treasurer; and Jimmy Dolan, John Hudgens, Samuel R. Corbet and a Mr. Hubbs, directors.

Like everyone else, Redman staked out mining claims. With partners Will Hudgens and a man named Sweet, he located well-defined leads of gold-bearing quartz (free gold) in the Little Mack and Little Nell Mines.[16] In an 1889 letter to Maj. William Caffrey of the White Oaks *Lincoln County Leader* Jim uniquely recalled his early days in that boomtown:

MY DEAR Major:

You want me to give you some of White Oaks first history.

I came to what is known as White Oak's Spring in the fall of 1879, in company with William H. Hudgens and a bottle of whiskey. We found camped there Jas. Allen, his son Harry, and old Livingstone. We visited the now famous Baxter Gulch and found imbedded in the side of a hill Chas. Starr, Geo. Gaines, Geo. Gay, Tom Walters, Dick McGinnis and Jack Winters, better known as "old blue skin." There was very little work done there at the time.

In the spring of '80, Mr. Hudgens and myself put up a saloon on the flat where White Oaks stands now, and just back of where Will Littell's stables are, and stocked it with Dowlin & DeLany's tanglefoot and pure Havana cigars, imported from Arkansas in the same box with the lost Charley Ross, and I tell you we did a rushing trade. But the town commenced to grow and we concluded to build the old Pioneer Saloon.

Along about the summer of '80 a man by the name of Joe Fowler, and Jim Calhoun made their appearance with a nice little woman, and also a dog with no hair on—I mean the dog—and they built a house where Peter Mackel lives.

About that time we made arrangements to have the mail forwarded from Lincoln and Stanton to the Pioneer Saloon, and there if we were sober enough, the people got their mail. But everything has its end and as our postoffice was developed by Uncle Sam sending us an old gosling of a man from the Grasshopper State, named McCutcheon, and his little daughter. I put on my best clothes, which consisted of a pair of overalls and suspenders, hat and boots, and calling on Mr. Jim Calhoun, I invited him to go with me and call on the pestiferous post-master and make some arrangements about moving the gunny-sacks, with the mail in them. We found the old sinner in a little log house somewhere in the neighborhood of Ed Bonnell's stone house. He said, "How are you?" and of course we replied, "Bully" and talked about moving the sacks, the danger of robbery, and finally concluded to darn up the holes in the sack and label it "Official survey of White Oaks."

So James and myself strutted down to the saloon, sacked the mail, took a drink, shouldered the sacks, took two more drinks, started, and stopped in Fowler's saloon, took three drinks and canned a yellow dog, then started up the street, and arriving in front of Whiteman's store we concluded that the mail was drunk and in no condition to be delivered to a sure-nuff post-master, so the wind being favorable, blowing in the direction of the P.M. to give her a little air, in the middle of the street, and you bet they sported over to the post-master's office. We opened some of them and some of them were never opened. We saved about 15 letters and papers out of the wreck and sidled up to the P.M.'s window.... He received the mail and kept the gunny-sacks, and James and I went down town, got blind drunk, canned two dogs (at least we canned one dog, the other was a stumpy tail and the can came off).

Yours respectfully,
W. J. Woodland or
Jim Redman[17]

Billy the Kid and Redman Cross Paths Again

The night of November 22, 1880, the Kid, Tom O'Folliard, and others of the Kid's gang tried to steal some of J. B. Bell's horses in the southwestern part of White Oaks. The next morning, they were rumored to be camped at Blake's sawmill near White Oaks on the trail to Coyote Springs. Deputy Sheriff Hudgens made up a posse consisting of Redman, John Longworth, George Neil, John Hudgens, James Carlyle, James Bell, and others and trailed the gang to Coyote Springs about six miles north of White Oaks.[18] In the ensuing gunfight, two horses were killed and the rustlers escaped, leaving the posse more interested in collecting the loot left by Billy the Kid instead of pursuing the thieves.

The following night, some of the gang, including the Kid and O'Folliard, boldly rode into the Oaks. The Kid aimed a "friendly" potshot at Deputy Redman, who happened to be lounging in front of Will Hudgens's saloon, but missed.[19] (This was likely intentional; O'Folliard was Redman's friend from Texas, and the Kid knew it.)

Christmas of 1883

Christmas of 1883 saw the Oaks settled down a bit and Redman still there, enjoying the season's festivities. The town ladies had fashioned evergreen and mistletoe wreaths into crosses on the side walls of the town hall. A magnificent tall spruce was the center attraction in the hall. Musicians furnished entertainment by organ, violin, guitar, and banjo. After the children performed their pageant, the most unique Santa Claus ever seen was ushered in—the irrepressible Jim Redman: "With profuse remarks he stripped the beautiful and glittering tree of its joyous fruit and handed it down to the eager, happy little hands. The boom of the Christmas guns, fired by the boys upon the street, meantime held a loud accompaniment."[20]

Later Years

In 1904, writer Emerson Hough revisited New Mexico in a nostalgic trip to the place he always referred to as "Heart's Desire." His account of that visit, "The West After Twenty Years," appeared in a *Field and Stream* magazine. Hough recalls his friends from yesteryear, bringing the reader up to date. He asks after them one by one, "Where is he now?" Of Redman, he answers himself, "Jim Redmond, of the Delaware? Working on the Pecos, somewhere. Ah, how the boys have changed."[21]

After more years of wandering, by 1908 Redman finally hung his hat in the railroad town of Carrizozo twelve miles southwest of White Oaks. Reverting to the name "S. J. Woodland" in his newspaper ads, he advertised himself as a contractor and builder in Carrizozo and farther south at Three Rivers.[22] It is as Jim Woodland that he was buried in Evergreen Cemetery in Carrizozo after his death October 26, 1922, at the age of seventy-two.[23]

NOTES

1. U.S. Bureau of the Census, *United States Census of 1880*, White Oaks.

2. Notes on James B. ("Jim") Redmond, n.d., Mullin Collection.

3. Earle, *Capture of Billy the Kid*, 70.

4. Notes on James B. ("Jim") Redmond.

5. Ibid.

6. Nolan, *Lincoln County War*, 201, 552 (chap. 22, note 2). Gilbert had a ranch at the lower crossing of the Peñasco River south of what is now Artesia.

7. Rasch, "Billy Wilson," 80.

8. Notes on James B. ("Jim") Redmond.

9. Nolan, *Lincoln County War*, 91.

10. Rasch, "Murder of Huston I. Chapman," 252.

11. Nolan, *Lincoln County War*, 381.

12. Ibid., 390.

13. Fulton, *History of the Lincoln County War*, 349.

14. Ibid.

15. *Las Vegas Daily Optic*, May 17, 1880.

16. Stanley Crocchiola, untitled article, *Santa Fe Register*, September 30, 1949.

17. Lincoln County Historical Society, "W. J. Woodland or Jim Redman," 23–24.

18. Nolan, *Lincoln County War*, 398; Rasch, "Gunfire in Lincoln County," 237.

19. Rasch, "Gunfire in Lincoln County," 237; Keleher, *Violence in Lincoln County*, 283; Nolan, *Pat F. Garrett's*, 117.

20. *Golden Era*, December 27, 1883.

21. Hough, "West after Twenty Years."

22. *Carrizozo Outlook*, January 31, 1908.

23. Blakestad, *Gravesite Directory*, 57.

JOHN BURCHEM SLACK
From International Diamond Hoaxer to Coffin Builder and Undertaker

John Burchem Slack was born to a staid old Hardin County family near Elizabethtown, Kentucky, on June 4, 1820. Abraham Lincoln's stepmother, Sarah Johnston, was his great-aunt.[1] Although he was raised as a farmer, Slack also learned the carpenter's trade that he would follow for much of his life.

Slack's difficulties in later life seem to have stemmed from his association with a cousin, Philip Arnold, who was ten years younger. The pair, known as the "Diamond Twins," salted a diamond field and scammed some of the world's greatest financiers, not to mention experts in business and law. Phil was a born con man and the complete opposite of the quiet, methodical Slack, who preferred to melt into the background. There was, however, a brave and adventurous side to his personality, especially when he was a young man.

Off to the Mexican War

At age twenty-six, Slack enlisted as a private at Greensburg, Kentucky, three weeks after the Mexican War was declared on May 11, 1846.[2] Five days later he was mustered into the infantry in Louisville for a twelve-month term. He served in Company A, Second Regiment of the Kentucky Foot Volunteers Infantry, commanded by Col. W. R. McKee.

If in fact Slack walked to war in Mexico, as he later claimed, it must have been a long, hard slog. In Mexico, he took part in many engagements, including the battles of Buena Vista and Monterey. He was honorably discharged at New Orleans on June 9, 1847. Slack returned home and got a job as constable for the Elizabethtown District. He served about a year.[3]

Cousin Phil had entered the war with Mexico on August 9, 1847. He started out as a teamster in the Quartermaster Department, spending six months in Tampico, one month in Mexico City, twenty days at Pueblo, and four months with the Indiana Regiment of Volunteers. According to his own account, he saw actual fighting for only about two weeks. Phil was discharged July 23, 1848, soon after the signing of the Treaty of Guadalupe Hidalgo that ended the war.

Slack and Phil led vagabond lives before finally joining forces to hatch their diamond-salting scheme. After they became embroiled in the scam, their activities became rapidly complex.

1848 and 1849

Gold was discovered at Sutter's Mill in California on January 24, sparking a frenzied gold rush. After hearing glowing reports of the gold discovery, in mid-1848 Slack set off from Elizabethtown by way of Independence, Missouri, hoping eventually to join the flood of westward-bound forty-niners. In July 1848, Phil moved to the goldfields directly from his point of military discharge in Mexico.

John B. Slack, March 19, 1892. This computer-enhanced image was produced by Pamela Clauss from a very small photo in a composite image of all the charter members of White Oaks Masonic Lodge No. 20. Photographer unknown. Courtesy John and Wesley Mottinger.

Slack joined an emigrant train on May 14, 1849, composed of a quasi-military force whose members called themselves the Morgan County and California Rangers of Illinois; he was named fourth sergeant for the group. After the Rangers had grown to a force of about two hundred men, the wagon train left over the Santa Fe Trail. Their numbers made them confident that they could repel any Comanches or Apaches who might have attacked on their overland journey.

The Rangers reached Santa Fe on August 9. Then they traveled south to the Gila River and crossed the desert to California. Reaching Los Angeles, the Ranger train headed north in December for San José along the El Camino Real of the old Catholic missions.

The 1850s

After finally reaching the goldfields, Slack stayed in California for almost a decade but met with little luck. Early in this decade, he and Phil found each other and began partnering in mining ventures. During the late 1850s, they did some hydraulic mining in Yuba County. They made the acquaintance of a man named Asbury Harpending, who would become one of California's foremost financiers and a major player in the Great Diamond Hoax of 1872.

Phil returned in 1854 to Kentucky to see his sweetheart, Mary May. They married in January 1855 and left for San Francisco.

The 1860s

Slack and Phil left their mining ventures in 1860 and went their separate ways until 1870. Phil bought a farm, moving his family from California back to Elizabethtown in 1862. Always restless, and tiring of soldiers from both sides of the Civil War tramping back and forth across his land, he left Mary and their four children and trekked westward once more. After mining at the Emma Mine in the Wasatch Mountains near Salt Lake City, Phil became involved with Harpending again in a project called the Lincoln Gold Mining Company near Lincoln in Placer County, California. This venture ended in the late 1860s.

Meanwhile, after leaving California in 1860, Slack wandered around the Southwest for a decade. He worked the Potosí Mines southwest of what is now Las Vegas, Nevada, and made his way to La Paz, Arizona, after hearing of the discovery of gold there in January 1862. In Arizona, Slack found work with the Walnut Grove Gold Mining Company, which owned the Bully Bueno Mine sixteen miles south of Prescott. He helped to build a road to the mine and later contracted to haul timber for the stamp mills. He spent another four years working for the company, harassed all the while by Apache raids.

Slack's companions at Walnut Grove held him in high esteem and, in 1866, sent him to the House of Representatives of Arizona Territory. By 1867, he had also joined Aztlán Lodge No. 177 of the Masons in Prescott, Arizona. The *Prescott Daily Arizona Miner* caught a snapshot of him in September 1867: "Hon. J. B. Slack, a member of the Third Legislature from this county called upon us yesterday. Mr. Slack . . . has worked hard in the mines and we hope he will yet make a fortune."[4]

In July 1869, the Apaches stepped up their raids, finally burning the mine's twenty-stamp mills and other buildings late that year and completely destroying the mining operation. Slack was free to roam once more.

1870

In March of 1870, Slack and Phil met up again. They were both working for Harpending and his business partner, George D. Roberts, with a prospecting party in Arizona and New Mexico. Harpending & Roberts Exploration Company, under the direction of San Franciscan William C. Ralston, founder and owner of the Bank of California, discovered promising silver deposits at what were first called the Burro Mines. The mines were in the Pyramid Mountains of southwest New Mexico near a town called Grant (the present-day ghost town of Shakespeare). Harpending & Roberts called its mines the Mountains of Silver and spearheaded the overnight growth of Grant. Harpending resurfaced in this venture as the senior business associate of a group of men who formed the Pyramid Range Silver Mountain Company Mines, absorbing the Mountains of Silver mines as part of a new company. Harpending and Roberts set about in San Francisco to raise money for the project.

Slack, who had traveled the Southwest for a decade and knew much of the area, handled the building and real estate end of this business while Phil superintended the mines.

Slack and Phil settled into life in Grant, briefly renamed "Ralston City" in honor of William Ralston, a heavy investor in Harpending's Pyramid company. On October 19, 1870, Slack and Phil wrote a letter to the editor of the *Tucson Arizonan*:

Ralston City, N.M.
Oct. 19, 1870

Having noticed that many of your Republican subscribers have withdrawn their support from your paper in consequence of your manliness in expressing honest sentiments, we the undersigned enclose you $30, and request that you forward your paper to us at this place until requested to change our address,

>Very respectfully,
>[signed] J. B. Slack
>P. Arnold [with several others][5]

About this time Harpending's partner, George Roberts, bought out Phil's share in the Pyramid Mines for $25,000.

The first mention of diamonds appeared in an article published in the *Albuquerque Republican Review* quoting the *Tucson Arizona Citizen* of November 19: "Messrs. J. B. Cooper [a diamond-drill salesman] and P. Arnold, recently from the Burro mines . . . called on us. . . . The latter is a pioneer of these mines, and interested with Harpending, Roberts & Company. . . . They also bring specimens of crystals which they believe to be diamonds and which were found at the mines."[6]

Writer Fern Lyons speculated that Roberts and Harpending used Cooper, the diamond-drill salesman, to get diamonds to Ralston City, and then cut him out of their plans as soon as they could. Cooper and Phil took their specimens from Fort Defiance in Arizona to San Francisco, where they met with Roberts and showed him their crystals (diamonds and rubies). Roberts was elated. Slack is not mentioned as a third party on this trip, but he may well have been.

Harpending was in London in November 1870 seeking financing for his Pyramid Range Silver Mountain Company Mines. Soon after, Cooper dropped out of sight. Slack and Phil—each now owners of an interest in the Harpending & Roberts Exploration Company—made a second trip to the site of the "crystal" discovery in the Burro Mines. It is logical to assume that on the way Slack and Phil refined the details of the Great Diamond Hoax.

1871

Harpending's Pyramid company failed as a result of an article published by the *London Times* attacking the integrity of the promoters. Continued raids by the Apache Indians on the Pyramid Mines had also made further mining unsafe.

With cash in hand, Slack and Phil went to London to buy gems. Phil had received $25,000 from Roberts for his share of the Pyramid silver mines in New Mexico. Slack had received $50,000 as his first payment toward a total of $100,000 from William Lent (a friend of Roberts's and president of the San Francisco and New York Mining and Commercial Company formed to exploit the diamond fields the cousins had supposedly discovered). Lent kept a 108-carat diamond for which a company had offered $96,000, to be divided among seven men according to their varying contributions.

On July 7, Slack (under his middle name of Burchem) and Arnold (under the name of Aundle) bought low-grade diamonds and rubies in London from Leopold Keller and Company. They returned on July 12, 19, and 21 to purchase a total of $19,000 worth of gems. Then they somehow got the stones back into the United States and began a train ride west to California.

From Reno, Nevada, Phil telegraphed ahead for someone to meet their train at the junction where the Central Pacific would head west for San Francisco. Harpending met the two men and accompanied them to Oakland. Along the way, the cousins told Harpending that they had struck a spot rich in gems at a location they refused to reveal; they claimed that they were also forced to cross a river in flood stage. To do so, they told him, they built a raft and were nearly swamped, losing one bag of diamonds in the river.

At Oakland, the cousins handed over one bag of stones to Harpending, who wrote a receipt and shepherded his eager gaggle of prospective investors by ferry to San Francisco. At his home on Rincón Hill, he spread a sheet on top of his pool table and, in a moment of high drama, dumped the gems onto the sheet. The bedazzled group decided to send some of the gems to Charles Lewis Tiffany of New York for appraisal. In a further show of caution, they also chose a mining expert to accompany the cousins in the spring for a personal tour of the field of discovery.

In New York, Tiffany and his lapidary appraised the diamond samples sent him at $150,000 (at maybe ten times their true worth, but without doubt "worth a rajah's ransom.")[7] No one knows how Tiffany arrived at this inflated figure, but it may be lack of experience with South African diamonds and the fact that the stones were uncut. (Neither Tiffany nor his lapidary had much experience with uncut stones.) American ignorance of recent South African finds and of diamond mining set the stage for the salting[8] of diamonds on a grand scale.

When he received Tiffany's appraisal, Harpending boosted the $150,000 appraisal to $1.5 million for all the stones. Samuel Latham Mitchill Barlow, a leading lawyer of the day and an expert in business affairs, was chosen as the legal representative of the diamond corporation newly formed by agreement between Phil and Harpending on October 31, 1871. This corporation, the Golconda Mining Company, was launched on the basis of Barlow's twenty-year friendship with Ralston and his slight knowledge of Harpending.

Phil received $100,000 in bank notes as an advance payment. In December, he took the cash and made a second trip to London to buy more gems from Leopold Keller and Company. He also purchased diamonds and rubies from the Paris branch of the same company.

1872

Back in the United States, Phil made several trips to the diamond fields to salt them in early 1872. The spot he finally chose was a remote three-thousand-acre mesa on the edge of Vermillion Creek at the foot of Diamond Peak (named as such after the hoax) in northwest Colorado. The Union Pacific Railroad had recently arrived a short distance away.

Phil traveled to nearby Laramie, Wyoming, and on the basis of a hard-luck story, pawned a single diamond with a jeweler. He did this several times, always leaving a different diamond with the jeweler to create the illusion of a large diamond field nearby.

Slack had taken his payment and gone to Saint Louis, where he got a job with Wilson, Warner & Company. The company manufactured wooden coffins and caskets and dealt in coffin hardware and trimmings. For the carpenter in Slack, here was a lot of wood to be fashioned into "furniture," and he returned to his comfort zone for a time.

The first part of the hoax was now complete: the initial salting of the diamond fields and the creation of rumors of a huge find. The second part of the plan required finding a highly respected mining engineer to accompany the Diamond Twins to the fields to look at their find. Harpending went to New York and selected Henry Janin, one of the most reputable mining engineers in the world. In early May, Harpending asked Janin to dinner in New York and sounded him out to see if he was interested. Janin accepted, a decision he would regret to his dying day.

On May 13, a party including Harpending and Janin headed west by train to meet the Diamond Twins in Saint Louis. (By then, Slack had been paid his second $50,000[9] and had apparently agreed to go with the party to the diamond fields.) Phil lost his temper when informed that the party planned to meet three other businessmen in Denver. He claimed that he would not expose his hand to the entire world.

One of the partners, William Lent, traveled ahead to Denver and talked to the waiting businessmen, then telegraphed Harpending, urging him to let them go with the Saint Louis party. Harpending refused. So Lent returned to San Francisco with his three men, and left Janin to verify the authenticity of the diamond fields. In June, the New York group got off the train near Rawlins, Wyoming (in order to make the horseback ride to the site seem longer than it really was).

The cousins pointed out where they had located gems before. Spreading out, the group found diamonds mixed with rubies, emeralds, and sapphires (a combination that never happens in nature). Harpending departed with his expedition, including Janin, leaving only Slack and another man named Rubery at the fields. At Cheyenne, Harpending fired off a telegram to New York with a glowing account of the find. On reaching New York, Janin also confirmed diamond fields rich beyond calculating. After only two more days at the site, Rubery and Slack set off for the railroad. Rubery headed west and Slack east, back to Saint Louis.

Baron Rothschild of London contacted Harpending immediately on hearing of the find. He created a European financing group on the strength of details from Harpending. Ralston, head of the Bank of California in San Francisco, offered to finance the entire diamond development project. Phil collected the rest of the $150,000 owed him after the successful trip to the diamond fields. Since the corporation had said he was also entitled to $1,875,000 in shares, Phil went to New York, visited Harpending at his home on the Hudson River and collected *another* $300,000. His take then totaled $550,000. Slack received only $100,000.

Back in San Francisco, twenty-five important men of their day were permitted to subscribe for up to $80,000 each for stock, paid into escrow in Ralston's Bank of California. Rumors of the true location of the diamond fields ran wild. It was said to be in Arizona—or perhaps in New Mexico, Colorado, or Wyoming. Santa Fe became one of the principal embarkation points in the great diamond rush of 1872.

Janin woke up to the extent of the diamond furor and lived in fear of being followed to the discovery site. Phil and Janin started for Laramie, and Janin retreated to San Francisco. But even as the craze escalated, the bubble burst.

The Hoax Exposed

In November 1872, a geologist and government surveyor named Clarence King exposed the hoax. In earlier years, as the first transcontinental railroad neared completion, King had thought a geological survey should be made along the line of the new railroad between the Rocky Mountains and the Sierra Nevada. Congress passed a bill appointing King geologist in charge of the exploration of the fortieth parallel in February 1867. While on this survey, King heard of the new diamond fields. On the basis of his knowledge of the terrain, he correctly deduced the location of the diamond fields.

He soon found the spot where the gems had been salted. King and his party gathered about a hundred rubies and four small diamonds. He devised a plan to test over the whole mesa and discovered the trail of Phil and Janin, even tracking their footprints. The placement of the gems and their association led King to conclude that these minerals could not occur this way in nature.

King arrived in San Francisco and, after an all-night conference with Janin on November 10, convinced the mining engineer that the site had been salted. On November 11, King telegraphed the parent company, San Francisco and New York Mining and Commercial Company, that it had been the victim of a hoax. On that same day, Janin, accompanied by King, called the executive committee of the offshoot diamond corporation together and broke the bad news. An expedition that included Janin was sent out directly from San Francisco on the Union Pacific Railroad to Black Buttes Station in Wyoming. They reached the diamond fields on November 18 and within twenty-four hours confirmed King's findings.

On November 25, the board of directors convened in Ralston's office in the Bank of California to hear Janin's report. Janin tried to explain how he had been fooled, blaming the short time he had been allowed to canvass the area. The company folded on November 27, 1872, before any stock had been sold to the public. Many prominent men had been fooled, but the big losers were Ralston, Harpending, and Roberts. Ralston was forced to pay off investors from his personal funds. Later, he hung a framed certificate of his company's failed stock on the wall of his office where he could see it every day as a reminder always to investigate thoroughly before investing.

In December, a committee was formed to present a written report to the board of directors of the San Francisco and New York Mining and Commercial Company. The report named Slack, Phil, and Cooper as the guilty parties. A grand jury issued indictments for misdemeanor offenses against the Diamond Twins. Phil, knowing time was running out, had sought refuge back in Elizabethtown around the spring of 1872. There his fellow citizens looked upon him as a hero who had outsmarted all those Yankees. On December 9, a suit was filed in Kentucky, and the local sheriff attached Phil's property. He could not be found and served with a summons.

1873

Tired of dodging the law, Phil negotiated a compromise to settle the joint suit against him and Slack. He agreed to repay William Lent $150,000 if all further litigation would be dropped and he received back his attached property. Slack, hiding in Saint Louis, apparently repaid none of the money he received. He continued in the coffin-making business in Saint Louis. He bought into partnership with Edward Wilson, P. H. Warner, and Robert Sweet. Almost certainly he invested most of his $100,000 from the hoax in this business.

1874

Slack rose to become president of the Saint Louis Coffin and Manufacturing Company. But the nationwide depression of 1873 and the bank failures it caused spelled the eventual end for Slack's enterprise.

On August 15, Phil, carefree and adventurous as always, journeyed to Louisville, where he was found and arrested at the Rufer Hotel in connection with two suits unrelated to the diamond hoax. He was hauled before Judge Stites on September 8. The judge ordered his release, and his friends immediately hustled him to a waiting carriage for a planned escape. The threat of extradition from Kentucky did not surface again. Left with about $400,000 from his scheme, Phil bought a farm on which stood a red-brick mansion. First, he raised Kentucky thoroughbred horses, then he sowed grass on which he put some two thousand head of sheep and got into the wool business. He raised hogs and built a store building. He even started a bank known as Arnold and Polk, Bankers.

1877–1879

In 1877, Phil became engaged in an argument with fellow townsman Harry Holdsworth—who owned a competing bank—and was wounded by a bullet in the right shoulder. The wound proved painful but not fatal.

After entering the hospital because of a bout of pneumonia in 1879, Phil worsened and died February 8 before reaching his fiftieth birthday. His funeral was the largest ever seen in Elizabethtown, even though in the short seven years since the Great Diamond Hoax his estate had dwindled to about $40,000. (True to form, he had lived to the hilt after his windfall from the hoax.)

1880–1887

Slack was now broke and saw no future in St. Louis. He looked west to the gold-mining hamlet of White Oaks in the New Mexico Territory. He began as a carpenter in White Oaks but was soon following his profession of making coffins. This profession led him into another, that of undertaking. His penmanship, marked by elaborate swirls and curlicues, began the White Oaks Record of Deaths on May 15, 1880, and ended with an entry on September 1, 1890.[10] A new script starts on that date, probably that of Max Koch, first his helper and then successor as undertaker. Slack officially sold his entire stock and undertaking business to Koch on May 30, 1895. Hiding in plain sight, Slack never changed his name or married; he lived out his years quietly and very privately in an ell attached to the neat frame building he had built for his undertaking parlors. A white picket fence ran along the front. To little Nettie Lemon, Maggie Kelt, and other children of the town, he was known as a pleasant, kindly old man.[11] His fellow citizens in White Oaks would have been amazed to learn that Slack was one of the greatest swindlers of the Old West.

In 1887, Slack applied for and received a pension as a Mexican War Veteran.[12]

1896

On July 26, 1896, John Burchem Slack died of heart failure at the age of seventy-six in White Oaks after sixteen years in the town.[13] His death is duly noted in the Record of Deaths.[14] His estate, administered by fellow Mason E. W. Parker, was valued at $1,611.14:

Cash in bank	$71.64
Cash on hand	2.90
Open Account—Thomas Link	5.00
Notes:	700.00
M. H. Koch	$350
W. E. Carmack	50
Mortgage	300
Personal property— principally carpenter's tools	231.60
Real property—house and lot in White Oaks	600.00
Total	$1,611.14[15]

Slack had lived more than seventeen years longer than his partner in crime and brash cousin, Phil Arnold. The *White Oaks Eagle* ran a long obituary eulogizing him in these words: "Mr. Slack was one of

Group of Masons in Roswell, 1891. John Slack is the Mason on the left end. Paul Larkam Photograph Collection, Archives and Special Collections, RG81-17,2300, New Mexico State University Library.

the oldest and most universally respected residents of White Oaks.... He was always honest and just in his intercourse and dealings with his fellow men and generous to those who gained his ready sympathy and confidence.[16]

Slack was a charter member of White Oaks Masonic Lodge No. 20, formed in 1892, and his fellow Masons also mourned him: "Not only this Lodge ... but the whole community today mourns the sad demise of him whose honest, sterling character, and kind heartedness made him not only a respected and honored brother and citizen, but whose virtues were proudly claimed by all who knew him as most worthy of emulation."[17]

Slack is buried in Cedarvale Cemetery outside White Oaks.

John B. Slack's headstone in Cedarvale Cemetery. Photograph by Bob Workhoven. Author's collection.

NOTES

1. Fern Lyon, notes for speech presented to White Oaks Historical Association, July 28, 1973, 1, in author's collection.

2. The lives and careers of Slack and Arnold are detailed extensively by Bruce A. Woodard in his well-researched book *Diamonds in the Salt*. Woodard refers to some accounts of the hoax written by Asbury Harpending, a major player in those events. Harpending wrote a book called *The Great Diamond Hoax* after the scandal to try and establish his own innocence (and that of his business associate, George D. Roberts). However, anyone following the long trail of Harpending's association with the Diamond Twins must be left with lingering suspicions that Harpending and Roberts, though unindicted, may not have been entirely innocent. Woodard's book was used to trace an abbreviated timeline of both the lives and events leading up to, and encompassing, the Great Diamond Hoax. Interested readers wishing a comprehensive account of this classic hoax should read Woodard's book.

3. Lyon, notes, 1.

4. *Prescott Daily Arizona Miner*, September 25, 1867.

5. *Tucson Arizonan*, October 1870.

6. *Republican Review*, December 3, 1870, quoting the *Arizona Citizen* for November 19, 1870.

7. Dunham and Dunham, *Flaming Gorge Country*, 130.

8. "Salting" is a mining term for placing valuable gems or minerals, usually in a mine, with the intent to deceive.

9. Wilson, "The Great Diamond Hoax of 1872," 71–79. Clarence King, the man who exposed the diamond hoax, attained international celebrity through his actions and later parlayed his fame into a two-year stint as first director of the U.S. Geological Survey.

10. Record of Deaths at White Oaks. The penmanship changes on September 1, 1890. This date probably coincides with the purchase of Slack's business by M. Koch.

11. Lyon, "The Reluctant Diamond Swindler," 12–13.

12. Woodard, *Diamonds in the Salt*, 171.

13. Birdsong, "John Burchem Slack."

14. Record of Deaths at White Oaks, July 26, 1896.

15. Woodard, *Diamonds in the Salt*, 172; Records of the Probate File of John B. Slack, Court House, Carrizozo, Lincoln County, New Mexico.

16. *White Oaks Eagle*, July 26, 1896.

17. Woodard, *Diamonds in the Salt*, 171–72.

LEVIN WASHINGTON STEWART
Mercantilist from Saint Louis

Levin Washington Stewart, born in 1849, was the son of Washington Levin E. Stewart, a descendant of Scots Stuarts and Dutch Stansbraughs.[1] Washington Stewart married Harriet Wheeler April 9, 1846, in Baltimore, Maryland. For some time, Washington was a business associate of Levin Jones, a ship chandler from Baltimore.[2]

In the 1850 census, the Washington Stewart family consisted of Washington, age twenty-five; Harriett, age twenty; and one-year-old Levin. Washington moved his family west to Saint Louis around 1855. The Stewarts are listed in the Saint Louis city directory of 1857, with Washington employed as a bookkeeper. The 1860 census lists the senior Stewarts (last name spelled as "Steward" in this census), son Levin, and daughters Emma, Laura (listed as "Louisa" in the census), and Eliza. Of the eleven children eventually born to the Stewarts, only five survived. Listings in Saint Louis city directories show the family living in Saint Louis from 1855 until around 1884.

Washington enlisted on September 2, 1862, in Company F, Seventh Enrolled Missouri Militia. He was discharged at Saint Louis in August 1865, at the close of the Civil War and returned home to his family.[3]

Levin Stewart Marries

Levin and Louella Vandevort were married April 9, 1874, in the Trinity Methodist Church at Saint Louis. Louella had been born in Pennsylvania to a Dutch family. Their only son, Eugene, was born in January 1876, in Saint Louis. The 1880 Saint Louis census shows Levin, age thirty-five, working for the Saint Louis Iron Works, wife, Louella, at home, and son Eugene, age five, at school. An 1883 Saint Louis city directory lists not only Levin Stewart's family living at 1205 Warren Street but also his father and brother William.

Levin Strikes Out for the Southwest

Levin worked as a clerk doing office jobs in Saint Louis until January 1883, when he decided to seek his fortune in the Southwest. Since the 1885 Saint Louis city directory shows information identical to that of the 1883 directory for Washington Stewart and son William, it is likely that Levin left home alone to scout out the Southwest while the other Stewarts stayed until a later date. First, Levin made his way to El Paso, Texas, where, hearing of the White Oaks gold town, he struck out for New Mexico alone on horseback. He liked what he saw and returned to Saint Louis to persuade his family to move to White Oaks.

Left to right: Louella, Eugene, and Mabel Stewart. Date and photographer unknown. Courtesy James Sorrells.

In Business in White Oaks

In late 1883, Levin partnered with A. J. Bond to establish a small grocery store before returning to Saint Louis to collect his family. In January 1884, the *Golden Era* noted that "Mr. Stewart, wife and family, arrived this week from San Antonio [New Mexico]. Mr. Stewart goes into business here in White Oaks with Arthur Bond."[5] Bond and Stewart moved to a new place of business on January 1, 1887. On the last day of 1889, Levin bought out his partner and continued alone.[6] He carried a large, well-assorted stock of goods and built a profitable business; by 1895, he numbered among the leading White Oaks merchants.

Levin was described as a person with a "sturdy will, steady application, tireless industry, and sterling integrity." Others said he was a "worthy representative of commercial interests and is a popular and highly respected man, well deserving of representation in [the] history of his adopted territory." Self-made, he was proud of his success in life, due not to any inherited fortune but to his own will and integrity.[7]

The Stewarts became active in political, social, and civic affairs in their community. A letter signed by Levin on December 15, 1890, shows him as justice of the peace for the Eighth Precinct.[8] Son Eugene acted in plays put on by the town's Dramatic Club. In 1890, the *New Mexico Interpreter* reported, "Levin W. Stewart is completing the lawn surrounding his handsome residence. With the arrival of grass it will be a beauty."[9] And

Levin W. and Louella Stewart. Date and photographer unknown. From *White Oaks*, by Morris B. Parker (Tucson: University of Arizona Press, 1971). Permission of Mrs. Louise Kelt, March 17, 2001.

A story from family descendant Harry Osell claims that Levin promised Louella he would buy an ornate secretary for their new home if she would move.[4] Louella agreed to join her husband, and the secretary, along with a matching sideboard, remains today in the family. Levin's father, mother, and at least two of his sisters joined the Stewarts in White Oaks not long after. Levin's daughter, Mabel, was born in July 1885, after the family moved. Washington died there at the age of sixty-six on December 11, 1888, and is buried in Cedarvale Cemetery.

Bond and Stewart store, ca. 1887–1889. Boy on the horse is Eugene Stewart; hatless man with white collar is Levin W. Stewart; elderly gentleman to right of horse is Washington L. Stewart; man with coat and holding hat is Mr. Bond. Photographer unknown. Courtesy Sara Jackson, My House of Old Things, Ancho, New Mexico.

258 *Gold-Mining Boomtown*

the mercantile business continued. The same newspaper reported Levin's advertisements for

- Preserves and jellies in bulk
- Flower pots, flower pot brackets, shelf brackets
- Evaporated and dried fruits, raspberries, blackberries, pitted plums, prunes, peaches, currants, grapes
- Base balls [*sic*] and bats
- Tops and marbles
- Croquet sets[10]

Good-hearted Levin allowed miners liberal credit, and when the mines began to play out he was often not paid back.

Operating a mercantile store in those days was not without danger. One family tale has Levin manning his store when a cowboy with two six-shooters entered, demanding to know where a certain man was. When Levin refused to give him any information, the cowboy accused him of lying. This angered the merchant, who jumped over the counter without a gun and chased the gunslinger down the main street of White Oaks. Another story has Levin physically rescuing and freeing a Mexican who was about to be strung up by a local mob.[11]

One of history's intriguing tidbits from the family has Levin befriending former Sheriff Pat Garrett. Sometime after Garrett shot and killed Billy the Kid in 1881, it is said that Garrett gave one of the Kid's guns to Levin.[12]

Levin and Eugene Invest in Angora Goats

In 1901, Levin and his brother-in-law Jesse Vandervort bought several goats in Lake Valley, New Mexico, to add to his son Eugene's Angora goat herd.[13] Angora wool (mohair) at the turn of the century was selling for high prices. Eugene had recently purchased several hundred quality Angoras from W. A. Hyde and owned two famous bucks: El Cid (valued at $250) and Argus, the half-brother of a champion buck named Lazarus. The Lake Valley goats were thought to be exceptional

Bond and Stewart's ad touting glass front in the *Nogal Nugget*, May 25, 1888. Author's collection.

Bond and Stewart's ad for groceries and merchandise in the *New Mexico Interpreter*, March 23, 1888. Author's collection.

since all were registered with the America Angora Goat Breeders Association of Kansas City. However, both gold and goats, it seemed, were fated to "go bust."

On December 20, 1905, Eugene Stewart married Maude Anna Grebles, a girl from Kansas, and two children were later born to them in Ysabel, Sonora, Mexico: Helen, March 16, 1908, and Howard Eugene August 26, 1909. On October 24, 1906, in White Oaks, Levin and Louella's daughter Mabel married Dr. Arvid E. Osell, a Swedish national born in 1870. Doctor Osell had settled first in Kansas City, Missouri, after coming to the United States.[14] At the time of his marriage, he was a resident of Cochise County, Arizona Territory, most likely because of the upsurge in mining in southeast Arizona at that time. A local newspaper reported, "Edith Parker, lifelong friend of the bride, acted as bridal matron; while the two little Misses, Esther Sager and Jessie Treat, ushered the bridal party to the altar, outlining the way with long streams of white satin ribbon.... After the ceremonies were all over, with many tender farewells the groom and his lovely bride departed under the silver rim of a slow declining moon for their future home at El Tigre [Sonora, Mexico]."[15]

Levin served as an agent for El Tigre Mining Company while living in White Oaks.[16] It was natural for the family to move to Mexico after White Oaks began going downhill. Thus, Levin followed son-in-law Osell to work for the El Tigre. The family may have lived in Mexico from 1906 to as late as 1915, though a 1910 census lists the entire family in Douglas, Arizona, where they could have come from Mexico to be counted in that year's U.S. census. (Some members of the family, including son Eugene, moved back and forth across the Mexican-Arizona border to work and visit.)

The Mexican Revolution

In Mexico, Doctor Osell began to drink heavily. Family member Harry Osell observed, "He drank the bar shut each evening."[17] He died at El Tigre sometime between 1912 and 1916. It is not known where in Mexico he is buried.

About the same time, Mexico erupted in a series of violent revolutions. The *insurgentes* wanted foreigners out of their country. The revolutionaries threw Levin, Louella, Mabel Osell, and Mabel's three children into a railroad boxcar and kicked them out of Mexico. When the train stopped across the border from Agua Prieta, Mexico, in Douglas, they were let out of the boxcar.[18]

The family decided to stay in Douglas and live there permanently. These events took place around 1916, considering that Levin Stewart's 1930 obituary noted that he had been a resident of Douglas for the past fourteen years.[19]

Life in Douglas

In addition to being a gateway to Mexico, Douglas is rich in western history. The town was founded just after 1900 when a copper smelter was built nearby. It also served as a roundup center for ranchers in southern Arizona and northern Sonora, Mexico. The impressive Southern Pacific Railroad Depot still stands today where the Sunset Limited once stopped on its way to California.

Levin owned and operated the Douglas Cigar Stand and Barber Shop at 938 G Avenue for some time. An ad in the 1920 Douglas city directory promoted the shop:

DOUGLAS CIGAR STAND AND BARBER SHOP

Largest and Best Line of Cigars, Tobaccos and Pipes in the City

Hot and Cold Baths—Everything Sanitary

Shoe Shining Parlor in Connection[20]

Deaths of Levin, Louella, and Eugene

Levin died April 7, 1930, in Phoenix and is buried at Calvary Cemetery in Douglas. Surviving him were wife, Louella, son Eugene, daughter Mabel Osell Eberhardt, and sister Laura Leighner of White Oaks. He was a member of the Episcopal Church. Louella died at age ninety-four on March 12, 1950, and is buried beside Levin at Calvary Cemetery. She was a member of Saint Stephen's Episcopal Church and past president of Saint Stephen's Guild in Douglas.

Eugene took over his father's cigar business, renaming it the Gadsden Cigar Store after the famous Gadsden Hotel in which the shop was housed.[21] The Gadsden Hotel was built in 1906, and opened on Thanksgiving Day 1907.[22] It drew its name from the Gadsden Purchase of December 1853, and the hotel soon gained a reputation as one of the finest in the Southwest. Today, the Gadsden remains a place of old-fashioned elegance and charm. It is both showpiece and social center of Douglas. The building's focus is its high-ceilinged lobby, which features a sweeping curved staircase of white marble with pillars inlaid with and topped by gold. At

Facade of Gadsden Hotel in Douglas, Arizona. Photograph by Bob Workhoven. Author's collection.

the head of the stairs are a large painting of a local scene and a series of magnificent Tiffany stained glass murals depicting desert vegetation. It must have been a pleasure to go to work every day amid such surroundings.

Sometime during the family's Douglas era, Zane Grey, the famous writer of Westerns, became a good friend of Eugene and modeled his leading character in *The Light of Western Stars*, published in 1914, after him.[23] Grey even retained the same name (Gene Stewart) for his hero. The family says that other characters in the book can be recognized as drawn from real-life Stewarts and their extended family. Although the real Gene Stewart did not wear six-shooters on his belt, some of the episodes in the novel were based on his life, especially his mining days in Mexico and clashes with Mexican guerillas.

Eugene retired in 1947 after serving more than ten years as justice of the peace. Called "Judge Stewart," he had previously worked for Phelps Dodge and was a member of the Elks for thirty-five years. On January 17, 1948, Eugene died at the age of seventy-one and is buried at Calvary Cemetery in Douglas.

Marble staircase in the Gadsden Hotel. The staircase steps are of white marble; the columns are inlaid with gold on the columns themselves, the bases, and the caps. Murals at the head of the stairs are Tiffany stained glass panels. Date unknown. Photographer James Sorrells. Courtesy James Sorrells.

NOTES

1. A brief biographical sketch of Levin W. Stewart is contained in *Illustrated History of New Mexico* (1895), 649–50.

2. James Sorrells of Waco, Texas, is married to a Stewart descendant and has done extensive genealogical research on the Stewart and allied families. He furnished a fifteen-page document titled "Stewart Genealogy" to the author on January 25, 2005. All of the census information in the article is taken from his research. The Sorrells document is on file with the Lincoln County Historical Society in Lincoln.

3. Ibid., 7.

4. Sorrells, Stewart Genealogy, 5.

5. *White Oaks Golden Era*, January 31, 1884.

6. "Levin W. Stewart," 650.

7. Sorrells, Stewart Genealogy, 2, 4.

8. Levin Stewart to Fransisca Marrnjo [sic], December 15, 1890, copy in author's collection.

9. *New Mexico Interpreter*, March 7, 1890.

10. Ibid.

11. Sorrells, Stewart Genealogy, 9, as related by Stewart descendant Harry Osell.

12. Ibid., 8, as related by Osell.

13. *White Oaks Eagle*, September 19, 1901.

14. Sorrells, Stewart Genealogy, 14.

15. Undated White Oaks newspaper clipping from Stewart Genealogy, 10, 11.

16. Ibid., 10, as related by Harry Osell.

17. Ibid., 10.

18. Ibid.

19. Sorrells, Stewart Genealogy, 11.

20. *Douglas 1920 City Directory*. (Douglas is a city in Arizona.)

21. Sorrells, Stewart Genealogy, 17.

22. American Automobile Association, *AAA Tour Book: Arizona and New Mexico*.

23. Sorrells, Stewart Genealogy, 10.

JONES AND STANLEY TALIAFERRO
Newspapermen, Politicians, Mercantilists, Miners

The Taliaferros first arrived in the United States in approximately the mid-1600s.[1] The first Taliaferro whose life can be traced was Richard (married to Sarah Jones). Their son, the first Jones Taliaferro, was born in Virginia on February 10, 1796. During the War of 1812, Jones, then sixteen, enlisted in the Fourth Ohio Regiment and was present at General Hull's surrender of Detroit to the British. Jones and others were sent in open boats down the Detroit River and then coasted along the southern shore of Lake Erie to the mouth of the Cuyahoga River and to what is now Cleveland, Ohio.

From there, Jones and another boy his age, Nicholas Walker, traveled on foot almost 250 miles through the forest to Cincinnati to deliver the first news of the surrender of Fort Detroit. Home at this time for Jones was Batavia in Clermont County, Ohio. In 1817, Jones married Eliza Townsley, also of Batavia, and the couple moved to Virginia within the year. They became parents of twelve children, the eldest of whom was a boy, Richard.

Dr. Richard Taliaferro, Father of Jones and Stanley

Richard was born in Virginia on August 11, 1818.[2] Alone, and while still a very young man, he left Virginia for Ohio, where he later studied medicine and graduated from the Medical School of Cincinnati. Then he moved to Indiana to practice medicine and finally went to Illinois, where he became a pioneer of Iroquois County. In 1848, Doctor Taliaferro settled in Old Middleport, Illinois, and began practicing medicine. Two years later, on June 22, 1850, he married Jennie Stanley, daughter of Micajah Stanley, a founding father of nearby Watseka, Illinois.

In 1850, Doctor Richard moved south to Clay County, where he and his wife lived for a time while he operated a general merchandising store. He became a staunch supporter of the Democratic Party and won the election for the circuit clerk of Clay County in 1861. He also held the office of justice of the peace and other official positions. In these years, the family moved back and forth between Clay and Iroquois Counties. Richard and Jennie became parents of Jones, born November 7, 1854 and named after his paternal grandfather; Stanley, born September 15, 1860, the first white child born in Iroquois County, and named after his maternal grandfather; and little sister Eliza "Lida," born in 1864.

During the summer of 1855, Richard published a Democratic newspaper in Old Middleport called the *Investigator*.[3] After six months, it died for lack of subscribers. However, the doctor's brief foray into the world of printing somehow rubbed off on sons Jones and Stanley—both fell in love with the smell of print early on. The family moved back to Watseka for good in 1867.[4]

Typhoid fever felled Doctor Richard in October 1872, leaving elder son Jones, at fifteen, to assume his father's role in supporting the family. Stanley was only twelve at the time; Lida, eight. Jones spent much of his boyhood in Watseka and was well known there as a dutiful son who tried to make life easier for his widowed mother. He was remembered as "industrious and painstaking" and a person of "great energy."[5] Stanley left school at fourteen to learn the printing trade at the local paper, the *Watseka Republican*.[6] Afterward, he ran a grocery business for a short time.

To Lincoln County

Somehow Jones heard of a gold strike in New Mexico Territory and headed west, arriving in White Oaks in May 1880. He spent his first year in White Oaks prospecting. Stanley joined his brother some time that same year.[7] The dry high desert must have seemed foreign to Stanley. He wrote home to Watseka in March of 1881, saying, "I notice you are having a little more than your

Jones Taliaferro and wife Ella. Date unknown. Photograph by Hentscher, Manitowoc, Wisconsin Territory. Courtesy Chris and Jack Harkey.

share of water. I don't see why the good Lord doesn't more equally distribute his water. We could stand several barrels full and not have an overflow either."[8]

In the beginning, the brothers maintained some kind of residence in both Lincoln and White Oaks and traveled back and forth to tend their business and mining interests. During these early times, the Taliaferro brothers also became acquainted with most of the survivors of the Lincoln County War, including Billy the Kid. In the early 1880s, Jones developed some lucrative mining claims that enabled him to build and furnish a home for his mother and sister.[9] In April 1882, he returned to Watseka to fetch them to live with him in White Oaks. Back in the Oaks, Jones met Ella Thomson, and they were married on November 15, 1883. Two sons were born: Richard on March 31, 1886, and Jones, Jr., on February 28, 1890. Stanley traveled back home to Watseka to marry Emma Louise Riggle on February 7, 1883, and brought her back to White Oaks.[10]

The Golden Era *Newspaper*

In 1882, the brothers bought the first newspaper in Lincoln County, the *Golden Era*, from its original owner, Jacob Wise.[11] (The *Golden Era*'s very first issue had rolled off the press in White Oaks on December 18, 1880.) In those days, anyone could go into the newspaper business with a light hand-set press and operate out of a small room.[12] In July 1884, the Taliaferros moved their paper from White Oaks to Lincoln.

The brothers shared publication of the newspaper, with Stanley acting as editor and manager. A registered Republican, Stanley proclaimed himself an independent but in 1884 supported Democratic candidates. The *Golden Era* was a vibrant newspaper, one of two in Lincoln County's earliest days. The competition was the *Leader*, a Republican paper that had been established by Lee Rudiselle in White Oaks October 1, 1882. In January 1885, the *Golden Era* reported hearing from the *Iroquois County Times* back in Illinois that "The Taliaferro brothers, Jones and M. S. [Stanley], of Lincoln, New Mexico, are congratulating themselves upon the success of their newspaper enterprise after two years of journalistic effort. . . . The *Golden Era* is certainly an excellent paper."[13]

Real Estate, Insurance, and Politics

Although his mining interests occupied him the rest of his life, Jones had other enterprises. Early on, he started a real estate and insurance business in White Oaks. In 1884, he placed an ad in the *Golden Era* listing houses to rent as well as a service he dreamed up: making collections and paying taxes for nonresidents in White Oaks,

Richard and Jones Taliaferro as children, ca. 1893–1894. Photographer unknown. Courtesy Chris and Jack Harkey.

TALIAFERRO & CO.,

Real Estate And Insurance Agents.

PROPERTY IN ALL PARTS OF THE CITY.

Houses to Rent, Collections Made And Taxes Paid For Non-Residents.

Warantee Title to City Property.

WE HANDLE PROPERTY IN THE FOLLOWING DISTRICTS:

OSCURAS, SAN ANDES,
NOGAL, RIO BONITO,
JICARILLA, GALLINAS,
BAXTER AND LONE.

(In the Post Office—Correspondence Solicited.)

No 25 IS A HORSE RANCH, near the Texas line. Good range. Plenty of water for twelve to fifteen hundred head.

No 27 IS A SHEEP RANCH, in Guadalupe mountains. Well improved ample range and water for 10,000 head of sheep.

No 28 IS THREE SPRINGS, forty miles from any other water. Splendid range; good winter protection. Will run from ten to fifteen hundred cattle.

No 29 IS AN ISOLATED RANCH. Abundant water and grass for 5,000 head of cattle.

Address: JONES TALIAFERRO & CO.,

White Oaks, - - N. M.

Taliaferro & Co. ad for real estate and insurance in the *Golden Era*, March 6, 1884. Author's collection.

Gallinas, Nogal, Jicarilla, San Andrés, the Oscuras, and the Río Bonito.[14] His ad revealed even further wide-ranging interests—acting as a real estate agent to sell a horse ranch near the Texas state line and a sheep ranch in the Guadalupe Mountains.

Another of Jones's interests was politics, and throughout the 1880s, he stayed active in those of Lincoln County.[15] He was elected clerk of Lincoln County in 1884 and served until 1888. At the Democratic county convention in Lincoln in September 1888, Pat Garrett and John Poe (formerly friends and political allies) had become bitter enemies and were each supporting different candidates for sheriff. A resourceful politician, Jones made an eloquent speech as county chairman of the Democratic Party. For the sake of party harmony, he urged the nomination of a controversial candidate for sheriff, Jim Brent. Brent, the current incumbent, was nominated by a majority of one vote. A few days later, the Republican county convention convened, with the Republicans naming candidates for all county offices except that of county clerk, for which the convention endorsed Democrat George Curry. Factional fighting on the part of the Democrats continued, with Jim Brent announcing his support of an independent candidate for county clerk and Jones Taliaferro backing for county clerk *his* candidate, James Parker.

It was the division of Lincoln County into smaller counties, however, that turned out to be the dominant issue in the contest for seats in the territory's 1889 legislature. Israel King of Silver City and Jimmy Dolan of Lincoln won seats in the senate; Frank Lesnet won a seat in the house. When the legislature convened in Santa Fe in January, a bill creating Cháves and Eddy Counties out of Lincoln County was hammered out and finally passed.

The General Merchandising Business

Jones Taliaferro was in the merchandising business in White Oaks continuously—sometimes singly, sometimes in partnerships—from 1880 until 1905.[16] In 1885, he bought the mercantile business of Robson, Young, and Bogard. In 1888, he joined forces with G. R. Young, fellow White Oaks old-timer who ran a dry goods store on White Oaks Avenue (and part of that same 1885 business trio).[17] The two men advertised themselves as Young and Taliaferro. In early 1895, the store renamed itself the Taliaferro Brothers. An ad in the *Old Abe Eagle* early in 1895 claimed,

> If you go to Taliaferro Bros., they will show you that celebrated White House coffee, some excellent old fashioned buckwheat, pure maple syrup; and tell you about that breakfast food; give you a list of their California canned goods and dried fruits, and ask you to see their line of fancy breakfast bacons and hams.... Their house is headquarters for Boots and Shoes and they sell more of each than any store in the county. It may be the price that sells them and it may be the quality. We suspect that it is both.[18]

By August 1895, as mining slacked off drastically, the White Oaks economy was pinching the townspeople hard. Taliaferro Brothers ran a special ad in an August newspaper announcing, "Beginning . . . September First

Wool wagons at the Taliaferro corrals. Date and photographer unknown. Courtesy John and Wesley Mottinger.

we will adopt the Spot Cash System. Absolutely no goods will be sold on time. Everything in our big stock of general merchandise will be sold at a small profit.... It is better business to sell at a smaller profit for cash in hand than for a larger profit with cash in the other fellow's hand. Yours for better times, TALIAFERRO BROS."[19] In the same month, they also ran an ad in the *Eagle* listing items that must have been hard to get in those days, declaring, "The first of the week we will have cranberries, oranges, figs, orange and lemon peel, nuts, oysters, celery."[20] Another ad in the *Eagle* in late 1895 claimed "Bed Rock Prices." Prices of staples were as follows:

Flour, $2.75
Irish potatoes, $1 for a hundred
Salt bacon, 10 cents a pound
Arbuckle coffee 3 pounds for 80 cents
Sugar, 4 pounds for $1
Rolled oats, 4 pounds for 25 cents
Sal soda, 4 pounds for 25 cents[21]

In 1896, Jones sold his general merchandising business to Charles Mayer but continued to operate a dry goods store until 1905. His store became Taliaferro Mercantile and Trading Company.[22]

Stanley Returns for Good to Watseka

After partnering with Jones in White Oaks in several enterprises, Stanley returned to Watseka for the final time in 1886 and soon afterward bought an interest in the *Watseka Republican*.[23] Stanley's first wife, Emma, died in 1889. He married again May 23, 1892, to Margaret Davidson in Chicago. In 1901, he became Watseka's first temperance mayor; he had never favored the liquor business. Stanley edited the *Watseka Republican* until his death at eighty-one in September 1941, only three months after he had vacated his desk.

Jones Taliaferro in front of his store. Jones stands next to the delivery wagon. Eusebio Cárbajal, sitting in the driver's seat, delivered merchandise for the Taliaferros. If Eusebio stayed too long on a delivery, the little mule would return to the store without him. Date and photographer unknown. Courtesy John and Wesley Mottinger.

Interior of Taliaferro store, ca. 1889. Photographer unknown. Courtesy Chris and Jack Harkey.

Lida Taliaferro

Jones and Stanley's little sister, Lida, married Charles Buford in White Oaks January 15, 1884. In November 1889, Charles lost his life in a fatal accident reported by the *Pecos Valley Register*: "He ventured into a well where he had just exploded a blast, and before reaching the bottom was overcome by the smoke and gas of the powder and fell about 20 feet to the bottom upon a rock. When brought to the surface he was badly bruised about the head and face and from every indication met his end at once, his skull was fractured, neck broke [*sic*], in fact everything but a mangled mass.... In 1878 he formed the acquaintance of Pat Garrett and John Poe in the Panhandle country and with these came to Roswell, where he clerked for Capt. Lea."[24]

Lida returned to Watseka and married a man named Cottrell. She, her mother (who had already returned to Watseka permanently), and her children traveled back and forth to White Oaks several times to visit. Tragedy entered her life once more when eldest daughter Kittie committed suicide by poisoning in White Oaks on April 27, 1902.[25] Uncle Jones paid for the casket. Lida died in Chicago January 2, 1947.

Later Years at White Oaks

Jones and his wife stayed on in White Oaks while raising their two boys, Richard and Jones, Jr. They sent Richard to the New Mexico Military Institute in Roswell.[26] Later Jones, Jr., was dispatched to Chicago to attend Bryant & Stratton Business College.[27]

The couple participated fully in the community life of White Oaks, including giving and attending parties. The *Golden Era* described Jones attending a dance in December 1885 dressed in "flowing pants embroidered around the edges with worsted ruchings and a Prince Albert coat polka-dotted with oleomargarine."[28] The *Lincoln County Leader* reported that the Taliaferros hosted a social at Young and Taliaferro's Hall in October 1890. The entertainment was literary exercises and a program dubbed "The Crowning of Autumn's Queen."[29] In 1899, Jones helped organize the Ancient Order of United Workmen, White Oaks Lodge No. 9. He was also a Mason and belonged to Baxter Lodge No. 9 of the Knights of Pythias.

Ever the mining man, Jones formed the Old Hickory Mining Company of the Gallinas Mountains of Lincoln County around 1900.[30] Jones shipped copper from his eight claims to the El Paso smelter after first hauling the ore fifty-five miles by wagon to the new railroad head at Carrizozo.

Of course, family matters surfaced from time to time. In 1907, Jones placed an ad in the *Carrizozo Outlook*: "$50 REWARD. Someone has poisoned our little bulldog 'Victor.' We do not know that he ever harmed any one. He would fight, but that was his nature. For his many good qualities, we loved him, and will pay the above reward for evidence that will convict the person that caused his death ... to learn who put out the poison."[31]

A Premature End

Besides his mercantile and trading business, Jones continued his mining interests and also dabbled in raising sheep. In May 1909, Jones became very sick, and when death appeared imminent, relatives wired Stanley in Watseka. Stanley took the first train to Chicago to pick up his nephew Jones, Jr., at school there, and they continued by rail in hopes of reaching Jones before he died. Boarding the Golden State Limited, the two traveled only as far as Kansas City when another telegram arrived with the news that Jones had died.

At the Carrizozo railroad terminal, a large delegation from White Oaks met brother and son and took them the twelve miles farther to White Oaks. The people "gave every possible evidence of the high esteem and respect with which Mr. Taliaferro was regarded, and the funeral was one of the largest ever held in that section."[32]

Jones died at the age of fifty-two. He is buried in Cedarvale Cemetery at White Oaks.

Jones Taliaferro's headstone in Cedarvale Cemetery. Photograph by Bob Workhoven. Author's collection.

NOTES

1. Taliaferro, *Taliaferro Family History*.
2. "Dr. Richard Taliaferro."
3. Beckwith, "Middleport and Belmont Townships."
4. "Funeral Rites Tomorrow for Prominent Man," *Iroquois County Times*, September 24, 1941.
5. "Jones Taliaferro Is Dead," *Watseka Republican*, May 7, 1909.
6. "Funeral Rites Tomorrow."
7. "The Newspaper Press," in Anderson, *History of New Mexico*, 755.
8. *Watseka Republican*, March 15, 1881.
9. *White Oaks Golden Era*, March 26, 1882.
10. "Funeral Rites Tomorrow."
11. "The Newspaper Press," in Anderson, *History of New Mexico*, 477.
12. Marc Simmons, "When Printers' Ink Was Yellow," New Mexico History, *Albuquerque Prime Time*, April 2004, 6.
13. *White Oaks Golden Era*, January 8, 1885.
14. *White Oaks Golden Era*, March 6, 1884.
15. Curry, *George Curry*, 62–63.
16. *Carrizozo Outlook*, November 30, 1905.
17. *New Mexico Interpreter*, January 9, 1891.
18. *Old Abe Eagle*, February 14, 1895.
19. *White Oaks Eagle*, August 15, 1895.
20. *White Oaks Eagle*, August 5, 1895.
21. *White Oaks Eagle*, November 28, 1895.
22. *Carrizozo Outlook*, November 30, 1905.
23. "Funeral Rites Tomorrow."
24. "A Fatal Accident," *Pecos Valley Register*, November 21, 1889.
25. Record of Deaths at White Oaks, 1902.
26. *White Oaks Eagle*, April 18, 1901.
27. "Jones Taliaferro Is Dead."
28. *White Oaks Golden Era*, December 31, 1885.
29. *Lincoln County Leader*, October 11, 1890.
30. Taliaferro, *Old Hickory Mining Company*.
31. *Carrizozo Outlook*, September 6, 1907.
32. "Jones Taliaferro Is Dead."

THE LEGEND OF MADAM VARNISH

Typical gold-mining towns began as boisterous places full of feverish energy. Prospectors swarmed the hills to stake claims, miners opened shafts, and settlers erected shacks and tents everywhere. Saloonkeepers, gamblers, and dance-hall girls arrived in droves. True to form, White Oaks developed in this manner and, along the way, inherited a story about a gambling madam that has persisted as White Oaks legend to this day. Not one iota of proof exists as to whether or not this incident happened, but there was a real saloon and gambling hall called the Little Casino.

Not all card sharps in gold-mining camps were men. Close on the heels of some of the first gambling men in the area in the early 1880s came a pretty blonde woman from Saint Louis. Improbably named Belle La Marr, she claimed to be a widow forced to make her living by dealing cards. Accompanying her were three charming women she referred to as her daughters.[1] Belle La Marr could deal: her way with cards was so slick that her cowboy victims soon nicknamed her Madam Varnish.

La Marr was casting about for a gullible, well-to-do man who would marry her and support both her and her so-called daughters when her eye fell on the treasurer from a neighboring county. This gentleman happened to be in White Oaks on business. In no time, the county treasurer proposed marriage to Madam Varnish, and he promised to take care of the daughters too.

A happy fivesome boarded the next stagecoach to Roswell for a proper wedding ceremony. That evening in Roswell the four ladies and the treasurer all enjoyed a late supper followed by drink after drink. The next morning the bride and young ladies caught the early stage for White Oaks, arriving back in town that night. Curiously, the bridegroom was nowhere to be seen. The good madam replied to all inquiries that her groom had to stay on a few days in Roswell to take care of some business. Before long, however, he would rejoin her and the girls in White Oaks.

The new husband stepped off the stage the following evening in a foul mood, immediately found the town

Little Casino in ruins, ca. 1942. Photograph by W. Edmunds Clausson. Courtesy Palace of the Governors Photo Archives NMHM/DCA, negative no. 55428.

Interior of Little Casino. The Little Casino was owned and operated by Capt. John Lee from 1885 to around 1888, when Lee sold it to A. Schinzing. Date and photographer unknown. Courtesy White Oaks Historical Association.

constable, and lost no time in swearing out a complaint for the arrest of his wife. He claimed that his bride had stolen $3,500 from under his pillow on their wedding night. The court released the lady under her own recognizance until a trial could be held. Dressed in high fashion, as always, the day after her arrest Madam Varnish and her girls called on the county treasurer and explained to him, "Honey boy, if you press this charge against your little sweetheart, she will have to have a charge sworn out against you, accusing you of stealing $3,500 of your county funds. And, sugar, I really don't want to do it."[2] The case was dropped, and the treasurer retreated to his home county.

Now richer by $3,500, the madam became part owner of a saloon built around 1883 called the Little Casino,[3] where she opened her own gambling hall. The one-story casino was a good-sized building. In front was the standard barroom with a long bar and a few tables. In back was a dance floor, and behind that were rooms for the dance-hall girls. Across from the barroom were tables for cards and roulette, where Madam Varnish dealt faro expertly as her daughters wheedled the customers into parting with the rest of their money for drinks and entertainment.

NOTES

1. Thorp, *Along the Rio Grande*, 56–57; James, *Madam Varnish*. James wrote a fictional account of the gold rush town of White Oaks as he imagined it to be during the 1880s in the mountains of the Southwest. It was James who used the name "Belle La Marr" for the fabled Madam Varnish who came from the Palais in Saint Louis. He also creates the character of Tom Short as Madam Varnish's love interest, and mentions that Varnish became part-owner of the Little Casino Saloon. A copy of this long out-of-print book may be accessed in the archives of the Lincoln County Historical Society in Lincoln.

2. This bit of fictional speech comes from Thorp, *Along the Rio Grande*, 57.

3. A description of the Little Casino layout appears in Birdsong, *1991 Lincoln County Pony Express Calendar*. The Little Casino was built on the corner of White Oaks Avenue and Placer Street. From about 1885 to 1888, the saloon was owned and operated by Capt. John Lee, who sold it to Augustus "Gus" Schinzing, who ran it until 1894. The saloon became vacant in the early 1900s and by 1930 was badly in disrepair. Eventually, it was torn down and the building material used in other locations.

WILLIAM H. WEED
Santa Fe Trail Trader and White Oaks Merchant, Dreamer, and Doer

William H. Weed was born May 22, 1823, in New York City.[1] He became one of the leading merchants of White Oaks and one of the most enterprising businessmen in southeastern territorial New Mexico. Having grown up in humble surroundings in New York City, he set about working his way up in the world to a hoped-for position of wealth by constantly moving.

The Making of a Western Mercantilist

Weed was sure that the West offered the best opportunities, and so in 1846 at the age of twenty-three, he headed for Independence, Missouri. From 1846 to 1848, he freighted over the Santa Fe Trail from Independence to Santa Fe. The Santa Fe Trail was first traversed by American trader William Becknell in 1821 and first mapped by U.S. surveyors in 1825.[2] An overland route now linked the United States and Mexico.

Trade remained slow with the United States in the beginning. It took the American "invasion" by General Kearny in August 1846 and the annexation of New Mexico Territory to open up the route to greatly increased commerce with the rest of the country. Wagons carrying manufactured goods from the East Coast and Europe poured into New Mexico and on down to Mexico, returning with mined gold and silver. Traders ventured as far as Saint Louis, New York, and even Paris to buy goods destined for Mexico.

Many a traveler lost his goods—and his life—to Indians along that trail in the late 1840s. Caravans asked for, and often received, an escort by U.S. troops. Weed freighted over the Santa Fe Trail before the days of huge profits. Since Santa Fe proved to be a small market, most American traders continued on down the Camino Real to Chihuahua City. Weed opted out in 1848, perhaps prematurely. In 1855, $5 million in goods were being moved, a figure that would balloon eightfold to $40 million annually by 1862.

In 1848, Weed settled briefly in Saint Louis, the "Gateway to the West" and the largest city west of the mighty Mississippi. There he continued to freight for a year before hearing of the fabulous California gold strikes. He started for the California gold rush from the jumping-off town of Saint Joseph[3] some three hundred miles west of Saint Louis in the far northwest corner of Missouri. Saint Joseph was founded in 1826 as a fur-trading post in the Blacksnake Hills and would become the future headquarters and home base of the Pony Express, which began its runs from the frontier town to Sacramento in 1860.[4] The discovery of gold in California transformed Saint Joseph into a major wagon-train staging area and supply depot—the perfect place for a young man who already had hard-won experience on the Santa Fe Trail. Weed once again hoped to make a fortune overnight.

From Saint Joseph, Weed traveled first to Fort Laramie, an important fur-trading post near the eastern border of Wyoming. This fort played an important role in aiding migrations to Oregon and California. Then he pushed on by way of the Humboldt River west of the Great Salt Lake Desert to California and the Pacific Slope. Weed mined on the North Fork of the American River some fifty-plus miles northeast of Sacramento near present-day Lake Tahoe, but soon realized he could pocket more gold as a mercantilist than as a miner. He targeted Sacramento to set up business and stayed until 1852.

A Boat Transportation Business Based at the Isthmus of Panama

In 1852, the always-restless Weed headed to the Isthmus of Panama and began transporting goods from Panama to Cruces in Cuba. He also took goods to Gorgona, an island due west of Cali, Columbia, in the Pacific Ocean

269

off the coast of South America. He invested heavily in the boating business and owned several boats that plied the Chagres River in Panama. Weed operated this business until 1854, around the time a railroad was built across the Isthmus, which caused boat transportation to decline. (It was decades before Teddy Roosevelt would build the Panama Canal.)

Abandoning his boat transport business, Weed returned to the United States and spent several years again in California, Oregon, and Washington. Sometime during this period, he began calling himself "Major Weed," or others bestowed on him that title,[5] despite there being no evidence to support any military service. Eventually, he moved out of the Northwest to Granada, Colorado.

Mobeetie in the Texas Panhandle

By 1873, Weed had traveled south to the Texas Panhandle and was trading with buffalo hunters at Mobeetie, then in their brief heyday before finally decimating the southern herd between 1874 and 1878. He also traded with cattlemen and cowboys who drove herds of cattle up the Western Trail to Dodge City, Kansas, from 1875 to 1880.

Weed opened a saloon, personally laid out the town of Mobeetie, and was instrumental in organizing the county.[6] He was elected county commissioner and spearheaded construction of the county courthouse. *Dodge City Times* newspapers from 1881 recorded some of the extent of his business:

> "W. H. Weed, a dealer in general merchandise at Mobeetie, Texas, was in the city [Dodge City, Kansas] this week and purchased goods to the amount of $20,000. He is doing a big business in the Pan Handle."

> "W. H. Weed has returned to Mobeetie. He intends establishing a store at White Oaks, and look on the mountains when they are high."

> "W. H. Weed, the big merchant of Mobeetie, Texas, is in the city [Dodge City], having returned from the East where he purchased a large stock of goods."

> "W. H. Weed has sold to Van Horn and Co. This is a business change of great importance in our community. Mr. Weed's stock was immense, composed of general merchandise, amounting to $35,000."[7]

A "Big New Store" in White Oaks, New Mexico

Weed arrived in White Oaks in March 1882, bringing with him $20,000 worth of goods. His freighting outfit carried on the largest mercantile business in southeastern New Mexico, with freight trains of goods coming and going constantly. People called his store the "Big New Store"—it was larger than any other mercantile establishment around.[8] From the beginning, he enjoyed great success, partly because he wisely extended his trade throughout the Pecos Valley to include what are now Cháves and Eddy Counties.

By April 15, 1882, Weed had completed his log store building in White Oaks.[9] He stocked it with goods of all kinds, including queen's ware (a cream-colored, glazed English earthenware), groceries, dry goods, clothing, boots and shoes, and notions.[10] His first year's trading with Gross, Blackwell & Co. of Las Vegas amounted to $126,000.[11] In 1887, he was calling his place Weed's Emporium, as noted in this advertisement in a local newspaper:

> To arrive in a few days, a fine and select assortment of Ladie's, Gent's and children's shoes.
>
> RICH BRUSSELS TAPESTRY, and INGRAIN CARPETS
>
> Dress Goods, Silks, Satins and Velvets.
>
> Col W. H. Weed at the Emporium is in receipt of two car loads of Kansas Corn.
>
> Go to Weed's Emporium for fine canned fruits, jellies, and provisions.[12]

Weed dreamed up the following ad for an 1892 White Oaks newspaper: "The place to buy your Weed is from Weed. If WE'D buy more from Weed, WE'D be more satisfied and so would Weed."[13] Weed's bookkeeper skipped town, and Weed called in an outside auditor to check the books. The auditor duly reported that the books indeed balanced. Still smarting from the defection of his bookkeeper, Weed bellowed, "The books balance, but does *WEED* balance?"[14]

Expanding His Interests

Weed also sold liquor. His store, a large log cabin with a lean-to shed on the back, had a side entrance away from the prying eyes of teetotalers and wives. He kept a barrel of whiskey on tap in that lean-to, and when miners quit work for the day they would drop by Weed's place, duck

THE LARGEST STOCK
Of General Merchandise
IN LINCOLN COUNTY.

Dry Goods and Groceries

AND MINERS' SUPPLIES,

AT RETAIL OR IN JOBBING LOTS.

A full and complete line of ladies fine dress goods, silks, worsteds, ginghams and satins, laces, ribbons and gloves, fine embroidered Cashmere shawls, ladies' collars and scarfs, ladies ready-made underwear. All at bottom prices.

W. H. WEED,
White Oaks and Agua Chiquito, N. M.

W. Weed's ad for dry goods, groceries, and miners' supplies in the *New Mexico Interpreter*, March 23, 1888. Author's collection.

in the side door and get a drink (or several) of whiskey. No matter how many drinks they consumed, Weed charged them for only one. He must have hoped that in this way he would corner all the trade of the miners, who numbered about two hundred at that time. Some of the miners complained that Weed billed them for that one drink which they never took. To that, Weed retorted: "It's *your* fault if you didn't."[15]

Besides his general merchandise business, Weed developed mining interests and owned considerable mining property in White Oaks. He established branch mercantiles throughout southeast New Mexico. In October 1885, Weed built a small general store at a tiny settlement on the Agua Chiquita in the Sacramento Mountains south of Fort Stanton, adding a post office soon after.[16] The hamlet was named Weed because the post office was granted to Weed's name on the application.

In 1889, he also opened a store in the old Spanish settlement of San Pedro about two miles east of San Antonio.[17] Not long after the San Pedro store opened, J. J. Kelly, a San Pedro butcher, went into the store and started toying with a revolver the clerk took out of the cash drawer. Kelly playfully tried to wrest the revolver from the clerk, when the pistol discharged. The bullet entered Kelly's stomach and exited the back near the spine. He died next morning at two, leaving behind a wife and eleven children.[18]

A Finnish Wife

Through most of his eventful life, Weed remained a dedicated bachelor. This changed somewhere in the mid- to late 1880s when he married a very young Finnish girl named Lena in White Oaks. Lena would have been only in her late teens; Weed, a man in his sixties. In 1890, a

Fred Mayer's account with W. H. Weed, 1888. This bill provides a glimpse of the goods Weed carried and his prices. Author's collection.

son named Johnnie was born. Young Johnnie, undoubtedly the apple of his father's eye, died of acute tonsillitis on September 10, 1897, at seven years and twenty-eight days and was buried in the Knights of Pythias Cemetery at White Oaks (present-day Cedarvale Cemetery).[19]

The Weed marriage went very sour in those years. William Blanchard of Louisiana came to Lincoln County in the early 1880s and lived in Texas Park a few miles from White Oaks. As an old man, he penned a lengthy, gossipy memoir in which he discussed many of the people and events of his youth. Of Weed, he wrote,

> The gold there [White Oaks] never was robbed by anybody on the road. At one time about Eight Thousand Dollars was missing. . . . It was all in one large lump as it comes out of the retort, and old man Weed was paying off the mining business and selling the gold, and paying all of the bills. [He means acting as a kind of financial officer for the mine owners.] He took it home with him one night and the next morning it was missing. They hunted for it quite a while and finally they found a meatsaw and an axe and some cold chisels that were all gold plated. They found these gold plated tools and found that his wife and her brother had stolen this gold and chopped it up into small pieces, and were preparing to sell it. Mr. Weed allowed his wife to go away and was very much relieved.[20]

The story of the theft of gold may have been true, but Lena Weed did not leave; she stayed on in White Oaks. However, the Weeds did divorce, according to information in the 1900 census.[21] In June 1889, the *Pecos Valley Register* reported, "Col. W. H. Weed will move his store building from Weed to Eddy [now called Carlsbad], Lincoln County, where he will establish a mammoth branch store."[22]

An Old Man Returns Home

Weed did not start up a new business in Eddy, or anywhere else in New Mexico. Instead, he returned to New York to live out his last days. The 1910 census shows him living in Schenectady as the single head of a household along with his widowed sister, Mary, and a nephew.[23] In 1910, he would have been eighty-nine. After 1910, there is no further evidence of him (he does not appear in the 1920 census). Most likely he is buried in his home state of New York.

NOTES

1. "William H. Weed," 585–86. The facts of Weed's early life are contained in these two pages and cover his days as a Santa Fe Trail trader, and his days later in Fort Laramie, the gold rush in California, Panama, and, finally, Mobeetie in the Texas Panhandle. This publication also gives the dates of Weed's later arrival in White Oaks to establish the "Big New Store" and his branching out to San Pedro, New Mexico, near San Antonio.

2. Flanagan, *Old West*, 18, 19.

3. "William H. Weed," 585.

4. Flanagan, *Old West*, 21.

5. The *Silver City Enterprise* of November 1, 1889, calls him "Major Weed."

6. "William H. Weed," 585.

7. *Dodge City Times*, January 29, 1881; April 14, 1881; October 6, 1881; and December 8, 1881. Texts of these newspapers are archived in the Panhandle-Plains Historical Museum, Canyon, Texas.

8. "William H. Weed," 586.

9. Ibid.

10. *Lincoln County Leader*, July 25, 1885.

11. "William H. Weed," 586.

12. *Lincoln County Leader*, October 29, 1887.

13. Undated notes by Chris Harkey of Carrizozo read, "From an advertisement in an 1892 White Oaks newspaper."

14. Ted Rayner, "Does Weed Balance?" Folklore Corner, *Lincoln County News and Carrizozo Outlook*, April 20, 1956.

15. Ibid.

16. *White Oaks Golden Era*, October 1, 1885; Postmaster Shirley Akers of Weed, New Mexico, to Haldane, October 22, 2002.

17. "William H. Weed," 586.

18. *Silver City Enterprise*, November 1, 1889.

19. Record of Deaths at White Oaks, 1902.

20. Blanchard, William E., untitled and undated memoirs of Lincoln County (the 143 pages are not numbered and it is not known if all pages have been included), from the collection of Robert and Dorothy Leslie of Carrizozo. A copy also exists in the archives of the Lincoln County Historical Society in Lincoln.

21. U.S. Bureau of the Census, *United States Census of 1900*, White Oaks Precinct. The resilient Lena Weed did not long remain a divorcée. According to an e-mail from Iginio Fitzpatrick to Paul Hile on April 12, 2004, sometime in 1900 she married Robert Forsythe, a former resident of White Oaks (and in mid-1900 a resident of Yavapai County, Arizona). In 1896, Forsythe had gunned down his sister's husband, George Fitzpatrick, on the road just outside White Oaks. Forsythe was tried in Lincoln for that murder, but the jury on April 16, 1897, returned a verdict of not guilty. This killing was considered a family affair, and juries of the day tended not to meddle with such outcomes. Forsythe did return to Lincoln County after his stay in Arizona and lived there with his wife, Lena. She died in 1928; Forsythe moved to Los Angeles, where he died in 1951.

22. *Pecos Valley Register*, June 27, 1889. Note that this newspaper now called Weed "Colonel" W. H. Weed.

23. U.S. Bureau of the Census, *United States Census of 1910*, Schenectady, New York.

SAMUEL AND MARTHA FRANCES WELLS
Not of the "White Oaks 400" and Proud of It

If ever there was an American story of the Old West on the verge of changing into the New West, it is that of the pioneering Wells family of White Oaks. Samuel "Sam" Wells was born in the midst of the Civil War on April 12, 1863, in Gainesville, Texas, north of Fort Worth and a scant seven miles from the border of Indian Territory.[1] His parents hailed from Illinois. The family moved to southwest Texas sometime in the early 1870s.

Martha Frances Forsythe, Sam's future wife of Irish extraction, was born the eighth of thirteen children in 1868 or 1869 in Belfast, Ireland.[2] Four of her siblings would eventually sail across the ocean and migrate cross-country to White Oaks. John, the first of the Forsythes to arrive in the Oaks, came to the United States some time before 1880 as the family's advance emissary; however, in his application for U.S. citizenship dated October 20, 1886, John swore that he had lived in the United States only for "at least five years."[3] He appears in the June 1880 census of White Oaks as a young man of twenty-three years along with a fourteen-year-old wife, Jane—who happened to be Sam Wells's sister.[4] According to the 1880 census, the Wells and Forsythe families lived in the same or adjoining households. The Wells household at that time consisted of seventeen-year-old Samuel, an older sister, and Sam's fifty-year-old mother.

Martha Forsythe would arrive in White Oaks in 1884 as a girl of sixteen or seventeen. She was better educated than many in that town and would teach

Martha Frances Forsythe Wells. Date and photographer unknown. Courtesy Norma Cinert.

Samuel Wells. Date and photographer unknown. Courtesy Norma Cinert.

school for a while in nearby Texas Park until her marriage to Sam Wells in 1888.⁵

A Witness of Indian Raids and Terror in Texas

Sam grew up in wild country near the head of the Nueces River in Uvalde County, Texas.⁶ Uvalde County lies in southwest Texas roughly one hundred miles west of San Antonio. The Wells were one of some twenty families who had settled in that part of Texas.

In those days, Comanche, Apache, and Kiowa Indians roamed the Southwest following the seasonal migrations of the buffalo that were at the center of their existence. The Native Americans resented any encroachment into their ancestral lands as well as the wholesale slaughter of their food sources and tried every means to prevent the newcomers from staying. They attacked isolated settlements of the whites seemingly at will. Only a two-company post at Fort Clark, thirty miles southwest of the Wells place, afforded meager assistance in helping the settlers ward off Indian raids.

Usually the Indians swooped onto the settlers' farms by the light of the moon. They stole horses, shot the work oxen, and sometimes killed everyone they could find. Under cover of darkness, they would hide nearby, then try to slip up silently and set the houses afire. The Wells family kept some fifty "bear" dogs around the place to provide an early warning system in case of attack.

The settlers had to be watchful every minute if they hoped to stay alive. Often they would hear of the massacre of one of their neighbors. If their own men or boys had to be a few miles from home, those left behind would be uneasy until the return of their loved ones. But the sheer beauty of the country that had drawn the settlers to Nueces country in the first place held them there. In his memoirs, Sam recalled,

> We could sit on the porch and see a hurd of deer grazing on the hill most any time of the day. Often we would have to shoo the wild turkeys out of our cornfield. With a light wind we would have to keep our hats on to keep the pecans from falling on our heads. The wild turkeys would roost in the trees over our house.
>
> We could hear that panther scream in nearly any direction. Any one that never hurd a panther scream cannot imagine the blood-curdling noise they can make. They—the best I can describe it is . . . say if a bear or a panther had a young girl down and was tearing her to pieces limb from limb.
>
> One could stand on the banks of the river running down near the house and count dozens of fish in all sizes, from 1 pound to 30 pounds.⁷

The Massacre of the Colston Family

The Colstons lived about twenty miles north of the Wellses on the head of the Nueces River by the side of a small stream called Blackberry Creek. A family with eleven children, the Colstons had a small house, a flock of sheep, and a few hogs and chickens. Behind their comfortable small cabin grew a very large red oak tree. In this tree the father and the Colston boys had driven some meat hooks on which to hang their fresh meat. Sam recounted,

> [One day] the old man and the oldest boy went out with the sheep. . . . Indians watched them from the hilltop. When the menfolks got well out of sight the Indians slipped up, and, as the door was open, they charged the house [and] ran in before anyone was aware of their presents, grabbed the mother and all the family, hog tied them, took them outside the house, [and] hung them on the meat hooks. Some they stuck the hooks in the back of there heads. Others they stuck the hooks under there chins, some they stuck in there feet hung them with there heads down. The old lady being a little taller . . . they stuck the hooks in her knees and hung her with her head down.
>
> Now then the Indians set the house on fire [and] while the poor family was roasting, the Indians held a war dance over them.
>
> When the old man and boy returned home and saw what had happened he put the boy on there old plug of a poney and started him off for Fort Clark for the U.S. soldiers. When he got to our house [the Wells place] his old horse was nearly gave out. Father put him on a fresh hourse and started him on his journey.
>
> Just before daylight some one shouted for Father. When Father went to the door he found it was Lieutenant [Henry] Lawton, General McKenzy's [Ranald Mackenzie] righthand man. He wanted a schout [scout] and trayler [trailer]. Father told him that I was the only boy at home. . . . When i got my cloathes on and went out I found that Lawton had brought me a nice government horse, gun and belt of cartridges. O but I was pleased to mount on that big government horse. . . . When Lieutenant Lawton found all he could he told me to hit the trail of the Indians. . . . Soon we saw that they was headed for the Río Grande.
>
> By the time we hit Devele River [Devils River] I found out that they were scattering. One Indian

would cut off three or four horses and turn off, then another one would cut off another bunch.... I saw that the maine buntch was heading for a shallow place on the Río Grande where they could cross into Mexico. So I made a bee line for that crossing. I led the troops to a high brushey hill where we could see with our field glasses all over the country. I took the field glasses [and] climed up the tallest tree I could find. ... I finley spied the Indians slipping thru the rough brushey country.... I told Lawton we had them.

We slipped up as close as we dared to ... and waited until dark.... Lawton and I got on a high hill and watched them cross the river. We knew that they were very tired and as soon as they got across into Mexico they would camp for the night.... They knew that a U.S. soldier would not dare to cross into Mexico.

We watched until about midnight.... Then we slipped acrost the river quite aways from there camp. ... We charged there camp. They were all found a sleep.... I don't think a one of them got away. We scalped every one of them, slipped back acrossed the river and straight toward Ft. Clark. I had two scalps for to show.... Lawton told us boys to say the Indians crossed the river and escaped.

The very next time I saw Lawton he had on a right new caption [captain's] coat.[8]

On to a New Home at White Oaks Spring

It is not known what prompted the Wellses to leave their Nueces home for White Oaks in New Mexico Territory, but Sam Wells arrived at the Oaks before June 1880 with his mother and two sisters.[9] With households so near, it was inevitable that the Wells and Forsythe youngsters would notice each other. John Forsythe married Sam Wells's sister Jane in 1880.[10]

Soon after arriving, Sam staked out a homestead in a choice valley two and a half miles southeast of town that included a Southwest commodity more precious than gold: water.[11] White Oaks Spring, and its surrounding white oak trees, gave the town of White Oaks its name. That spring and the many nearby oak, juniper, cedar, and piñon trees made the Wells homestead a place of great beauty, enhanced by the nearby Patos and Carrizo Mountains.

In 1888 Sam further cemented the Forsythe-Wells connection by marrying Martha Frances Forsythe when he was twenty-five years old and she was twenty. Sam and his wife would have seven children.[12]

The Only Family Pond

The first order of business for Sam on his homestead was to build a comfortable log house.[13] Next came the planting of gardens and an orchard, which was irrigated by White Oaks Spring. Because water in the area was scarce, Sam also sold the water from the spring to the miners.

Sam dug a pond that was fifteen feet deep in the middle and kept it filled with spring water. The large carp with which he stocked the pond were difficult to catch, but when they swam to the edge where plants grew, Sam would shoot them. Because water from the spring was plentiful, Sam decided to build an icehouse, covering the floor with several inches of sawdust obtained from a local sawmill. Every winter the family chopped ice in six-to-seven-inch-thick blocks from the pond, stored it in the icehouse, then sold it to the townspeople in summer. It was daughter Edith's job to float the blocks of ice to a dock, where they were hefted by large tongs into the icehouse.

Southern Baptists immersed their converts in that same pond. Twenty-five people were baptized one Sunday. All of them backslid and had to be baptized again when a new minister came to town. The icy cold water required a wash boiler constantly full of hot coffee to revive the baptized converts while they changed from wet clothes to dry in a small room of the icehouse.

When the gold at White Oaks played out, valuable tungsten was discovered. A New York company bought the mine and set up headquarters in the Hewitt Building. The wife of the mine owner told Martha Frances that because it was God who had given the Wellses water, she should let the people from New York have it for free. Martha Frances replied that since God had given the New Yorkers their tungsten, they should exchange some of it for the water. Over the years, as the family grew, Sam would build a second log house and, finally, a large adobe house.

Daily Life

To help feed the family, Sam would hunt the plentiful wild turkey for Thanksgiving, Christmas, and other holidays. He would find a turkey roost and wait until dark to shoot. He also killed deer, and venison hung in the meat house year-round. Sam cut off strips of venison, salted them, and hung the strips on a line to make jerky. Daughter Edith kept a cowchip-stoked fire burning under the meat to cure it; beef was also jerked in this manner.

In partnership with a man named George Treat who owned the local meat market, Sam bought, butchered, and delivered meat to the shop. The Wells also milked a few cows. Again, it fell to Edith to drive her horse and buggy to town to deliver milk and cream, ice, and vegetables to townspeople. For a barter worth $500, Martha Frances acquired the first piano to arrive in White Oaks. Urbain Ozanne, proprietor of the Ozanne Hotel, had run up his meat bill and could not pay, so Martha Frances, who loved music and dancing, took his piano in exchange. Thereafter, the Wells house was filled with music.[14]

As young marrieds, Sam and Martha Frances regularly drove to Las Vegas for supplies in a light covered wagon. The round-trip, including a stay for reprovisioning, took more than a month. Once they also drove south to El Paso, Texas, in the wagon. Martha Frances had her own little saddle horse called Dunny. As a pastime, she rode her horse with several other White Oaks women, always sidesaddle as befitted proper ladies. All the women wore long riding habits and big hats tied on by long veils or scarves. Besides horses, the family raised cattle. Their first herd of cattle carried the KV brand. After this herd was sold to the Bar W Ranch, the replacement herd was branded FAW.

Because cash was chronically scarce, Sam sometimes worked in the coal mine a short distance from his house. He had a blind horse that he rode to work; the miners used that same horse to bring cars of coal to the surface from the mine. Like many others, Sam was deafened in a mine explosion, making communication with him quite difficult. And like most other men in White Oaks, he always prospected off and on for gold, iron, and other minerals.

George Edward Fitzpatrick and John Forsythe. George stands with his arm on the shoulder of his friend and brother-in-law John. Date and photographer unknown. Courtesy Evelyn Fitzpatrick Dobbs.

According to the memoirs of daughter Edith, there were two "classes" of people in White Oaks: the first—bankers, grocery store owners and mercantilists, bookkeepers, lawyers, ranchers, mine owners, and sheepmen—called themselves the "400" and lived in large, lovely homes on the east side of town. The other class was made up of those like the Wells family, but Martha Frances never considered herself inferior, telling her children that *true* aristocrats could live on good terms with anyone.[15]

The Baptists and Congregationalists shared the same church for their services. Eventually, the Baptists met in a hall in the Hewitt building while the Congregationalists built a new church. The Wellses were Congregationalists, and Sam donated the first one hundred dollars to build their church.

The Later Years

The Wells children grew up and scattered. Daughter Lina Cherrille married William Coe, son of George Coe of Glencoe. Edith and Cherrille became schoolteachers like their mother. After high school, Edith attended Normal College in Las Vegas, New Mexico, before teaching at San Patricio, White Oaks, Glencoe, Capitán, and Encinoso. She married Clarence Palmer of Texas Park in 1915 and moved to Oklahoma. Cherrille also taught in the White Oaks Schoolhouse for a time. Sam made a brief foray into politics; he ran for Lincoln County Commissioner as a Republican.[16]

Along the way, Sam and Martha Frances divorced (nobody knows why, and divorce was unusual for those days) but eventually got back together and remarried.[17] With the passage of the years, Martha Frances began traveling to California to spend the winters with one or another of her daughters; Sam stayed in White Oaks year-round.

Martha Frances died in 1938 at daughter Cherrille's house in Tularosa. Found by her daughter and granddaughter on the front-room couch, she had died peacefully while combing and arranging her long hair, which she always wore in a pompadour. Sam followed on December 13, 1941. Both are buried in Cedarvale Cemetery in White Oaks.

Robert Forsythe Murders George Fitzpatrick

On May 3, 1894, Martha Frances's youngest brother, Robert, furnished the family some unwelcome notoriety by gunning down his brother-in-law George Fitzpatrick on a road outside White Oaks.[18] Robert's accomplice was Oliver Peaker. The exact motive for the killing is unclear but seems to have centered on Fitzpatrick's wife, Elizabeth Forsythe (another of Martha Frances's sisters), and on Fitzpatrick's decision to move his new family to a ranch in Mexico owned by his brother.

The Forsythes hired for Robert the very able and well-known defense attorney Col. George Prichard. The case was prosecuted by district attorney George B. Barber. In keeping with the code of the times—that a man always took care of his own affairs—the jury found Robert not guilty on April 16, 1897. In a strange twist, the widow Elizabeth Fitzpatrick afterward married Oliver Peaker, codefendant in the murder of her husband. Oliver and Elizabeth eventually left White Oaks to live in England.

NOTES

1. Edith Wells Palmer, Sam Wells's daughter born in 1890, wrote a five-page undated biography of her life in the 1930s. The biography was written for a government-funded WPA project during the Depression years. Norma Coe Cinert, Sam Wells's granddaughter (also the granddaughter of George Coe of Lincoln County fame) furnished a copy to the author.

2. Forsythe family genealogy, n.d., copy in author's collection.

3. John Forsythe, Copy of Application for Citizenship made October 20, 1886, before the Lincoln County Clerk for the Third Judicial District, County of Lincoln, Territory of New Mexico, Lincoln County Courthouse, Carrizozo.

4. U.S. Bureau of the Census, *United States Census of 1880*, White Oaks.

5. Palmer, biography. Missing from the biography was Martha Frances's account of coming West and riding the Ozanne stagecoach from Carthage to White Oaks. Only one other passenger happened to be riding on the coach at the time, a very drunk Indian. The young girl's apprehension grew on the rough trip but evaporated once the coach rolled into White Oaks and she spotted a doll buggy in the yard of one of the homes. She decided that where there were dolls, civilization could not be far away.

6. Samuel Wells, Sr., three-page undated memoir told to (and typed by) daughter Edith Wells Palmer at her Oklahoma home. This is a fascinating account of life in the wild and remote southern "horn" of the Texas frontier during the days before the Plains Indians were finally forced onto reservations.

7. Ibid., 1.

8. Ibid., 2, 3.

9. U.S. Bureau of the Census, *United States Census of 1880*, White Oaks.

10. Ibid.

11. Palmer, biography.

12. A list of the names of the seven Wells children was furnished by Norma Coe Cinert.

13. Mrs. S. F. Wells, from an article written for the Glencoe Woman's Club in 1936 and reprinted in Paul Baker, "Mrs. S. F. Wells," Ramblin' Around Lincoln County, *Lincoln County News*, April 25, 1958. The article was sent by Mrs. Wells to Mrs. Sydney Bonnell as her contribution to a county history being compiled by the Glencoe Woman's Club. Most of the description of life in White Oaks in the 1880s and 1890s came from Mrs. Wells's article.

14. Martha Frances owned an extensive collection of records by the opera singers Enrico Caruso and John McCormack. Some of the songs she enjoyed included "The Rose of Tralee," "Mother McCree," "An Old Guitar and An Old Refrain," "I Hear You Calling Me," "Remember Me," "When Irish Eyes Are Smiling," and "The Last Rose of Summer."

15. Palmer, biography.

16. According to an unidentified, undated clipping from the Carrizozo newspaper, copy in author's collection.

17. Norma Coe Cinert, telephone conversation with author, December 29, 2006.

18. Transcript of Case No. 1433, *Territory of New Mexico vs. Robert Forsythe*, filed in the State Records Center and Archives, State of New Mexico, Santa Fe.

MARCUS WHITEMAN
Russian Jew and Founder of the Pioneer Store

Marcus Whiteman joined the White Oaks rush looking for gold of a different sort—the kind found in the pockets of miners who needed picks and shovels and basics like clothes, food, ammunition, lumber, and wash basins. Whiteman was born in Russia in 1819.[1] As a young man keen for adventure, he toured the world twice. On the second tour, he was taken captive by South Sea Islanders. During his captivity, Whiteman learned much about the characteristics and customs of the islanders and complained that his captors did not compare favorably with people of other foreign countries he had visited. After his world-roaming days ended, he decided that his fortunes lay in the exuberant young country of America.

Whiteman set sail for New York, where he worked hard and had the honor and distinction of becoming a charter member of the Masonic Lodge of New York City. It was also in New York City that he met and married Mary Levy, another Russian Jew, early in 1868.[2] By this time, he was almost fifty years old; she was twenty-nine.

Out West

After his wedding, Whiteman wasted no time heading west for ever-greener pastures. The couple landed first in Colorado, where their first three sons were born: Joseph in 1868, Louis in 1870, and Charles in 1871.[3] One of Whiteman's daughters, Flora, claimed that some of the years between his marriage to Levy in 1868 and his arrival in White Oaks in 1880 were spent in Alaska.[4] This would have been twenty years before the more famous Alaskan gold rush of the late 1890s. He discovered gold in Alaska, Flora maintained. The possibility that he spent some of the mid- to late 1870s in Alaska or elsewhere away from his family is bolstered by the fact that a second wave of children was born to the Whitemans in New Mexico with the births of son Abraham Lincoln in 1879 and daughters Carrie in 1882 and Flora in 1884.

It is thought that the family came to New Mexico in 1879 with the all-important advent of a railroad to Las Vegas in that year. Hearing of the discovery of gold in White Oaks, Whiteman moved there with his family sometime in the latter part of 1880.[5] He appears in Emerson Hough's novel *Heart's Desire* as "Whiteman the Jew, ever a Greatheart."[6]

The Rise of the Pioneer Store in White Oaks

Whiteman maintained a corral for his patrons to the side of his store, located next to the post office. The corral became a favorite rendezvous for the young men of the gold camp. They gathered to discuss the affairs of the day. Prominent businessmen and Whiteman's brother Masons also joined in awaiting the never-ending stream of freighters who drove in their teams and unloaded sundry boxes, bales, and packages at Whiteman's emporium. Everyone enjoyed Whiteman's stories of adventures around the world.

Whiteman's store, the Pioneer Store, was one of the first, and largest, in town. It had a long front porch reached by steps from either end. There were small windows, barred against nighttime theft. A rough board wall divided a small office in the front from the rest of the store. The young men called their merchant "Mr. Viteman" after his own pronunciation of his last name. In his novel, Hough describes him as "squat, stocky, bristling, blue shirted like the rest."[7] Standing in the doorway of his store, he surveyed the long counters and shelves piled with clothing and hats, boots and gloves, pickaxes, shovels, saddles, spurs, wagon bows, flour, bacon, and all manner of things packaged in tin cans. He also stacked wood in his storerooms for sale around the region.[8] Dust hung everywhere in town, but the air inside the store was not one of decay but of expectancy amid the hustle-bustle. In short, the ever-ebullient storekeeper Whiteman was a man who fervently believed in the future of

M. WHITEMAN,
—:OF THE:—
Pioneer Store,
Keeps His Stock
Up With The Increased Demand of the Country.
WHITE OAKS AVENUE,
White Oaks. **N. M.**

M. Whiteman's ad for Pioneer Store in the *Golden Era*, April 24, 1884. Author's collection.

White Oaks: "'Keep your eye on Viteman,' he said. 'Der railroat may go . . . der saloon may go, but not Viteman. My chudgment is like it vas eight years ago. Dis stock of goots is right vere I put it. If no one don't buy it, I keeps it. I know my pizness. Should I put in twenty thousand dollars' vort of goods, and make a mistake of der blace vere a town should be? I guess not!'"[9]

In 1884, Whiteman built a new building with walls of adobe and a forty-four-foot storefront of glass and renamed his store Whiteman and Sons. Between the iron roof and rafters, six inches of dirt were placed in hopes of fireproofing the structure.[10] In the cornerstone of his new store, Whiteman placed a sealed glass bottle inside a small wooden box. In the bottle he placed several U.S. coins, Mexican coins, local newspapers from Las Vegas and other towns, and a list of the names of every man, woman, and child then living in White Oaks. He also enclosed a list of all the mines showing pay ore. Over the years, Whiteman bought up the land adjoining his store until his store and property came to be known as the "Whiteman Block."

Mary Whiteman

Mary Whiteman was loved and widely known as "Mother Whiteman."[11] Never happier than during the first hard years of settlement in southeast New Mexico, she opened her doors to all newcomers and operated a rooming house for travelers as well.

As a small boy, Wilbur Coe accompanied his father from their orchard at Glencoe over the mountains to peddle a wagonload of apples in White Oaks.[12] After stopping to sell apples at the Ziegler Brothers grocery, Wilbur and his father drove on to Whiteman's store and rooming house. Mother Whiteman, delighted to see the pair, led Wilbur by the hand to the store's soda stand and candy counter. She pulled out a strawberry soda from under a wet blanket used to cool the bottles and handed it to the boy. Then she reached into a large glass drum and emerged with enough hard candy to fill one of his pockets.

Wilbur had never in his life seen so much candy. When the good lady turned her head, he could not resist snatching up a luscious big red gumdrop and cramming it into his mouth. At that moment, Mother Whiteman inquired after Wilbur's mother and sisters, and the boy found to his dismay that he could not answer; the filched gumdrop had locked his jaws shut. Seeing his plight, Mother Whiteman popped her finger into his mouth and wrapped it around the gumdrop. With a sharp tug she brought it out—along with a loose tooth. Taking pity on the boy, now a little pale from the blood in his mouth, Mother Whiteman reached up on a shelf and brought down a bottle of sugar pills, saying, "Now, you're all right again, Wilbur." In the meantime, Marcus Whiteman bought ten or fifteen bushels of apples at a good price, saying that if they could not sell the rest from house to house he might buy them at a reduced price.

After peddling more apples around town, the Coes returned to the Whitemans for supper and to spend the night. But first everyone enjoyed a toe-tapping evening of music furnished by "Dad" Coe; the Whiteman sons, Abe and Charley; and a Whiteman daughter. (Wherever the Coes went, music followed.) Dad Coe tuned up a borrowed fiddle, one of the Whiteman girls sat at the piano, and Abe and Charley played their banjo and guitar. They struck out with favorites like "Irish Washerwoman" and "Arkansas Traveler" and played on until bedtime. As a last tune, Dad Coe fiddled the "Home Sweet Home Waltz."

Thanking Mother Whiteman for her hospitality, the pair were led to a couple of cots in a hallway (the rooming house was full). She put on clean pillow slips, kissed Wilbur good night, and said that she hoped his tooth would not bother him. The next morning the Whitemans gave the Coes breakfast before they set out for home.

Roswell Beckons

On May 10, 1889, Whiteman moved his family to the fast-growing town of Roswell, New Mexico, where he established the M. Whiteman Grocery in a store he built on south Main Street.[13] Roswell received him with open arms: "Mr. Whiteman, the new merchant, is a stem-winder to push things, he has raised and completed a 60-foot building since our last issue."[14]

Most of the Roswell business houses were then on Main Street between Walnut and Third.[15] Those first business houses were of adobe or lumber sawed in the mountains and hauled to Roswell; the Whiteman store was also a lumber building. As they had in White Oaks, the Whitemans lived in back of the store.

The White Oaks Store Burns

In spite of all the dirt and its famous iron roof, in January 1890 a disastrous fire burned the Whiteman Block in White Oaks.[16] The loss included the Whiteman store building, household goods, and dressed lumber, amounting to a total of about $2,000. It was 8:45 on a

Mary Levy Whiteman with grandson Dick. Date and photographer unknown. Courtesy Historical Center for Southeast New Mexico, Roswell, negative no. 109.

Carrie and Flora Whiteman posing with flowers. Date and photographer unknown. Courtesy Historical Center for Southeast New Mexico, Roswell.

Sunday morning when the fire broke out in the ceiling over a stove in a barber shop in the Whiteman Block. Not long afterward, the *Lincoln Independent* noted that Whiteman had sold his "burnt" district property in White Oaks to Tackman of Roswell.[17]

Ray Lemon, longtime White Oaks and Carrizozo resident, told the following anecdote about that Whiteman and Sons business partnership in White Oaks: "Whiteman and Sons operated a general merchandise store. Whiteman had three sons. Once a son did something that didn't suit Whiteman, and he painted out 'and Sons' on the sign, dissolving the partnership."[18]

Roswell Fires

In 1893, much of Roswell's business district burned in a fire, the major hazard for buildings in early days. The Whiteman business and residence was luckily spared in this first fire. But several years later there was another serious fire in which almost all the businesses between Walnut and First Streets burned; this time the Whiteman store did not escape. Worse, though many of the store's goods were carried out of the burning building by friends and neighbors, much that had been rescued was stolen by onlookers. Whiteman's losses of goods in this fire totaled well over $20,000, a large sum of money for those times. Nevertheless he—or, more accurately, his sons—continued business in Roswell.[19]

Whiteman himself had pushed on alone to the new, bustling railroad town of Alamogordo to start another business around 1898. Son Abraham took over as owner and manager of his father's old grocery business in Roswell and operated it successfully for decades under the name Whiteman Brothers Store.

Whiteman Fails to Make His Subscription to the Railroad Cause

In 1895, though Whiteman's store had long since burned and he had relocated to Roswell, the *White Oaks Eagle* severely chastised him. He had failed to send in a promised subscription to a committee formed a year before to induce a railroad to build from El Paso to White Oaks. The paper reported, "Last winter our committee requested Mr. Whiteman, who owns some of the most desirable property in the town . . . to subscribe towards its . . . enterprise. . . . This he promised to do but has not yet sent in the subscription. . . . Others who would reap far less advantage from the road . . . are subscribing from $1,000 to $5,000, while he writes slush to the newspapers and neglects to contribute anything."[20]

Marcus Whiteman "Suicides" in Alamogordo

Eight years after Whiteman left his family in Roswell to open his Alamogordo store, the *Alamogordo News* of May 26, 1906, carried the following news item:

M. WHITEMAN SUICIDED [*sic*]:
Ending of a Long and Eventful Life by Shooting Himself

Tuesday morning our people were shocked by the report of the suicide of M. Whiteman. The doors of his store were not open Tuesday morning. . . . Upon investigation it was discovered that he was dead in his bedroom in the rear of the store. Evidence showed that he had shot himself with a Winchester, and that he had made two attempts, first time missing his aim and the second time he placed the gun under his chin and by the discharge of the gun tore a great hole through his head, thus ending his long and eventful life.

M. Whiteman was 85 [88] years of age and a native of Russia. He had lived in the West for many years and was a noted character as a Jew merchant. . . . He was a man of superior intelligence and had been noted for business shrewdness.

He had lived here since the founding of the town and conducted a store on Pennsylvania avenue. Most of the time he had lived here alone while his family lived at Roswell.

The remains were . . . held by Undertaker Buck till the arrival of a son, Chas. Whiteman of Roswell, who had been notified of the death, and upon his arrival the body was shipped to Roswell for interment.[21]

NOTES

1. M. Whiteman, interview by Georgia B. Redfield. In this WPA interview, Flora Miller, daughter of M. Whiteman, recounts her early life in White Oaks and Roswell.

2. Ibid.

3. New Mexico Quinquennial Census for 1885.

4. Whiteman, interview.

5. Ibid.

6. Hough, *Heart's Desire*, 262–63.

7. Ibid., 262.

8. *Lincoln County Leader*, November 11, 1882.

9. Hough, *Heart's Desire*, 262.

10. *White Oaks Golden Era*, April 17, 1884.

11. Whiteman, interview.

12. Coe, *Ranch on the Ruidoso*, 238–41.

13. John Kelt to Paul Mayer, January 30, 1971, copy in author's collection.

14. *Pecos Valley Register*, May 30, 1889.

15. Shinkle, *Reminiscences of Roswell Pioneers*, 133.

16. *New Mexico Interpreter*, January 3, 1890.

17. *Lincoln Independent*, February 7, 1890.

18. Rayner, *Gold Lettered Egg*, 15.

19. Whiteman, interview.

20. *White Oaks Eagle*, May 30, 1895.

21. *Alamogordo News*, May 26, 1906.

JOHN E. WILSON AND SON-IN-LAW JOHN WAUCHOPE
Bit Hard by the Gold Bug

John E. Wilson, "Uncle Johnny," was born in New York on May 17, 1807, and later lived in Illinois.[1] Two daughters were born to him and his first wife before she died. He had five children from a second marriage, including John Jr. The elder John was described by people who knew him as a man of an "energetic and pushing disposition."[2] He moved west to Colorado to prospect and then on down to the Jicarilla district, placer mining before working his way to nearby Baxter Gulch at the age of seventy-one in 1878.

John E. Wilson. Date and photographer unknown. Courtesy Ina Hunter Dow.

Wilson Buys Out George Wilson, First Discoverer of Hard-rock Gold at White Oaks

It had long been known that there was gold in Lincoln County. Spaniards, and, later, Mexicans, explored the area for gold early in the nineteenth century, and the first American prospector encountered gold sometime before the Civil War.[3] In the 1860s, prospectors around Jicarilla began to mine gold by sluicing placer gravel near the head of Ancho Gulch. Gold in solid rock was first discovered in the region in 1866 at what later became the American Mine near Nogal, sixteen miles southeast of White Oaks and inside the reservation set aside for Jicarilla Apache Indians.[4] The gold discoveries caused Congress to abolish the Jicarilla reservation and to relocate the Jicarillas to northern New Mexico, leaving the land open to prospectors and settlers.

A dozen or more prospectors including John Baxter, a veteran of the California gold rush, arrived in 1879 and were struggling along in Baxter Gulch. A native of Missouri, Baxter had first passed through New Mexico on his way to California in '49.[5] He returned in 1862 as a soldier in the California Column under General Carleton. Their mission: to guard against a Confederate reinvasion of the Southwest. Mustering out after the war, Baxter found good placer diggings beyond the malpais in a gulch running from what is now known as Baxter Mountain east to the arroyo bounding the west side of present-day White Oaks.[6]

In 1879, Baxter, John Wilson, Jack Winters, Jimmy Dolan of Lincoln County War notoriety, and others were still working the gulches around that region when George Wilson (no relation to John) joined the group. While exploring some foothills, George came upon a blowout boulder and had sat down on it for a while to rest and eat lunch when his prospector's eye spotted fine gold sticking to an outcrop on the rock. He hurried back to report his find to his companions.[7] The lucky

John Wauchope. Date and photographer unknown. Courtesy Ina Hunter Dow.

Ione Wilson Wauchope. Date and photographer unknown. Courtesy Ina Hunter Dow.

discoverer, George Wilson, sold his interest to John Wilson and Jack Winters for forty-five dollars in gold washings, three silver dollars, a bottle of whiskey, and a six-shooter.[8] He then rode off on his pony and was not heard from again.

Origins of the South Homestake Mine

Jack Winters and John Wilson divided their claim, with Winters taking the north half and Wilson the south half—hence the North and South Homestake Mines.[9] Standard claims were 1,500 feet long by 600 feet wide. After the split, Wilson added another 750 to his holdings, to cover 1,500 feet south of the original partition line.[10] A stock company was organized, a hoist and other machinery put in place, and a Fraser, Chalmers and Company twenty-stamp mill installed to work the ores. John did not hold his mine very long; a little over a year later, the *Las Vegas Daily Optic* reported, "Old Uncle Johnny Wilson has sold his Homestake to the Homestake Gold Mining Company for $300,000."[11]

The new company consisted of Saint Louis investors who would become prominent in White Oaks: Erastus Wells Parker (president), V. W. Parker, William Watson, and Will Patterson.[12] Unfortunately, Uncle Johnny took payment in stock for the South Homestake. He would eventually realize only about $30,000 in cash out of the property.[13]

Ione Wilson and John Wauchope

John Wilson's daughter, Ione, was born July 10, 1842, and married William John Wauchope (pronounced "walkup") in 1863. They lived in Illinois and had five children—Mary, Ada, Ina, Samuel, and George—by 1874.[14]

In the spring of 1873, John (as he was always known) Wauchope went to Kansas and bought a farm in Belle Plaine in Sumner County twenty-five miles from Wichita. That fall he sent back to Illinois for his wife and children, who arrived in Kansas by train. For six years, the family farmed in Kansas, raising mostly wheat and corn. The bumper crops they raised in 1874 were destroyed by an enormous cloud of Rocky Mountain grasshoppers. The next year Wauchope planted a big wheat crop, and just as it was ready to harvest, he took the whole family in a covered wagon to Wichita to fetch supplies. They arrived home after a terrible rain and hailstorm to find their wheat plastered flat to the ground, ruined by hail as large as goose eggs.

Discouraged, Wauchope left the family in Kansas and went to Chicago to find work. The hired man he left back on the farm turned out to be worthless, and sunflowers took over the land. In the fall of 1876, Wauchope came home and the next year planted mostly corn. He made a bumper crop, but when it was ready to harvest he found he could not sell it at any price. Corn overflowed in the yard and the fields. The resourceful Wauchope bought hogs, fattened them on all that corn, and sold them—but he wanted out of the farming business.

Wauchope went to Fort Dodge in Kansas looking for a different kind of work. There he met Charles Siringo, who had just arrived with a herd of cattle he had driven from Texas. Siringo gave Wauchope the idea of going to Texas and getting in the cattle business. (A few years later in White Oaks, Wauchope's daughter Ina and her husband, Charles Mayer, became good friends with Siringo—by then a well-known Pinkerton detective and writer.)

In a Covered Wagon to Texas

In the fall of 1879, Wauchope sold the farm, and the family of nine (seven children by that time) set out for Texas in a covered wagon drawn by six horses. Ione's Singer sewing machine was tied on the wagon along with the water kegs. The Wauchopes headed straight south through dangerous Indian Territory, and they had two Indian scares along the way. Once when the family camped for dinner near an Indian village, Indians flocked around the camp. To placate them, Ione cooked a pot of sweet potatoes and set it down in their midst. They ate all they could hold and put the rest in their blankets to take away.

Later on, the family camped for the night in dense sagebrush. Two Indians rode up on horses and looked over the camp while talking and laughing loudly. After they rode off, the Wauchopes could hear whooping and yelling. Wauchope broke camp in the night and got out of the area fast. When the wagon arrived at the Red River, Wauchope decided to continue south to the Wichita River. On the road, the family ran into a big snowstorm and stopped for a week. The search for warmer weather took them even farther south to Fort Worth, Texas, then on to San Antonio, where they camped near town and rested a month.

Wauchope bought thirty cows at Castroville just south of San Antonio. The family broke camp, drove through Castroville to pick up the cattle, and started for west Texas and Fort Davis. On the way, two men, one riding a horse and one a burro, caught up with the family and traveled along, helping with the cattle. They were welcomed since Indians had attacked a wagon train and killed everyone just a month earlier. It took the family and their new riders three days to cross a dry stretch of country sixty-five miles long with no water holes. The team almost gave out. One of the cows they were driving had a baby calf, so the children took the calf in the wagon, and the cow provided them with much-needed milk. On reaching a watering hole that the military had dug, the cows and horses finally drank their fill and the family continued to a beautiful canyon with tall grasses to camp for a week. Then they went on to Fort Davis, where they stayed for a month before striking camp again. Reports of Indian uprisings and the slaughter of more travelers unsettled Wauchope, who sold all his stock except for the horses and turned back to San Antonio. They wintered nearby.

One day Wauchope picked up an old newspaper and brought it home. As Ione was reading it, she noticed an article describing a rich gold strike in White Oaks made by a John E. Wilson. Although Ione had not heard from her father in twenty years, she said, "I just know that is my father."[15] She wrote him a letter; Wilson answered immediately, inviting them to come to White Oaks.

On to White Oaks

On March 21, 1881, the family left San Antonio, Texas, for White Oaks. They had to outfit again and bought a covered wagon with four horses. During one night as they camped along the way, some commotion at a nearby army fort caused the horses to break loose and run away. Wauchope had to look for them on foot; he found them twenty miles below where the family was camped. It took him four days to round them up. In New Mexico, the family traveled west toward the mountains by way of Hondo and Lincoln. On April 28, 1881, near Hondo the wagon was stopped by a man who warned them, "Billy the Kid broke out of jail today. He killed his jailers and you had better not go on toward Lincoln right now."[16] Wauchope replied that it was doubtful the Kid would want anything from them, so they continued on to Lincoln. The first night in this area they camped within sight of Fort Stanton.

After five weeks of traveling, the Wauchopes finally arrived in White Oaks on the Sunday of May 1, 1881.[17] As they entered the town, they met a woman and her children coming from Sunday School, which allayed their fears about the roughness of the mining camp. The Wauchopes lived with John Wilson in White Oaks while Wauchope freighted from Las Vegas and Socorro to White Oaks.

Off to the Mines in South America

In 1886, after five years in White Oaks, John Wauchope was restless. He left to prospect in South America, leaving Ione and their eight children (a baby girl had been born in 1883) behind with Grandfather Wilson. Wauchope described his trip from White Oaks to Bolívar, Venezuela, in a letter to John McMurchy, a mining contractor at White Oaks:

> On arriving at El Paso I concluded that the proper way to reach my point of destination was via [rail] of N.Y.; $37.85 paid for the ticket. At 1 o'clock on the morning of the 9th my journey began. The next day brought to view nothing but the dry barren plains of Texas.
>
> Long before reaching the eastern line of the state [Texas], the way was through a timber country. Night overtook us just as we crossed the line. Next morning we looked out upon Arkansas.
>
> Feb. 26—Upon arriving in the city of N.Y. we went to the steamship ticket office, and were informed that the steamer would leave for Trinidad on Monday night.... At 4 o'clock we were on board.
>
> Feb. 27—On shore today for a stroll here at St. Lucia. Here our boat puts on coal and water. The coal is carried from the coal yard to the deck of the steamer by black girls. The market was crowded with people; everything is carried on the head; not an animal to be seen.
>
> Feb. 28—Looking out on the bay at Barhad where our steamer anchored in the night one beholds a beautiful country.... Beautiful fields of cane can be seen, the outlines of which are marked by a border of trees. Here the water is still, and in this harbor I count 62 large ocean sail ships and steamers.
>
> Bolívar, March 13—After leaving Barbados, the steamer was soon at Trinidad, where I had to stop three days waiting for the Oronoca steamer. Three days spent in Port of Spain. Here one sees many things that are very strange, one of which is the different classes of people, including those of all nations. Fine featured, fair looking ladies from the East Indies with rings around their wrists and ankles, rings in their nose, and on their toes, and in their ears, and heavy beads around their necks.... Again we halt at Port Establos. Now the river is four times as wide as it was in the morning. On leaving here it is dark, and in the morning we are at Bolívar, which is a city of 8,000. The ticket from Port of Spain cost $10.00. The journey is made in one day and two nights.
>
> From here the journey is by burros.... The prospect seems good.... I have to leave my rifle in the customhouse, none are allowed rifles without permit from the President of Venezuela.
>
> Wm. J. Wauchope,
Bolivar, Venezuela,
South America[18]

The Rest of the Story

In 1890, Wauchope returned to the United States. On the way back, he became very ill while visiting his mother in Iowa, and Ione went there to nurse him. He died after a short illness at the age of forty-nine, and Ione took his body back for burial at their old home in LaSalle, Illinois.[19]

John Wilson outlived his son-in-law. Wilson's death in 1893 at the age of eighty-six was noted in the *Old Abe Eagle*: "In this city, in which he was a pioneer, his character has been irreproachable and unassuming. He was not a business man in the strict sense of the term, but in all his dealings was fair and open and was warmly esteemed by all who knew him."[20]

During John's last years, John Wilson, Jr., looked after his father and his father's White Oaks interests. At the time of his death, the elder Wilson owned two valuable patented claims: the Silver Cliff and the Miner's Cabin. He also owned a gold milling plant and considerable town property. Ione returned to White Oaks from Iowa in 1890 and lived with son George in El Paso. She died at the age of sixty-six in May 1910.[21]

John and Ione's daughter, Ina, who first set eyes on White Oaks in May 1881 as a girl of thirteen, married Charles Mayer in 1888 in White Oaks. This couple made their home in White Oaks until 1921, when they moved their merchandising business to Carrizozo.[22]

NOTES

1. *Old Abe Eagle*, August 31, 1893.
2. Ibid.
3. Sinclair, "Little Town."
4. Parker, *White Oaks* 4.
5. *Albuquerque Journal*, April 6, 1881.
6. James Abarr, "White Oaks and Faded Dreams," *Albuquerque Journal*, July 26, 1998.
7. Sinclair, "Little Town."
8. *Las Vegas Daily Optic*, April 24, 1880.
9. Sinclair, "Little Town."
10. Parker, *White Oaks*, 5.
11. *Las Vegas Daily Optic*, February 22, 1881.
12. Ibid.
13. Obituary of John E. Wilson, *Old Abe Eagle*, August 31, 1893.
14. Ina W. Mayer, interview by Edith Crawford, 1938. In this WPA interview, Ina recounts her early life in Illinois, Kansas, Texas, and White Oaks.
15. Ibid.
16. Mary Nell Taeger, story about Wauchopes in *Ruidoso News*, November 26, 1948.
17. Mayer, interview.
18. John Wauchope to John McMurchy, published in *Lincoln County Leader*, May 1, 1886.
19. Mayer, interview.
20. *Old Abe Eagle*, August 31, 1893.
21. Mayer, interview.
22. Ibid.

JOHN V. "OLD JACK" WINTERS
Owner of the North Homestake Mine

In the fall of 1879, Jack Winters was one of the first prospectors to file a mining claim in White Oaks. His strike came late in life when he was already sixty-two years old. He had tried his luck before as a prospector in the California gold rush of 1849 and had failed miserably. When the Civil War broke out, he enlisted as a saddler in the First California Cavalry under Gen. James Carleton. Winters's service record states that he was born in Little York, Pennsylvania, and lists his trade as "shoemaker."[1]

Winters was one of some 1,350 soldiers who reached the Río Grande in 1862, a unit that became known in New Mexico as the California Column. Many prominent (and notorious) New Mexican pioneers came to New Mexico as part of the Column. Winters stayed on in New Mexico searching for gold. The 1870 census lists him as living in Santa Fe County, and lists his occupation as "miner."[2] By 1877, he was running a ranch two miles north of the Jicarilla mines where he boarded miners for half-interest in their claims instead of cash.

The Homestake Mines

In 1879, Old Jack joined forces with John Baxter and John Wilson in setting up camp at Baxter Gulch on the slope of a nearby mountain. George Wilson (no relation to John) drifted into camp in August 1879 and discovered a vein of gold ore on the lower east slope of what would come to be called Baxter Mountain. The inevitable stampede to the White Oaks goldfields followed.

George Wilson and Baxter moved on,[3] and in the fall of 1879, Winters and John Wilson, the remaining partners, staked a claim to the gold lode that they named the Homestake. It was the first gold-bearing quartz mine to be located in the Jicarilla Mountains and the first rich hard-rock claim at White Oaks. To celebrate the Homestake strike, Winters immediately bought a wagonload of whiskey, passed around tin cups, invited everybody to help himself, and, according to lore, got much of the town drunk.[4]

Winters and Wilson dissolved their partnership and divided the property into north and south sections, with Winters taking the North Homestake and Wilson taking the South Homestake. On December 23, 1879, Winters deeded 350 of his 750 feet in the North Homestake to Caroline Fritz Dolan, wife of Jimmy J. Dolan of Lincoln County War notoriety. On the same date, Jimmy and Caroline transferred one-half of their interest to Joseph A. LaRue in return for several thousand dollars from LaRue's silent partner, Marcus Brunswick, to work the mine.[5] The two Homestakes generated much excitement among developers. In February 1881, the *Las Vegas Daily Optic* reported that the Homestake Gold Mining Company had "bonded" (secured an option on) Jack Winters's interest in the North Homestake for $100,000.[6]

Old Jack Chews up a Glass of Whiskey

Winters did not live to enjoy his bonanza. He died March 21, 1881, at the age of sixty-four. A myth surrounds his cause of death. A local character named Frank "Big Foot Wallace" Clifford, a friend of Winters, told a bizarre story. He said that Winters had been boarding with the Bolton family, when one morning one of the Bolton girls, Ella, called out to Clifford for help. Winters—called "Old Blue Face" by Bolton—was trying to get out of the house to go buy whiskey, without clothes on. Clifford ran over to the Boltons to find Winters just inside the door and Mrs. Bolton trying to force him back to bed. Clifford took over and wrestled Jack back into bed. Then Clifford asked Mrs. Bolton for a drink of whiskey. When she brought it in a glass, Clifford held it to Jack's mouth and let him drink. After he had gulped all the whiskey down, Jack bit a large piece out of the glass, chewed it up, and swallowed that as well. Clifford choked Old Jack a bit trying to keep him from swallowing the ground-up glass, with no luck, so he asked Mrs. Bolton to bring another big drink in a tin cup—this time, water. After drinking the water, Winters fell asleep. He died the next day. The main cause was probably pneumonia. Clifford did not think the glass had anything to do with it, because Winters was "pretty far gone from whiskey" long before that.[7]

Another version has it that while at the Homestake, Old Jack was trying to winch the ore by hand while awaiting shipment of a steam hoist to White Oaks when he caught a bad cold and got so sick that he took to his bed. (Neighbors told him that it would help if he would also quit talking to "John Barleycorn.")[8] The old man's cold quickly turned into pneumonia, and he was taken from his mine shack to board with the Bolton family. An unnamed friend came to visit and brought a bottle of whiskey along. Jack got roaring drunk, leading to visions of snakes crawling up the wall that night. The next morning the Bolton daughter caught Clifford outside chopping wood and got him to help put the howling and shouting prospector back to bed, where he died the next day.

Jack Winters has the only grave in Cedarvale Cemetery whose occupant reclines in a north-south direction, feet to the north, instead of the traditional east-west. He wanted to face his strike on Baxter Mountain forever, the place where he had finally found his El Dorado.

Winters died intestate and without any known wife or relative in New Mexico. A notice of his death was published in Pennsylvania, where he was thought to have relatives.[9] Several false claimants immediately sprang out of the woodwork to claim the estate. Harvey Fergusson, a well-known lawyer from Alabama (and future New Mexico senator), arrived in White Oaks in 1882 to represent Jack Winters's heirs in a case called *Brunswick v. Winters' Heirs*.[10] Fergusson was a member of the law firm of Jacob, Cracraft, and Fergusson, located in Wheeling, West Virginia. He argued his case before the territorial bench and lost at district court level but won on final appeal to the territorial supreme court. The legal heirs represented by Fergusson then sold the North Homestake for $40,000 to Dr. W. G. Hunter of Kentucky.[11] After winning the case, Fergusson took possession of the North Homestake and sank a shaft on it to a vein about five feet wide. Free gold was visible on most of the pieces of quartz.

John V. Winters's tombstone in Cedarvale Cemetery. Winters's date of death is erroneously given as November 21, 1881; it should be March 21, 1881. Dorothy Leslie designed the crossed pickax and shovel for the tombstone. Photograph by Bob Workhoven. Author's collection.

James A. Sigafus Buys the North Homestake

In the fall of 1884, the North Homestake was bought by James A. Sigafus of Colorado, an experienced mine operator.[12] Sigafus brought in a capable manager named Frank Lloyd. Lloyd sank a shaft to one hundred feet, with drifts at the fifty-foot and one-hundred-foot levels, then fitted the shaft with a fifty-horsepower mine hoist. Shortly afterward, Sigafus and Lloyd also built a mill consisting of two five-foot Huntington stamp machines. Sigafus operated the mine successfully from the time he started early in 1885.

In 1892, Sigagus sank the vertical shaft down to 1,066 feet, with stations every 40 feet and a total of about one mile in drifts on the north end of the mine. A 450-foot inclined shaft brought the total vertical depth to roughly 1,400 feet.

The mine shut down about the time of World War I. When WWI broke out, a Mrs. Whiteman and her daughter came to White Oaks from New York City. They leased the North Homestake for its tungsten, then much in demand for the war effort. The tungsten operation caused the mine to cave. Before mining could be resumed, miners had to dig a tunnel below the cave-in from the Rita Mine. Later that tunnel also caved in. Later efforts to mine the North Homestake were made, with no success. The North Homestake produced about $1 million in gold, most of it before 1904.

NOTES

1. Miller, "Carleton's California Column," 29–30.

2. Ibid., 30.

3. Keleher, *The Fabulous Frontier*, 36. Keleher says that after abandoning his claim, Baxter drifted down to Silver City, where in 1885 he was shot and wounded in an Indian raid on the Gila. In the late eighties, Baxter returned to White Oaks to find a rip-roaring mining camp and his claim now an important mining property.

4. Ibid., 36.

5. Miller, "Carleton's California Column," 29–30.

6. *Las Vegas Daily Optic*, February 22, 1881.

7. Clifford, "Winter at White Oaks," 7–9. (Frank Clifford's real name was John Menham Wightman.)

8. Birdsong, "The Old Prospector."

9. Miller, "Carleton's California Column," 30.

10. Keleher, *Fabulous Frontier*, 38.

11. Miller, *California Column*, 61.

12. Parker, *White Oaks*, 60.

GEORGE RICHARD "DICK" YOUNG
Mercantilist from Mississippi and Klondike Prospector

Dick Young was born on July 15, 1854, at Abbeville, Mississippi, not far from the Tallahatchie River, and lived through the Civil War but was too young to enlist.[1] At age fifteen, Dick Young became a member of the Methodist Church and carried the deep religious convictions of that faith throughout his life. When he was about twenty-one, he married his first wife in Mississippi. She died in childbirth in 1877, leaving a baby boy named Robert.

The death of his young wife left Dick grieving, restless, and eager for a new life far away from his memories and the woes of the Reconstruction South. Once little Robert was old enough to travel, Dick crossed the South and the Texas plains to Fort Worth, as far as the railroad could take him. It is thought that he worked for a time in Fort Worth before deciding to try his fortunes in the territory of New Mexico.

Young arrived in Fort Sumner, New Mexico, at a time when lawlessness defined Lincoln County to the extent that historian Frederick Nolan has dubbed it the "Violent Paradise."[2] It was a land of pitched battles, outlaws, murders, and turmoil of all kinds. Dick worked briefly for John Chisum's Jinglebob cattle outfit in southeast New Mexico and somehow escaped being killed like so many others during the Lincoln County War.[3]

On to White Oaks

Dick Young does not appear in the June 1880 census for White Oaks. He probably arrived later that year to start his business called "Hides and Wool." On April 7, 1882, he partnered with a man named Robson to apply for a business license.[4] Joe Lea, deputy sheriff, signed the application for Sheriff Pat Garrett. The application was for three months, and it does not appear that the partnership survived long. By 1884, Dick was again sole proprietor of his business, now a mercantile. A *White Oaks Golden Era* in that year lists the following goods:

George Richard "Dick" Young, December 1897, taken in Salt Lake City. Photographer unknown. Courtesy Dr. G. Richard "Dick" Wheelock.

Dry goods
Staple groceries
Hats, caps
Smoking and chewing tobaccos
Smokers' outfits
Boots, shoes
Fancy notions
Hardware, doors
Sash, etc.
MINERS' SUPPLIES A SPECIALTY[5]

An 1888 *San Francisco Chronicle* listed Young as "general merchant," one of five in White Oaks and still sole

Application for license by Robson and Young, 1882. Courtesy J. N. "Snooks" McDaniel.

owner of this business.[6] A *Lincoln County Leader* ran the following ad, also in 1888:

G. R. YOUNG
—Dealer In—

Dry Goods, Groceries, Miners' Supplies, Boots and Shoes, Cigars and Tobaccos, Pipes, Fancy Goods, Notions, Fresh and Canned Fruits, Corn, Oats, and other Feed Stuff.

Prices 'way down. Goods good. Square dealing. No house in White Oaks can or will do better with and by you, and "don't you forget it."[7]

Dick Marries Kate Thomson

Not long after Dick arrived in White Oaks he met and married Catherine "Kate" Thomson. Kate's sister Ella married Jones Taliaferro in White Oaks in November 1883, and Kate married Dick around the same time. The newlyweds began married life in White Oaks with Dick's son Robert already part of their family. In

Taliaferro building, ca. 1883. On left is G. R. Young, Hides and Wool. Photographer unknown. Courtesy Chris and Jack Harkey.

Robert Young with pets. Date and photographer unknown. Courtesy Chris and Jack Harkey.

the June 20, 1885 New Mexico Quinquennial census, thirty-year-old George R. is listed as a merchant living with wife, Kate, twenty-one; son Robert, eight years old; and new baby Ella, born in 1885 (and named after Kate's sister).[8]

Partnering with Taliaferro

Dick continued as sole proprietor of Hides and Wool until early 1888, when an ad in the *New Mexico Interpreter* listed the new partnership of Young & Taliaferro.[9] Most likely Dick Young continued to specialize in miners' supplies and hides and wool, while Jones Taliaferro carried general merchandise. This partnership lasted until at least March of 1892, as evidenced by a large ad in a local newspaper:

> Young & Taliaferro
> Dry Goods, Groceries, and General Merchandise
> Highest cash price paid for country produce,
> WOOL, HIDES, AND PELTS.[10]

Dick Young became active in White Oaks politics along with his brother-in-law. He began on the town level, first serving as city treasurer of White Oaks while Taliaferro was school board president. Widening his political ambitions, he is listed in 1887 as one of three county commissioners of Lincoln County. Two years later the *Pecos Valley Register* lists G. R. Young of White Oaks as county treasurer of Lincoln County.[11]

Besides his civic and political duties, in everyday life Dick quietly carried on his good deeds as a devout Meth-

Little Ella Young around three years old, ca. 1888. While playing with Billy the Kid's IOU for a $25 hat, Ella tossed the paper into the fireplace. Billy, a frequent visitor to White Oaks, had bought the hat and paid with a promissory note. Ella had gotten into her mother's purse where the note was always kept. Photograph by Max H. Koch. Courtesy Chris and Jack Harkey.

Young & Taliaferro's ad in the *New Mexico Interpreter*, March 23, 1888. Author's collection.

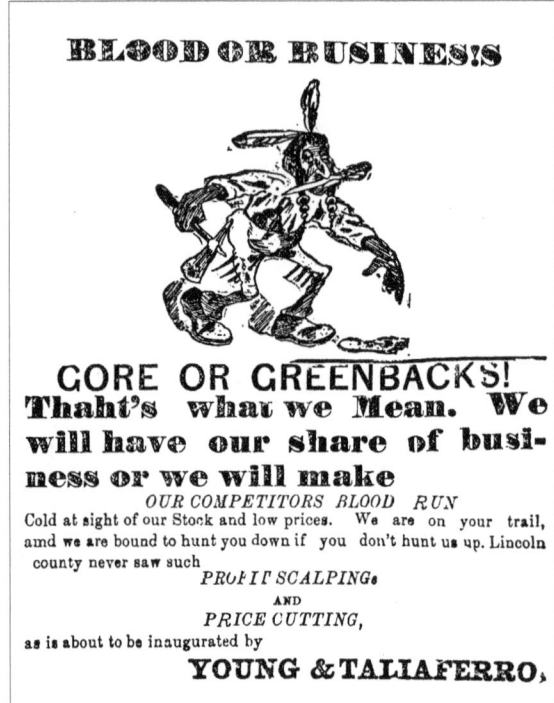

Young & Taliaferro's "Blood or Business" ad, *New Mexico Interpreter*, January 9, 1891. Author's collection.

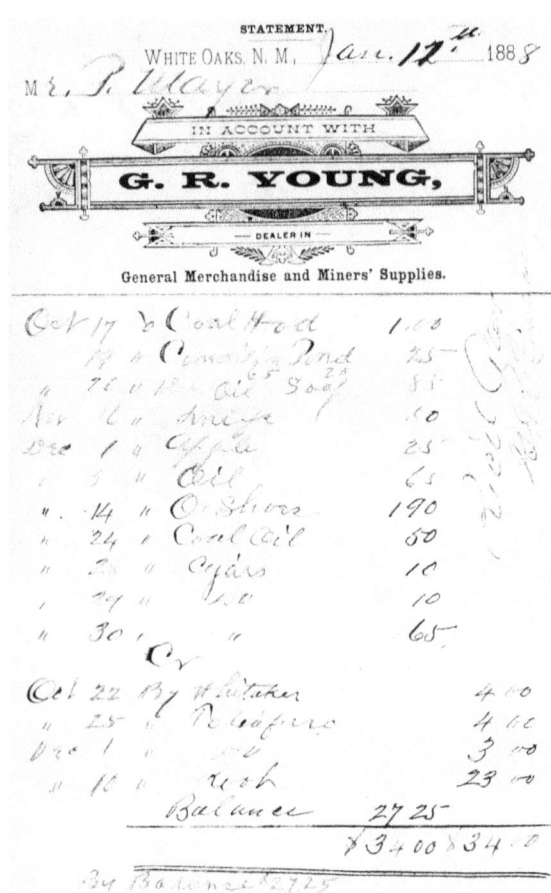

P. Mayer account with G. R. Young, January 12, 1888. Author's collection.

odist. The young schoolmarm Olive Rencher came out in 1887 from Mississippi to Lincoln County on the new railroad as far as San Antonio, then by stage to White Oaks and on to Nogal to take a teaching position (as she had been misled to believe).[12] She soon found herself in a dangerous situation working as a virtual slave for a Mr. and Mrs. Hughes, who had placed an advertisement for a teacher in the *Home and Farm Magazine* but who really wanted to snare an unpaid servant; there was no teaching assistant job. After a few weeks, Olive fled to shelter with a nice couple in Nogal, and while trying to decide what to do next, she received a letter from White Oaks advising her of an opening there. At once she went to inquire and found that one of the school trustees was from her home state of Mississippi. The relieved young woman wrote, "God had found me friends. Mr. Young was like a brother, took me right into his home and made me feel welcome."[13] Not long after, Olive married Alfred Ozanne of the Ozanne Stage Line.

Measles Kills Two of Dick and Kate's Children

In the days before routine inoculation against childhood diseases, epidemics periodically swept through towns and entire families, often killing several children in a family at one time. On March 5, 1892, the *Lincoln County Leader* reported,

> On Saturday the 20th ult., the grim monster, death, entered the house of our respected townsman, G. R. Young, and departing took with him the bright, scarly [sic] 5 year old boy Dick, whose mortal remains were laid away in the cemetery on Sunday 21st.... On Sunday the 28th, [death] again stalked into the stricken household, snatching therefrom the precious little life of the fourteen months old daughter, Mary Florida.... But one more little flower was left in the garden of Mr. and Mrs. Y., and sickness threatened it. Believing that the Young home had become permeated with fever, W. H. Weed . . . threw open the doors of his residence and invited the survivors to enter and sojourn until such time as the measel [sic] scourge could and would be extirpated, which invitation was gratefully accepted.[14]

Another paper wrote words of praise for the boy: "Bright little 'Dick,' as he was affectionately known by many of the older residents, was a pupil of Mrs. J. O. Parker's kindergarten school and although less than five years of age had exhibited remarkable talent, occupying a position in the second reader grade."[15]

To Roswell

With business in White Oaks declining, and heartsick at the loss of their children, the Youngs thought a change would be good. They moved to Roswell in June of 1893, where daughter Myra was born March 5, 1896. Sometime during their years there, another child was born and died. They buried this child in White Oaks with their other two.

In Roswell, Dick served as registrar of the U.S. Land Office during President Cleveland's second term, which began in 1893. He must have had to travel extensively in this job; the Pecos Valley Railway Company issued a three-year pass to G. R. Young as "Registrar, U.S. Land Office."[16]

The Klondike Beckons

The Alaskan Klondike gold rush began in 1898 and generated a fever pitch of excitement in the United States, which had been gripped throughout most of the 1890s in an ongoing deep depression. The Klondike was America's last big gold rush. At first, Dick merely grubstaked a few miners he knew to follow the Trail of '98, but soon he longed to be part of the action himself.[17] Unlike many prospectors, Young took his family along with him north to the large port of Dyea, Alaska, the staging point for the Klondike.

The Youngs planned to spend five years in Alaska in a place where winter temperatures often reached fifty degrees below. This time Dick did not get into the miners' supplies business but instead prospected for gold himself. To reach Alaska, the Youngs went through Salt Lake City to Seattle, where they boarded a steamer to Dyea.[18] The family lived in a tent at first while Dick was searching for gold. Dyea, no longer in existence today, was at that time the main port and entry over Chilkoot Pass to the Klondike. (The cemetery and the remnants of the docks are the only surviving structures in that once-bustling town.)

In May 1898, the first rails of the White Pass and Yukon Route Railway were laid at Skagway, and the line to White Horse opened July 1890.[19] Chilkoot Trail, originating near Dyea at sea level, wound over the 3,246-foot Chilkoot Pass before dropping down to the Yukon River. It is an arduous thirty-two mile trek of three days. Chilkoot Pass was very steep and dangerous; a landslide once killed nearly one hundred miners trying to reach the top.[20] In the late 1890s, an endless stream of pack-laden men struggled up Chilkoot Pass. Eventually, even equipment of massive proportions was also packed over the pass. Machines such as dredges were disassembled and carried piecemeal by manpower to be reassembled at the gold camps. To increase the miners' odds of survival, the Canadian Northwest Mounted Police (Mounties) required each man who wanted to work the Klondike to take one ton of food, supplies, and survival gear—enough to last a year—across Chilkoot Pass.[21] Then the cargo had to be floated over the rapids of the Yukon River. It is said that the local Native Americans (who in the North call themselves "First Americans") stood by and silently watched this operation before offering the observation that across the next hill was a pass much easier for people and animals to climb.

By spring 1898, more than sixty thousand men and women had passed through Alaska's Chilkoot and White Passes on their way to the Klondike. From 1899 to 1904, the Klondike produced $100,000,000 in gold.[22] As always, only a few would find riches; most were lucky just to survive. Those who did would forever remember the experience as a high point in their lives.

To Salt Lake City, Northern Arizona, and Safford, Arizona

In 1902, the Youngs returned to the United States, stopping over for a couple of years in Salt Lake City, where youngest daughter, Olga, was born September 13, 1902. They lived briefly in northern Arizona in 1904 or 1905, then pushed on to Safford. Safford would be home to Dick and Kate for the rest of their lives. Dick opened a grocery store in an old adobe building and did very well as one of the town's pioneer merchants.[23] Dick built up the grocery business from a small beginning until the firm had to find larger quarters, prompting Young to purchase a building that was enlarged and remodeled. He took son-in-law David Ridgway into partnership with him, and the name of the firm became Young & Ridgway.[24]

Daughter Ella married David L. Ridgway in 1907.[25] He was a deputy sheriff in 1910 before becoming half of Young & Ridgway. Daughter Myra married Gilbert Wheelock.[26] Gilbert had accompanied his Boston family as a teenager to Riverside, California. After high school, he worked for the railroad and transferred to a Southern Pacific branchline that ran from Fort Bowie, east of Willcox, to Globe-Miami, copper-mining towns in southeast Arizona. The branchline ran through Safford north to Globe, and it was in Safford that he met and married Myra. The couple moved to Coronado, California. Youngest daughter, Olga, grew to womanhood and graduated from college. She taught school for two years before what is now known as Lou Gehrig's disease hit. She deteriorated within three years and died November 2, 1929.[27]

Throughout life, Dick maintained his interest in politics. In 1928, Dick advised Cordell Hull of the Tennessee delegation at the Democratic National Convention in Houston, Texas, via a Western Union telegram, "Use all your power to defeat Governor Smith. [It would] Put Arizona in the Republican Column if he is nominated. I helped to elect Samil [sic] J. Tilden in 1876. I am a Mississippian. Would be pleased to hear of your nomination. Give us a dry man."[28]

The day after Christmas 1929, in a decision no doubt brought on by the death of youngest daughter, Olga, in November, the Youngs returned to White Oaks to supervise exhumation of the bodies of their three children buried in Cedarvale Cemetery. The many friends of the Youngs, and the undertakers, thought it would be impossible to remove the bodies because of the number of years they had been buried. However, Dick—believing at the time of his children's deaths almost thirty-seven years before that there was a possibility he would some day want to remove them to another resting place—had encased the caskets in boxes of zinc, which were in turn placed inside wooden coffins.

After the bodies were exhumed, the Youngs found that, although the outer wooden boxes had decayed, the inner encasings of zinc were still intact, with the caskets in almost the same condition as when they were first lowered into the grave. The three boxes were put in new hermetically sealed containers by the New Mexico undertaker and shipped by train to Safford for final burial in the family plot.

Heatstroke Claims Dick

Dick was home on Stafford's Pioneer Day, July 25, 1930.[29] At about noon, he decided to irrigate his pecan orchard. When he failed to return to the house for lunch, Kate (thinking he was busy with the irrigation and could not leave it) waited some time before going to look for him. She found him lying in the shade on the canal bank stricken from the heat of an extreme Arizona day. He was taken to the house, and a doctor was called immediately. All that could be done was done, but at 5:30 the next afternoon he died at seventy-six years of age.

A Baptist minister preached a rousing eulogy at the funeral, relating how Dick had lived his religion throughout life and had helped the needy in many ways unknown even to his closest friends. He was buried in Union Cemetery near the graves of the four children who preceded him in death. Kate died in 1935 and is buried next to him. After Dick Young's death, Ridgway and Wheelock went into partnership in the business Dick had started. They kept the original name of Young & Ridgway for their firm.

NOTES

1. From an undated newspaper obituary in the possession of Dr. Wheelock. The name of the newspaper is not listed.

2. Nolan, *Lincoln County War*, 60.

3. Wheelock, telephone conversation with author, January 7, 2005.

4. Robson and Young, Application for License, filed by Robson and Young with the Clerk of Lincoln County, New Mexico, April 7, 1882, Lincoln County Courthouse, Carrizozo.

5. *White Oaks Golden Era*, January 10, 1884.

6. "New Mexico—Victoria—White Oak [sic]," *1888 San Francisco Chronicle*, 497.

7. *Lincoln County Leader*, October 27, 1888.

8. New Mexico Quinquennial Census for 1885.

9. *New Mexico Interpreter*, March 23, 1888.

10. *White Oaks Eagle*, March 3, 1892.

11. *Pecos Valley Register*, November 21, 1889.

12. Rencher, "Ups and Downs of Eighty Years," 104.

13. Ibid.

14. *Lincoln County Leader*, March 5, 1892.

15. Undated newspaper obituary of Wheelock.

16. The Pecos Valley Railway Co. Pass No. 1896, good until December 31, 1896, copy in author's collection.

17. Wheelock, telephone conversation with author, March 7, 2005.

18. Ibid.

19. American Automobile Association, *Western Canada and Alaska TourBook*.

20. Wheelock, telephone conversation with author, January 7, 2005.

21. AAA, *Western Canada and Alaska TourBook*.

22. Ibid., 266.

23. David Smitheran to Haldane, October 2004.

24. Wheelock, telephone conversation with author, March 7, 2005.

25. *Arizona Days and Ways Magazine*, April 28, 1963, 36.

26. Wheelock, telephone conversation with author, March 7, 2005.

27. Undated newspaper obituary of Wheelock.

28. Dick Young to Cordell Hull, telegram, June 23, 1928.

29. From an undated Safford newspaper article in the possession of Dr. Wheelock. The name of the newspaper is not listed.

ALBERT AND JACOB ZIEGLER
German-Jewish Merchants Chasing the American Dream

Albert Ziegler was born June 20, 1862, in Kaiser's Esch, Germany. On January 2, 1881, at the age of nineteen, he sailed from his home on the Rhine River in Koblenz, Germany, to join his brother Jacob "Jake" in America. After crossing the ocean in thirteen days on the Furnesia of the Anchor Lines, he arrived in New York City.[1] Albert thought Jake was in Pittsburgh, where he had been six years before. On arriving in New York City, however, Albert got a wire from Jake saying that he had just left for the territory of New Mexico and the new town of Albuquerque.

Two days later Albert bought a ticket on an immigrant train for Albuquerque and, after a week's travel, arrived with only thirty-five cents in his pocket. Jake already had a job clerking for the Jaffa Brothers; he was happy to see his brother and tried to get work for him. But Albert had no luck—maybe because he could not yet speak a word of English—and decided instead to travel to San Francisco and live with relatives where he could work during the day and might attend school to learn English.[2]

A couple of weeks after landing in San Francisco, Albert got a job with Lusk Canning, overseeing Chinese workers packing salmon. When the season ended after two months, his cannery job played out. He bought a small stock of notions and set out as a peddler. Albert was a born salesman; he made quite a success of this new enterprise until he was arrested for not having a peddler's license. Discouraged with California, Albert returned to New Mexico after a year and landed a job as a clerk with Price Brothers in Socorro, which was located at that time in Lincoln County.

While Albert was struggling in California, Jake had relocated to the tiny village of Manzano, south of Albuquerque. He and his partner Herman Goodman ran a small store selling dry goods, groceries, and liquor. The partners were unaware that Manzano was one of the

Albert Ziegler. Date and photographer unknown. From *White Oaks*, by Morris B. Parker, (Tucson: University of Arizona Press, 1971). Permission of Mrs. Louise Kelt, March 17, 2001.

toughest towns in the territory of New Mexico.[3] Jake, Goodman, and a man named Charles L. Kusz were the only Anglos in the village. Kusz published a newspaper called the *Gringo and Greaser*, a title not calculated to win many friends in this all-Hispanic community. Kusz disliked Hispanics and constantly criticized them in his paper. One night as he was eating supper, someone shot him in the back through a window and killed him instantly.

On one occasion, when Albert was visiting Jake, a group of local men stood around in the store drinking. When Jake refused to sell them any more whiskey, the men complained loudly but eventually left. Later that

Invoice from Goodman, Ziegler & Co., February 9, 1888. Author's collection.

Goodman, Ziegler & Co.'s ad for new goods in the *New Mexico Interpreter*, March 23, 1888. Author's collection.

evening, a mob of twelve men returned to the store armed with Winchester rifles, looking for Jake. He had been tipped off and hid out in the hayloft. After failing to find Jake, the men left. The next day, the men obtained a warrant, arrested Jake, and hauled him before a justice of the peace, who not only fined Jake but made him pay court costs as well.

Fortunately, Jake was still alive, but he had soured on Manzano. He and Goodman moved farther south to White Oaks and put in a dry goods store in a small log cabin, taking in a third partner named Joe Freudenstein.

Freudenstein did not like White Oaks and left after three months. Meanwhile, Albert's employers, the Price Brothers, became very much interested in White Oaks and encouraged Albert to join his brother in business. In late 1886, the town was booming, with about two hundred people at work in the mines.

Albert Partners with Jake in White Oaks

Albert left Socorro for White Oaks around Christmas of 1886 soon after filling out a Certificate of Alien's Intention to become a citizen of the United States in

Goodman, Ziegler & Co.'s emporium ad, *Nogal Nugget*, May 25, 1888. Author's collection.

Ziegler Bros.' ad for spring suits in the *White Oaks Eagle*, May 23, 1901. Author's collection.

November.[4] It is thought that he finally became a citizen in April 1889 when the Socorro County Clerk filed a copy of his original declaration of intention.[5]

Albert's winter trip to White Oaks was by so-called stagecoach—actually a buckboard drawn by two Spanish mules. Leaving Socorro about ten o'clock in the morning, the travelers reached Ozanne's ranch halfway between Socorro and White Oaks about six in the evening. After supper at the ranch, the drivers changed the team and began the last half of the journey. It was a bitter cold night, the road was difficult, and at times the mules could make only two miles an hour while struggling through the many sandy places before the passengers finally arrived at their destination.[6]

With Albert's arrival, the three partners renamed the White Oaks business Goodman, Ziegler & Company. All the company's merchandise was freighted from San Antonio, New Mexico, mostly by ox team. The trip took a week and was rarely without incident. In one shipment, Albert had ordered ten gallons of fine wine. When the keg arrived at last, the partners eagerly opened it to sample the contents and discovered that someone had removed the wine and refilled the keg with water.

In 1888, after the Ziegler brothers and Goodman had been in White Oaks two years, the partners decided there was not enough business for all three.[7] They opened a branch store at San Pedro, New Mexico, an old Spanish settlement on the Río Grande one mile east of San Antonio. Goodman drew the straw to operate this store. The move turned out to be a bad one, with the Zieglers having to buy Goodman out at the end of two more years and assume all debts of the store. In 1890, their dry goods company was reorganized in White Oaks as the Ziegler Brothers.

Peddling Goods around Lincoln County

The White Oaks store was quite small at first. As the brothers' business grew and prospered, Albert often traveled back East for weeks at a time to buy fresh stock. He would go to Saint Louis, Chicago, and sometimes New York before returning. To further increase business, Jake often made forays into the countryside, at times far from home. The brothers had a good pair of horses and a stout lumber wagon on which Jake would load big stocks of merchandise. He usually did very well. However, the country in those days was not safe, and Jake often took a man along to help drive the team and watch out for trouble.

On September 3, 1891, Jake left White Oaks with a load of dry goods bound for the Peñasco in the Sacramentos. He had hired Ike Smith, an old-timer who knew the country well, to accompany him.[8] After a week of peddling, the two men left the Peñasco, intending to reach Jesse Brazel's ranch on Eagle Creek and stay the night. Jake rode ahead of the wagon and arrived at Brazel's before Ike, who was driving the team. When Ike

Ziegler Brothers.
ARE WELL PREPARED TO SUPPLY YOUR WANTS.

Their Stock of General Merchandise is Complete.

Dry Goods, Clothing, Groceries, Gents Furnishings.
A COMPLETE STOCK
of Boots and Shoes.

Give Us a Chance and See How Cheap We Can Sell You a Bill of Goods.
Ziegler Bros.

Ziegler Bros.' ad for general merchandise in the *Carrizozo Outlook*, March 23, 1905. Author's collection.

entered the house, he noticed a sitting stranger who got up and left immediately. He asked the children who the man was, and they told him it was Jim Connor.

Jake and Ike stayed the night and left early the next morning, driving as far as the Angus VV Ranch, where they lingered awhile before leaving around nine that evening. As they neared Henley Hill leading downhill to the Río Bonita, Jake got out of the wagon and walked behind. Halfway down, a man riding a bay horse and wearing a gunnysack as a mask jumped in front of the team. He waved a pistol and commanded Ike to stop. The bandit ordered Ike out of the wagon and told him to stand beside Jake. Ike said he had no money, and the man replied that he wanted nothing to do with Ike but ordered Jake to hand over his cash. The amount was so small that the robber said he should kill them anyway, but instead demanded Jake's watch. All the time Ike was observing the man carefully so that he would be able to identify him. After pocketing Jake's watch, the man ordered the pair to drive on and not look back.

The two drove straight to Nogal and reported the robbery. A posse was formed consisting of William and Charles Henley, O. Davis, and Ike Smith. Ike and Jake had become convinced that the robber was the man they saw at Brazel's ranch and told William Henley of their suspicions. The posse rode toward the scene of the holdup and separated at the VV Ranch, with Smith and William Henley traveling one way and Davis and Charles Henley another, and agreed to meet at Brazel's ranch. Ike and William arrived first at Brazel's. Jim Connor was sitting down when they entered the house. Ike recognized Connor as the robber and made a sign to William. When Ike told Connor that he wanted to talk to him, Connor sprang up with a pistol in his hand and jumped behind a partition, exchanging shots with Ike. Ike was shot in the head and fell, blinded by blood. Connor then ran out the back door and William out the front, leaving Ike bleeding and badly hurt. William was able to arrest Connor.

As soon as Ike could see through the blood in his eyes, he went to the creek with Davis's help and washed his face. The men then left for the VV Ranch so that Ike could get his wound treated, thinking that William Henley would secure Connor as a prisoner. The next day Ike went on to Nogal and waited for Connor to be brought in. At about 4:00 P.M. word came that Connor had escaped. Jake and Ike later discovered that, rather than disarm or arrest Connor, William Henley had set him free to hunt up his witnesses, with Connor promising to return. Jake was astounded that a highway robber and attempted murderer was allowed to travel around finding his own witnesses. Back in White Oaks, the outraged Ziegler brothers decided not to take this injustice lying down. They posted a one-hundred-dollar reward notice in the *New Mexico Interpreter*: "For arrest of Jim Conner [*sic*]; he is about 5 feet 11 inches high, 30 years old, light red hair and beard, sandy complexion, inclined to be freckled, large thin nose, spare built, keen blue eyes, had on boots with ridged legs, striped pants. Committed highway robbery in this county on the 10th day of September, 1891. The above reward will be paid on the delivery of his body to the Sheriff of Lincoln County."[9]

Connor was never arrested and brought to trial. This incident had two consequences: Jake ended his trips around the country to peddle merchandise, and a year later Ike Smith died from the effects of the gunshot wound.

Albert Finds a Pretty Bride in Trinidad, Colorado

Albert claimed the best move he ever made was to marry Carrie Leon of Trinidad, Colorado, whom he called "one of my greatest helps in business as well as socially. From that time I commenced to live right."[10]

Although White Oaks was about a hundred miles from the nearest railroad at the time, its social life compared favorably with that of much larger towns in other parts of the country. Jake hosted the post-wedding festivities for Albert and Carrie on Friday evening, August 29, 1891. The newlyweds were guests of honor at a formal reception that began the ceremonies. At half past nine, the orchestra struck up the grand march, led by the bride and groom. Dancing lasted until dawn, with everyone happy and dressed to the nines:

> The bride wore white silk, "en traine" with corsage trimmed in lace and completed by diamond jewelry. She is described as a "lovely brunette, petite of figure and possessing a charming grace of manner which has already won her a high place in the esteem of her husband's host of friends."
>
> Susan Barber, the Cattle Queen of New Mexico and always a fashion plate, wore a gown of pink satin and white lace with her trademark diamonds. Mrs. Colonel Heman wore black silk India mull trimmed with cherry-colored ribbons. A Medici collar with a "V" opening in front showed to advantage her "exquisite curves of throat and chin." Black silk gloves reached nearly to the shoulders.
>
> Mrs. Whiteman, a dress of black lace with handsome jewelry. Mrs. Sam Wilson, pink satin trimmed with chantilly lace. Mrs. Lloyd, pink silk and cream lace. Mrs. Emil [sic] Ozanne, white China silk. Mrs. Dr. Paden, brown silk Francaise. Mrs. Sidney Parker, black Indian silk with velvet trimming.
>
> Mrs. Woodland, white nun's veiling trimmed with satin. Mrs. Stewart, figured old rose silk trimmed in black velvet.
>
> Mrs. William Watson . . . a gown of cream challis with shirred vest and full sleeves of Gobelin chine silk. Miss Rice, an elegant gown of black grenadine with garnet jewelry. Miss Green, satin-striped cardinal silk corsage decollete trimmed with black lace. Miss Edith Lane, electric blue nun's veiling trimmed with cream lace and ribbons. Miss Jennie McCourt, old rose satin with sleeves and corsage trim of nile green surah. Miss Cooper, light-blue satin trimmed with cream lace. The Misses Nabours and Moor in white Swiss muslin.[11]

To please his bride, Albert freighted her large square piano to White Oaks. By some miracle, it arrived in fine shape.[12] In the next few years, the couple became the parents of two daughters, Effie and Selma.

Albert's Memories of White Oaks

Wilbur Coe told a story of Albert Ziegler when Wilbur was a young boy and he and his father, Frank, were peddling apples in 1903 from the Coe orchards at Glencoe.[13] They drove up to the Ziegler Brothers grocery in White Oaks with their wagon heaped with apples. Frank went inside and returned with Albert, who was carrying a wooden candy bucket estimated to hold a bushel. Albert looked at the apples and said that he would take five bushels if they were cheap enough; he was proud of being a good trader. He warned Frank not to do to him as other peddlers had and sell him a big order at a high price and then go around to people's houses and sell the same thing cheaper.

Ziegler house in White Oaks. Date and photographer unknown. Courtesy Larue Lane Wetzel.

Selma Ziegler, daughter of Albert and Carrie Ziegler, ca. 1886. Photograph by Max H. Koch. Courtesy Historical Center for Southeast New Mexico, Roswell.

Willie Gallacher (a former resident of old White Oaks) related in another story that a man once came into the store and was trying on a coat when two more men burst in with pistols drawn. They must have been law officers looking for the man trying on the coat. Jake began warning them in a loud voice (tinged with a heavy accent), "Don't shoot him in the coat—it's still ours and he hasn't paid for it yet!"[14]

Life could be funny, too. Carrie and Albert were invited to a dinner party at the house of their neighbor, Mrs. John McCourt. Her five-year-old son Ben enjoyed the meat course very much. He heard the adults talking about the kind of meat served, which happened to be kid. Ben inquired, "What did they kill the poor little fellow for?" During the Cleveland administration, the Ziegler brothers had quite a few light grey stovepipe hats in the store. A group of Apache Indians came over from their reservation to buy some dry goods. Jake sold each of the men a stovepipe hat, after which the entire group paraded through the streets of White Oaks showing off their blankets, moccasins, and new tall hats.[15]

A Short Move to Trinidad, Colorado, and Relocation to Carrizozo

Albert and Jake held on to their store in White Oaks as the town declined. They finally sold in 1906 to Solomon Wiener and moved to Trinidad to try the hide and wool business. They stayed with this business for about two years until the market fell drastically to almost nothing. They quit hides and wool after a huge loss and decided to get back in the mercantile business.[16]

In 1908, the Ziegler brothers relocated to Carrizozo to become the first dry goods store in that new railroad town.[17] Eight years later, Albert and his wife built a handsome two-story Italianate home on Tenth Street in Carrizozo—the only home in town of the many built by Frank English that was designed by an architect. There he and Carrie hosted many social events.

Later Years

Jake died in El Paso, Texas, April 16, 1919. He never married; his half of the Ziegler mercantile business went to Albert.[18] In 1928 or 1929, a prosperous Albert, Carrie, and daughter Effie (who had married a Judge Guteknecht) took an extended pleasure trip to Europe, which included visiting Albert's former homeland of Germany.[19] They spent their first week in Paris, and then traveled to Cologne. Every day they went sightseeing, mostly on the Moselle River, where they could best view the many castles on the river. Then they traveled along the Rhine to Frankfurt. Albert's son-in-law Judge Guteknecht cabled the Zieglers to meet him in Baden-Baden in the Black Forest. He had rented a car in Paris and would drive the family on a tour of Switzerland and the Alps, then to Hungary and back to southern Germany.

Keenly interested in the country's economy, Albert observed the high inflation plaguing Germany following World War I. One of his sisters-in-law in Germany showed him a tin box of the size kept for papers. It was stuffed with German marks, paper money that before World War I had represented real wealth but was now worthless. By this time, Albert considered himself an American through and through, was proud of it, and was eager to return home. He recorded his thoughts as he entered New York harbor aboard the steamship *Europa*: "We were very happy to see the Statue of

Liberty in New York harbor and to get back home to our own free country and all our friends."[20]

Albert continued to do business in Carrizozo, always careful of the quantity and quality of his goods. He would carry local ranchers all year on credit; then when they sold their calves and paid their bill for the year, he would usually give them a new Stetson hat.[21] At last, at the end of August 1940, he closed his doors for the final time. A "Quit Business Sale" rang down the curtain on an enterprise that had been a source of pride to every citizen in and around Carrizozo. It had weathered every depression and never failed its friends.

On September 2, only days after the store closed, Albert passed away at the age of seventy-eight.[22] He had been in business in Carrizozo over thirty-two years and had become an institution. As a business and civic leader, he had made hundreds of friends from all walks of life. Albert had also been a Master Mason since 1919 and followed faithfully the order's teachings of loyalty and helpfulness in everyday life.

A funeral address for this Jewish immigrant who had started out in America with thirty-five cents to his name was delivered at the Carrizozo Masonic Temple and was followed by Masonic funeral rites. The temple overflowed with flowers, which in life Albert had loved. Then the body was taken to El Paso for funeral services at Temple Mount Sinai, presided over by Rabbi Wendall Phillips. Burial was at Mt. Sinai Cemetery in El Paso.

The store passed on to new owners, Mr. and Mrs. Joseph Franklin Petty.[23]

NOTES

1. Albert Ziegler, interview by Edith Crawford, March 1938.

2. Albert Ziegler, interview by Nettie Lee Lemon, n.d. (Nettie Lemon was a former White Oaks resident.) Furnished by her granddaughter Marilyn Lemon Ingley of Falls Church, Virginia, by letter to Haldane, April 20, 1999.

3. Ziegler, interview by Crawford.

4. Albert Ziegler, Certificate of Alien's Intention filed in Socorro County, Second Judicial District Court of the Territory of New Mexico, November 18, 1886.

5. Albert Ziegler, Socorro County Clerk's certification of original declaration of intention to become a citizen, Second Judicial District Court of the Territory of New Mexico, April 26, 1889.

6. Ziegler, interview by Crawford.

7. Ziegler, interview by Lemon.

8. *Lincoln County Leader*, September 19, 1891.

9. Choate, "History of the Capitan Community," 77. Choate quotes the *New Mexico Interpreter* reward advertisement.

10. Ziegler, interview by Lemon.

11. *Lincoln County Leader*, August 29, 1891.

12. Ziegler, interview by Lemon.

13. Coe, *Ranch on the Ruidoso*, 239–40.

14. Johnson Stearns to Haldane, March 6, 2004.

15. Ziegler, interview by Crawford.

16. Ziegler, interview by Lemon.

17. *Lincoln County News*, August 23, 1940.

18. Last will and testament of Jacob Ziegler, filed by the Judge of the Probate Court September 7, 1920, at the Lincoln County Courthouse, Carrizozo, copy in author's collection.

19. Ziegler, account of his trip to Europe.

20. Ibid.

21. Stearns to Haldane, April 3, 2004.

22. *Lincoln County News*, September 6, 1940.

23. *Lincoln County News*, September 13, 1940.

AFTERWORD

White Oaks seemed touched with magic in that golden era between 1879 and 1904, from boom to bust. No mining town in the Southwest offered more excitement, color, and political intrigue or produced more citizens of note in so short a period.

These days White Oaks is usually referred to as a ghost town, but it has never completely "gone ghost." Some seventy people still live in the area, including a small colony of artists. Today most structures are gone, destroyed by fire or dismantled for use elsewhere. But some have survived. The Brown building to the north on the main road at one time or another housed a school, dance hall, and offices. The Taylor house, originally of logs with an adobe portion added, is across from the Brown building. East of the Brown building is the old Watson-Lund law office, now home to the No Scum Allowed Saloon.

Larue Lane Wetzel, granddaughter of the Oaks' first physician, Dr. Alexander Lane, continues to live across the valley to the south in her beloved brick-and-stone masterpiece, the Hoyle House. The beautiful Victorian Gumm House owned by Anne New graces the hill across the valley from the Hoyle House. The two-story schoolhouse, home to the White Oaks Historical Association, contains many photos and artifacts of mining days and is open most weekends in the summer to hundreds of visitors.

Many families who left long ago to find work elsewhere settled at first in New Mexico locations—the nearby railroad town of Carrizozo, then Tucumcari,

Old Cedarvale Cemetery. Date and photographer unknown. Courtesy Chris and Jack Harkey.

Cedarvale Cemetery as it looks today. Photograph by Bob Workhoven. Author's collection.

Roswell, and little villages in Lincoln County like Nogal and Capitán where some descendants can be found even today. Those who continued in the mining tradition moved to Mexico and throughout the West. Carol Queen Watt of the four-generation Queen mining family died in 2005; her cousin Donald, the last Queen in Lincoln County, died in 2009.

Stirling Spencer, great-grandson of William McDonald, continues to successfully manage the Bar W Ranch just north of Carrizozo. Farther north is the J + H Ranch run by Gallacher descendants as one of the best cattle-working outfits in New Mexico. Roy Dow still serves up ice cream treats to customers from the original backbar and soda fountain at Dr. Melvin Paden's Drug Store in Carrizozo, one of the few remaining historic drugstores in New Mexico.

Cedarvale Cemetery a mile west of town is on the National Register of Historic Places. It is home to many notables, including the first governor of the state and the Cattle Queen of New Mexico. Longtime White Oaks Historical Association volunteers at the schoolhouse museum, Robert and Dorothy Leslie, are among the most recent citizens to be buried there in this century.

Cedar trees and piñons still waft their scent down the mountain slopes and through the tiny village. And White Oaks remains, as Emerson Hough once characterized it, as his fictional town of Heart's Desire, "A spot lovely, lovable. Nothing in all the West is more fit to linger in a man's memory than the imperious sun rising above the valley of Heart's Desire."[1]

NOTES

1. Hough, *Heart's Desire*, 12.

BIBLIOGRAPHY

Unpublished Sources

Account ledgers for business of Ed Bonnell. Collection of Eleanor Shockey. Ruidoso, New Mexico. Copies in author's collection.

Ball, Eve. Interview with Amelia Bolton Church, December 3, 1951. Roswell, New Mexico. Copy in author's collection.

Barber, Susan. Letters to Bertha Mayer Hunter, March 23, 1926, and May 9, 1926. Collection of Ina Hunter Dow. Salinas, California. Copies in author's collection.

———. Letter to Governor Miguel Otero, June 12, 1930. Copy in author's collection.

———. Letter to Paul Mayer, July 18, 1928. Copy in author's collection.

Blanchard, William E. Untitled memoirs of Lincoln County, n.d. Collection of Robert and Dorothy Leslie. Carrizozo, New Mexico. Copy in author's collection.

Bonnell, Ed. Letter to Charles A. Siringo, February 27, 1893. Copy in author's collection.

Bonnell, Erva. Letter to "Dear Aunt," November 11, 1886. Copy in author's collection.

Brimberry, J. S. Notes on the origin of the Aguayo name, n.d. Copy in author's collection.

Brimberry, Lorraine Aguayo. Aguayo family biography, n.d. Ruidoso, New Mexico. Copy in author's collection.

Brinkman, Ronald A. Letter to R. E. Lemon, August 3, 1971. Copy in author's collection.

Charles, Mrs. Tom. Telegram to Associated Press, n.d. Copy in author's collection.

Coe, Frank. Diary. In Eleanor Shockey (Coe's granddaughter) collection. Ruidoso, New Mexico.

Collinson, Frank. "Jim Greathouse or Whiskey Jim," n.d. Panhandle-Plains Historical Museum, Canyon, Texas.

———. Letter to Bruce Gerdes, January 4, 1938. Panhandle-Plains Historical Museum, Canyon, Texas.

Corn, Betty. "Mama Donnally, the First 'First Lady' of the State of New Mexico," April 29, 1995. Collection of Betty Corn. Las Cruces, New Mexico. Copy in author's collection.

Davis, Edna Koch. Letter to Bertha Mayer Hunter, April 7, 1926. Collection of Ina Hunter Dow. Salinas, California.

"Descendants of Nathaniel Bonnel," n.d. Collection of Eleanor Shockey. Ruidoso, New Mexico. Copy in author's collection.

Dow, Ina Hunter. Letter to Mary Jo Jamison, August 10, 1994. Copy in author's collection.

———. Obituary for Mrs. Alfred T. Davis (Edna Alice Koch Davis), n.d. Collection of Ina Hunter Dow. Salinas, California. Copy in author's collection.

Dow, Roy. "This Is the Story." The Life and Times of Roy N. Dow. Accessed September 28, 2011. royndow.tripod.com/index.html.

Early Pawnee County Collection. Santa Fe Trail Center. Fort Larned Historical Society. Fort Larned, Kansas.

Eastwood, Rich. Genealogy of the Lacey and LaLonde families. Collection of Rich Eastwood. Enseñadas, Mexico. Copy in author's collection.

Forsythe family genealogy, n.d. Copy in author's collection, furnished by Norma Coe Cinert.

Galusha, J. Letter to Andrew Hudspeth, November 10, 1916. Copy in author's collection.

Gard, Warren. Letter to Andrew Hudspeth, September 26, 1916. Copy in author's collection.

Gillett, Arizona Vera Leslie. Early history of the first Robert Leslie family memoirs, n.d. Collection of Robert Leslie. Carrizozo, New Mexico. Copy in author's collection.

Guck, Dorothy. "Coalora," n.d. Copy in author's collection.

Guevara, Arturo "Bud," and Pat Guevara. Genealogy of Guevara family. Copy in author's collection.

Gumm, Benjamin. Diary of Benjamin Gumm. Collection of Evelyn Lynch. Paso Robles, California. Copies in author's collection.

Haley, J. Evetts, Collection. Haley Memorial Library and History Center, Midland, Texas.

Haley, John A. Nominating Speech of William C. McDonald before Democratic State Convention, Santa Fe, October 2, 1911. Collection of Aileen Haley Lindamood. Carrizozo, New Mexico. Copy in author's collection.

Harkey, Chris. Notes from an advertisement in an 1892 White Oaks newspaper, n.d. Collection of Chris Harkey. Carrizozo, New Mexico.

Hart, Robert. "Ozanne Family Summary," March 8, 1994. Tularosa, New Mexico. Copy in author's collection.

———. "The Ozanne Stage Line." Speech presented to the Carrizozo Women's Club, September 23, 1995. Transcript in author's collection.

———. "The Ozanne Stage Line and a Brief History of the Ozanne Family." Speech presented to the Lincoln County Historical Society, September 28, 2002. Transcript in author's collection.

Hayes, Charles. Genealogy of Guevara family. Collection of Charles Hayes. Morro Bay, California. Copy in author's collection.

Hill, Benjamin. Descendants of James Alexander Hill, n.d. Collection of L. A. Brimberry. Ruidoso, New Mexico. Copy in author's collection.

Hough, Emerson, Collection. Iowa State Department of History and Archives, Des Moines.

Kelley, Troy. Hudgens family genealogy, 2007–2008. Collection of Troy Kelley. Johnson City, New York. Copy in author's collection.

Kelt, John J., Sr. Genealogy Records, Family Group Sheet. Copy in author's collection, furnished by Jane Gallacher Shafer.

———. Letters to Amanda Treat, May 14, 1969, and January 1, 1970. Collection of John Mottinger. Santa Teresa, New Mexico.

———. Letter to Paul Mayer, January 30, 1971. Collection of John Mottinger. Santa Teresa, New Mexico.

Key, Ellen. Records of Green and Gilmore Families. Copy in author's collection.

Krattinger, Mary. Interview with Truman Spencer, Sr., and Truman Spencer, Jr., August 31, 1980. Transcripts in author's collection.

———. Interview with Truman Spencer, Sr., October 11, 1980. Transcripts in author's collection.

Kudner, Arthur. Letter to Andrew Hudspeth, February 27, 1931. Copy in author's collection.

Lee, John. Family Bible with pages listing births and deaths. Copies in author's collection.

Lemon, Nettie. Interview with Albert Ziegler, n.d. In notes furnished by Marilyn Ingley to Haldane, April 20, 1999. Copy in author's collection.

Leslie, Dorothy. "Tribute to John Haley," from the 1993 White Oaks Historical Association financial report, n.d. Copy in author's collection.

Leslie, Robert. Drawing depicting early-day brands of White Oaks, n.d. Collection of Robert Leslie. Carrizozo, New Mexico. Copy in author's collection.

———. Listing of occupants of the Gumm House, n.d. Collection of Robert Leslie. Carrizozo, New Mexico.

Lyon, Fern. Notes for speech presented to the White Oaks Historical Association, July 28, 1973. Collection of White Oaks Historical Association. Copy in author's collection.

Mayer, Gertrude. Letter to Edgar Turrentine, n.d. Copy in author's collection.

Mayer, Ina Wauchope. Letter to Ina Dow, March 14, 1948. Copy in author's collection.

McDonald, William C. Letter to Andrew Hudspeth, February 18, 1913. Copy in author's collection.

———. Thanksgiving Day Proclamation. November 28, 1912. William McDonald Collection. New Mexico State Records Center and Archives, Santa Fe. Copy in author's collection.

Merrill, Ruby, and Glen Merrill. Merrill family history, n.d. Collection of Loris Norris. Abilene, Texas. Copy in author's collection.

Mottinger, John. Papers about the Aguayo family, Lincoln County, New Mexico, n.d. Copies in author's collection.

Mullin, Robert N., Collection. Haley Memorial Library and History Center, Midland, Texas.

Newton, Howard. "Memories of a Boy Scout Adventure in Lincoln," n.d. Copy in author's collection.

Nolan, Frederick. E-mail to Karen Mills, August 2, 2004. Copy in author's collection.

Norris, Loris. History of Charles Madison Merrill family, July 2002. Collection of Loris Norris. Abilene, Texas. Copy in author's collection.

Owen, Garry. Collection of Ozanne memorabilia. Copies furnished to author by Robert Hart.

Ozanne Stage Line schedule, ca. 1888. E. V. Long Collection. New Mexico State Records Center and Archives, Santa Fe.

Palmer, Edith Wells. Biography of her life in the 1930s, n.d. In *American Life Histories: Manuscripts from the Federal Writers' Project, 1936–1940*. Washington, D.C., Works Progress Administration. Copy in author's collection.

Parker, Morris Brown. "The Parker Family: Ancestors, Relatives, Descendants." 1940. Lincoln County Historical Society. Copy in author's collection.

Queen, Lawrence. Letter to John Kelt, August 18, 1971. Copy in author's collection.

Reily-Branum, Barbara Jeanne. Interview with Robert Leslie, August 8, 1995. Transcription in author's collection.

———. Interview with Truman Spencer, Sr., October 11, 1980. Transcription in author's collection.

———. Interview with Truman Spencer, Sr.; Truman Spencer, Jr.; and Marion Spencer, July 24, 1981. Transcription in author's collection.

Roberts, Albert. Letter to Johnson Stearns, November 15, 1978. Copy in author's collection.

Sisson, Mrs. Lula. Letter to Cora Barber Dutton, March 12, 1945. Copy in author's collection.

Sorrells, James. Stewart Genealogy, January 25, 2005. Lincoln County Historical Society. Copy in author's collection.

Spencer, Steven. "History of the Bar W Ranch." Transcript of speech presented to the Lincoln County Historical Society, April 13, 1996. Lincoln County Historical Society. Copy in author's collection.

Spencer, Truman, Jr. "The Block and Bar W Ranches." Transcript of speech presented to the Lincoln County Historical Society, March 13, 1974. Copy in author's collection.

Stewart, Levin. Letter to Fransisca Marrnjo [sic], December 15, 1890. Copy in author's collection.

Tucumcari History Museum Collection. Tucumcari Public Library, Tucumcari, New Mexico.

Turrentine, Edgar. Genealogy research of the Mayer families. Copy in author's collection.

———. Letters to Mary Jo Jamison, August 31, 1987, and August 16, 1988. Copies in author's collection.

Wauchope, John. Letter to John McMurchy, May 1, 1886. Copy in author's collection.

Wells, Samuel, Sr. Memoir told to, and typed by, daughter Edith Wells Palmer, n.d. Copy in author's collection.

Wetzel, Charles. "Mining in White Oaks, New Mexico," n.d. Collection of Larue Lane Wetzel.

Wetzel, Larue Lane. Letter to Robert Leslie, n.d. Copy in author's collection.

Young, George Richard "Dick." Telegram to Cordell Hull, June 23, 1928. Collection of G. Richard Wheelock, Del Mar, California.

Ziegler, Albert. Account of trip to Europe in 1928 or 1929, n.d. Copy in author's collection.

Local, State, and Federal Records

Archdiocese of Santa Fe marriage records. Historic/Artistic Patrimony, Archives and Museum, Santa Fe, New Mexico.

Bonnell, Ed R. Application for membership in B. F. Larned Post No. 8, Department of Kansas Grand Army of the Republic, n.d. Santa Fe Trail Center. Fort Larned Historical Society. Fort Larned, Kansas.

Cháves County, New Mexico Articles of Incorporation No. 5722, Office of the Secretary, Territory of New Mexico, Book B, December 28, 1908. Cháves County Courthouse, Roswell, New Mexico.

Criminal Court Cases for Lincoln County. New Mexico State Records Center and Archives.

Douglas [Arizona] 1920 City Directory. Copy furnished by Rose Barton. Capitan, New Mexico.

Elbert County, Georgia, Deed Books H, K, N, W, and Five. County Clerk's Office. Elbert County Courthouse, Elberton, Georgia.

Frances J. McCourt, Final Divorce Decree No. 819 from Thomas B. McCourt, April 11, 1891. District Court of the Fifth Judicial District of the Territory of New Mexico for the County of Lincoln. New Mexico State Records Center and Archives, Santa Fe, New Mexico.

Lincoln County Clerk's Office. Lincoln County Courthouse. Carrizozo, New Mexico. Barber, Susan E., Probate File. Office of the Probate Clerk. Lincoln County Courthouse. Forsythe, John. Application for Citizenship before the Lincoln County Clerk for the Third Judicial District, Territory of New Mexico October 20, 1886. Grand Jury Indictment of William Wilson, ____ Rudebaugh [sic], and José Sánches, Territory of New Mexico at the County of Lincoln, signed by Simon Newcomb, District Attorney, Third Judicial District, August 8, 1881. Justice Docket, Lincoln County, New Mexico, Precinct No. 8, November 1883 to August 1885. Justice of the Peace Court Records for White Oaks, 1882. Criminal Action No. 2, *The Territory of New Mexico, W. C. McDonald, Plaintiff, vs. W. H. Hudgens, Defendant.* Justice of the Peace Court Records for White Oaks, 1883. Criminal Action No. 2, *The Territory of New Mexico, W. C. McDonald, Plaintiff, vs. W. H. Hudgens, Defendant.* Justice of the Peace Court Records for White Oaks, 1884. Criminal Actions No. 23; No. 41; and No. 42, *The Territory of New Mexico, Plaintiff, vs. W. H. Hudgens, Defendant.* Justice of the Peace Court Records for White Oaks, 1885. Criminal Actions No. 42, *The Territory of New Mexico, Plaintiff, vs. John N. Hudgens, Defendant*; and No. 59, *The Territory of New Mexico, Plaintiff, vs. James A. Alcock, et al., Defendants.* Lincoln County Deed Book C, 175–176. Transaction dated February 15, 1881. Payment by Territory of New Mexico, to Dr. M. G. Paden, December 20, 1886. Filed by Clerk Jones Taliaferro, January 7, 1887. Collection of Historical Records Clerk. Lincoln County Courthouse. Request to Jesús Lueras from the Constable of Precinct No. 1, filed April 29, 1881, by Ben H. Ellis, Probate Clerk. Office of the Probate Clerk. Lincoln County Courthouse. Robson and Young, Application for License filed with Lincoln County Clerk, April 7, 1882. Lincoln County Courthouse. Slack, John Burchem, Probate File. Office of the Probate Clerk. Lincoln County Courthouse. Taliaferro, Jones. Letter to Ed R. Bonnell, June 15, 1888. Office of the Probate Clerk. Lincoln County Courthouse.

Lincoln County Historical Society Recording TR771107 B. Interview of Urbano Carrillo and Rosa Montaño Carrillo, November 7, 1977. Lincoln County Historical Society, Lincoln, New Mexico.

New Mexico Quinquennial Census for 1885. New Mexico State Records Center and Archives, Santa Fe, New Mexico.

New York passenger list for the ship *Aurania*, 1888. Copy in author's collection.

Record of Deaths at White Oaks and Vicinity, 1880–1905. By J. B. Slack and (later) M. H. Koch. Collection of Robert Leslie. Carrizozo, New Mexico. Copy in author's collection.

Socorro County, Second Judicial District Court of the Territory of New Mexico. New Mexico State Records Center and Archives, Santa Fe, New Mexico.

    Ziegler, Albert. Certificate of Alien's Intention. November 18, 1886.

    Ziegler, Albert. Certification by County Clerk of original declaration of intention to become a citizen, April 26, 1889.

Timber Culture Act of 1873. U.S. Statutes at Large 17 (March 3, 1873): 605–606.

U.S. Bureau of the Census. *United States Census of 1840*. Washington County, Mississippi.

———. *United States Census of 1850*. Orleans Parish, Louisiana, Second Ward, Second Municipality. Terrebonne Parish, Louisiana, Bayou Caillou, Fam. No. 514.

———. *United States Census of 1860*. Carroll Parish, Louisiana, Fam. No. 568. Orleans Parish, Louisiana, Second Ward, First District.

———. *United States Census of 1870*. Lincoln County, New Mexico. Orleans Parish, Louisiana, Second Ward.

———. *United States Census of 1880*. Lincoln, Lincoln County, New Mexico. Christian County, Missouri. White Oaks, Lincoln County, New Mexico.

———. *United States Census of 1890*. Lincoln County, New Mexico. *Special Schedule*. Enumeration Dist. No. 36, "Surviving Soldiers, Sailors, and Marines, and Widows."

———. *United States Census of 1900*. Lincoln County, New Mexico. Teller County, Colorado. White Oaks, Lincoln County, New Mexico.

———. *United States Census of 1910*. Lincoln County, New Mexico. Schenectady, New York. White Oaks, Lincoln County, New Mexico.

Books, Articles, and Published Interviews

Adams, Clarence S., and Joan N. Adams. *In the Shadow of the Malpais*. Roswell: Old-Time Publications, 1983.

Adler, Rana, and Hy Adler. "How a Plucky Woman Foiled a Governor." *Albuquerque Journal Magazine* (February 10, 1981): 10–12.

American Automobile Association. *AAA TourBook: Arizona and New Mexico*. Heathrow, Fla.: AAA Publishing, 2003.

———. *California, Nevada TourBook*. Heathrow, Fla.: American Automobile Association, 1999.

———. *Western Canada and Alaska TourBook*. Heathrow, Fla.: American Automobile Association, 2005.

Anderson, George B. *History of New Mexico: Its Resources and People*. New York: Pacific States Publishing, 1907.

Avant, Bundy (as told to Arthur Clements). "The Bundy Avant Story." *True West* (May–June 1978): 13–14.

Ball, Eve. "David Jackson." *New Mexico Magazine* (July 1978): 34–35.

Bancroft, Hubert Howe. *History of Arizona and New Mexico, 1530–1888*. Albuquerque: Horn & Wallace, 1962.

Bartholomew, Ed. *The Biographical Album of Western Gunfighters*. Houston: Frontier Press of Texas, 1958.

Beckwith, H. W. "Middleport and Belmont Townships." In *History of Iroquois County*. Chicago: H. H. Hill, 1880.

Birdsong, Ruth. "E. W. Parker House." In *1991 Lincoln County Pony Express Calendar*. White Oaks, N.Mex.: Self-published, 1991.

———. "Gumm House." In *1991 Lincoln County Pony Express Calendar*. White Oaks, N.Mex.: Self-published, 1991.

———. "Jane Malcolm Gallacher." In *1998 Lincoln County Calendar*. White Oaks, N.Mex.: Self-published, 1998.

———. "John Burchem Slack." In *1992 Lincoln County Pony Express Calendar*. White Oaks, N.Mex.: Self-published, 1992.

———. "Leslie House." In *1992 Lincoln County Pony Express Calendar*. White Oaks, N.Mex.: Self-published, 1992.

———. "Mrs. Jane Malcolm Gallacher, Lincoln County Pioneer." In *1992 Lincoln County Pony Express Calendar*. White Oaks, N.Mex.: Self-published, 1992.

———. *1991 Lincoln County Pony Express Calendar*. White Oaks, N.Mex.: Self-published, 1991.

———. *1992 Lincoln County Pony Express Calendar*. White Oaks, N.Mex.: Self-published, 1992.

———. *1998 Lincoln County Calendar*. White Oaks, N.Mex.: Self-published, 1998.

———. "The Old Prospector Who Ate a Water Tumbler, or the Old Prospector's Last Days." In *1993 Lincoln County Pony Express Calendar*. White Oaks, N.Mex.: Self-published, 1993.

———. "Schoolhouse Tales." In *1996 Lincoln County Calendar*. White Oaks, N.Mex.: Self-published, 1996.

———. *Tracks North: A Story of Charles B. Eddy's Railroad from El Paso North*. White Oaks: Birdsong's Press, 1998.

———. "White Oaks School House." In *1991 Lincoln County Pony Express Calendar*. White Oaks, N.Mex.: Self-published, 1991.

Blakestad, Alice, comp. *A Gravesite Directory of Cemeteries in Lincoln County*. Lincoln County Historical Society Publications 8. Hondo, N.Mex.: Lincoln County Historical Society, 2001.

Bond, Marshall, Jr. *Gold Hunter: The Adventures of Marshall Bond*. Albuquerque: University of New Mexico Press, 1969.

Bonney, Cecil. *Looking over My Shoulder: Seventy-Five Years in the Pecos Valley*. Roswell: Hall-Poorbaugh Press, 1971.

Booker, Susan. "Did They Really Sing Opera in the Opera Houses? Public Entertainment in Oklahoma and Indian Territories, 1895–1907." *Chronicles of Oklahoma* 81 (Summer 2003): 132–53.

Branton, Katherine C., and Alice C. Wade, eds. *Early Mississippi Records: Boliver County*. Vol. I. 1836–1861, 1988.

Brown, George S. Interview by Edith Crawford. In *American Life Histories: Manuscripts from the Federal Writers' Project, 1936–1940*. Washington, D.C.: Works Progress Administration, August 2, 1938.

Bryan, Howard. "The Coming of the Railroad." In *Railroads and Railroad Towns in New Mexico*. Compiled by William Clark. Edited by Ree Sheck. Albuquerque: *New Mexico Magazine*, 1989.

———. *True Tales of the American Southwest*. Santa Fe: Clear Light, 1998.

Buckley, Eleanor. "The Aguayo Expedition in Texas and Louisiana, 1719–1722," *Quarterly of the Texas State Historical Association* 15, no. 1 (July 1911): 1–65.

Castel, Albert E. *William Clarke Quantrill: His Life and Times*. Norman: University of Oklahoma Press, 1999.

Chamberlain, Kathleen P. "Billy the Kid, Susan McSween, Thomas Catron, and the Modernization of New Mexico, 1865–1912." In *New Mexican Lives: Profiles and Historical Stories*, edited by Richard W. Etulain, 193–202. Albuquerque: University of New Mexico Press, 2002.

———. "In the Shadow of Billy the Kid: Susan McSween and the Lincoln County War." *Montana* 55 (Winter 2005): 36–53.

Chávez, Fray Angélico. *Origins of New Mexico Families*. Santa Fe: Historical Society of New Mexico, 1954.

Choate, Edward Burleson. "History of the Capitan Community, 1890–1950." Master's thesis, Eastern New Mexico University, 1954.

Clifford, Frank [Big Foot Wallace, pseud.]. "The Winter at White Oaks 1880–1881." In "Deep Trails in the Old West." Unpublished journal of Frank "Big Foot Wallace" Clifford. Self-published pamphlet edited by Ruth Birdsong. White Oaks: Ruth Birdsong, January 1991.

Coe, Wilbur. *Ranch on the Ruidoso: The Story of a Pioneer Family in New Mexico, 1871–1968*. New York: Knopf, 1968.

Crutchfield, James A. *It Happened in New Mexico*. Helena, Mont.: Falcon Press, 1995.

Curry, George. *George Curry, 1861–1947: An Autobiography*. Edited by H. B. Hening. Albuquerque: University of New Mexico Press, 1958.

"Dr. Richard Taliaferro." In *Portrait and Biographical Record of Iroquois County, Illinois*. Chicago: Lake City Publishing, 1893.

Dunham, Dick, and Vivian Dunham. *Flaming Gorge Country*. Denver: Eastwood Printing and Publishing, 1947.

Earle, James H., ed. *The Capture of Billy the Kid*. College Station, Tex.: Creative Publishing, 1988.

Eidenbach, Peter L., and Robert L. Hart. *A Number of Things: Baldy Russell, Estey City, and the Ozanne Stage*. Tularosa, N.Mex.: Prepared for White Sands Missile Range by Human Systems Research, 1997.

Erisman, Fred, and Richard W. Etulain, eds. *Fifty Western Writers: A Bio-Bibliographical Sourcebook*. Westport, Conn.: Greenwood Press, 1982.

"E. W. Parker." In *Illustrated History of New Mexico*. Chicago: Lewis Publishing, 1895.

Flanagan, Mike. *The Old West: Day by Day*. New York: Facts on File, 1995.

Fleming, Elvis E. *Captain Joseph C. Lea*. Las Cruces: Yucca Tree Press in cooperation with Historical Society for Southeast New Mexico, 2002.

Fleming, Elvis E., and Ernestine Chesser Williams. *Treasures of History III: Southeast New Mexico People, Places, and Events*. Roswell: Historical Society for Southeast New Mexico, 1995.

Fulton, Maurice G. *History of the Lincoln County War*. Tucson: University of Arizona Press, 1997.

Garcez, Antonio R. *Adobe Angels: The Ghosts of Las Cruces and Southern New Mexico*. Santa Fe: Red Rabbit Press, 1996.

Garcia Carraffa, Alberto, and Alturo Garcia Carraffa. *Diccionaria Heráldico y Genealógico de Appellidos Españoles y Americanos*. Salamanca, Spain: 1931. http://perso.wanadoo.fr/rancho.pancho/Ortiz.htm.

"George B. Barber." In *Illustrated History of New Mexico*. Chicago: Lewis Publishing, 1895.

"George Worth Prichard." In *Illustrated History of New Mexico*, edited by Benjamin M. Read. Santa Fe: New Mexican Printing Company, 1913.

Gillett, James B. *Six Years with the Texas Rangers, 1875–1881*. Lincoln: University of Nebraska, 1976.

Glenn, W. S. "The Hunters' War." In "The Recollections of W. S. Glenn, Buffalo Hunter." *Panhandle-Plains Historical Review* 22 (1949): 15–64.

Greer Society. *Cousins of Thomas Alfred Bragg: A Little Family Book*. Sparks, Nev.: Greer Society, 1994.

Griswold, George B. *Mineral Deposits of Lincoln County, New Mexico*. Bulletin 67. Socorro: New Mexico Bureau of Mines and Mineral Resources, a division of New Mexico Institute of Mining and Technology, 1959.

Hackett, Charles. "The Marquis of San Miguel de Aguayo and His Recovery of Texas from the French, 1719–1723." *Southwestern Historical Quarterly* 49, no. 2 (October 1945): 193–214.

Haines, Helen. "Biographical Sketches: Urbane [sic] Ozane [sic]." In *History of New Mexico*. New York: New Mexico Historical Publishing, 1891.

Haldane, Roberta. "What Became of Susan McSween Barber's Diamonds?" *Southern New Mexico Historical Review* 6, no. 1 (January 1999): 39–41.

Haley, John A. *1913 Year Book, Lincoln County, New Mexico*. Carrizozo, N.Mex.: Self-published, 1913.

Hall, Bruce Edward. "A Walk with My Great-grandfather through the Last Foreign Country in New York City, Chinatown." *American Heritage Magazine* (April 1999): 54–72.

Hamilton, Nancy. "El Pasoan Finds Kin Helped Build N.M.," *El Paso Herald Post*, September 11, 1973.

Harris, Charles H. *A Mexican Family Empire: The Latifundio of the Sánchez Navarros, 1765–1867*. Austin: University of Texas Press, 1975.

Hough, Emerson. *Heart's Desire*. New York: Grosset & Dunlap, 1905.

———. "In the Jewel Box." In *Let Us Go Afield*. New York: D. Appleton, 1916.

———. "The West after Twenty Years." *Field and Stream* 9 (June 1904): 178–89.

Hutchinson W. H., and R. N. Mullin. *Whiskey Jim and a Kid Named Billie*. Clarendon, Tex.: Clarendon Press, 1967.

Jackson, David L. "Hoyle House and Electricity Come to White Oaks." In *The Saga of the Sierra Blanca*, by Herbert Lee Traylor and Louise Coe Runnels. Roswell: Old-Time Publications, 1986.

James, Henry. *Madam Varnish and the Golden Era*. Hicksville, N.Y.: Exposition Press, 1979.

Johnson, Carole M. "Emerson Hough's American West." *Books at Iowa* 21 (November 1974): 26–28, 33–42.

"John Y. Hewitt." In *Illustrated History of New Mexico*. Chicago: Lewis Publishing, 1895.

Jones, Fayette A., *New Mexico Mines and Minerals*. Santa Fe: New Mexican Printing, 1904.

Kalloch, Eunice, and Ruth K. Hall. *The First Ladies of New Mexico*. Santa Fe: Lightning Tree Press, 1982.

Keleher, William A. *The Fabulous Frontier: Twelve New Mexico Items*. Albuquerque: University of New Mexico Press, 1982.

———. *Memoirs: 1892–1969: A New Mexico Item*. Santa Fe: Rydal Press, 1969.

———. *Violence in Lincoln County, 1869–1881: The New Mexico Feud That Became a Frontier War*. Albuquerque: University of New Mexico Press, 1957.

Klasner, Lily. *My Girlhood Among Outlaws*. Edited by Eve Ball. Tucson: University of Arizona Press, 1972.

Lavash, Donald R. *Wilson & the Kid*. College Station, Tex.: Creative Publishing, 1990.

Layne, Floyd B. *Layne–Lain–Lane Genealogy*. Los Angeles, Floyd B. Lane, 1962.

Lea, Frank H. Interview by Gertrude Lea Dills. In *American Life Histories: Manuscripts from the Federal Writers' Project, 1936–1940*. Washington, D.C.: Works Progress Administration, February 26, 1937.

Leslie, Elisha. Interview by Edith Crawford. In *American Life Histories: Manuscripts from the Federal Writers' Project, 1936–1940*. Washington, D.C.: Works Progress Administration, October 17, 1938.

"Levin W. Stewart." In *Illustrated History of New Mexico*. Chicago: Lewis Publishing, 1895.

Lincoln County Historical Society. "From Mr. G. Gauss." In *Voices from the Past*. Lincoln County Historical Society Publications 7. Hondo, N.Mex.: Lincoln County Historical Society, 2000.

———. "Jerry Hockradle." In *Voices from the Past*. Lincoln County Historical Society Publications 7. Hondo, N.Mex.: Lincoln County Historical Society, 2000.

———. "W. J. Woodland." In *Ramblin' Around Lincoln County*. Lincoln County Historical Society Publications 6. Hondo, N.Mex.: Lincoln County Historical Society, 1999.

Looney, Ralph. *Haunted Highways: The Ghost Towns of New Mexico*. New York: Hastings House, 1968.

Lyon, Fern. "The Reluctant Diamond Swindler." *New Mexico Magazine* (October 1975): 12–13, 38.

Marrin, Albert. *The Spanish-American War*. New York: Atheneum, 1992.

Mayer, Charles D. Interview by Edith Crawford. In *American Life Histories: Manuscripts from the Federal Writers' Project, 1936–1940*. Washington, D.C.: Works Progress Administration, July 19, 1938.

Mayer, Ina W. Interview by Edith Crawford. In *American Life Histories: Manuscripts from the Federal Writers' Project, 1936–1940*. Washington, D.C.: Works Progress Administration, 1938.

McCourt, Mary ("Butte Rat"). Ghost Towns & Characters. Butte, Mont.: Self-published, 1993.

McStay, Gary. "Beyond a Glimpse." *El Palacio Magazine* (Winter 2005): 64.

"Melvin G. Paden, M.D." In *Illustrated History of New Mexico*. Chicago: Lewis Publishing, 1895.

Metz, Leon Claire. *The Encyclopedia of Lawmen, Outlaws, and Gunfighters*. New York: Checkmark Books, 2003.

———. *Pat Garrett: The Story of a Western Lawman*. Norman: University of Oklahoma Press, 1974.

Miller, Darlis A. *The California Column in New Mexico*. Albuquerque: University of New Mexico Press, 1982.

———. "Carleton's California Column: A Chapter in New Mexico's Mining History." *New Mexico Historical Review* 53, no. 1 (January 1978): 5–38.

———. "The Women of Lincoln County, 1860–1900." In *New Mexico Women: Intercultural Perspectives*, edited by Joan M. Jensen and Darlis A. Miller, 169–200. Albuquerque: University of New Mexico Press, 1986.

Miller, Michael. "Colorful Governors." *New Mexico Magazine* (January 1987): 24–26.

National Park Service. *Andersonville National Historic Site, Georgia: Official Map and Guide*. Washington, D.C.: Government Printing Office, 2002.

Nolan, Frederick. *Bad Blood: The Life and Times of the Horrell Brothers*. Stillwater, Okla.: Barbed Wire Press, 1994.

———. *The Lincoln County War: A Documentary History*. Norman: University of Oklahoma Press, 1992.

———, ed. *Pat F. Garrett's "The Authentic Life of Billy, the Kid."* Norman: University of Oklahoma Press, 2000.

———. *The West of Billy the Kid*. Norman: University of Oklahoma Press, 1998.

Otero, Miguel Antonio. *My Life on the Frontier, 1864–1882*. New York: Press of the Pioneers, 1935.

———. *My Nine Years as Governor of the Territory of New Mexico 1897–1906*. Albuquerque: University of New Mexico Press, 1940.

———. *The Real Billy the Kid*. Santa Fe: Sunstone Press, 2007.

Ozanne, Olive R. "A Romance." *Home and Farm*. Reprinted in Eidenbach and Hart, *A Number of Things*, 98.

Parker, Morris B. *Mules, Mines and Me in Mexico, 1895–1932*. Tucson: University of Arizona Press, 1979.

———. *White Oaks: Life in a New Mexico Gold Camp, 1880–1900*. Tucson: University of Arizona Press, 1971.

Phillips, Charles, and Alan Axelrod, eds. *Encyclopedia of the American West*. New York: Macmillan Reference, 1996.

Poe, Sophie. *Buckboard Days*. Albuquerque: University of New Mexico Press, 1981.

Prichard, Paul Preston. *The Prichard Family: History and Genealogy of the Descendants, Antecedents and Collateral Relatives of Harmon and Nancy Purcell Prichard of Mason County, Kentucky*. El Paso: Self-published, 1976.

Queen, May Lee. Interview by Edith Crawford. In *American Life Histories: Manuscripts from the Federal Writers' Project, 1936–1940*. Washington, D.C.: Works Progress Administration, June 13, 1938.

Rasch, Philip J. "Alias 'Whiskey Jim.'" In *Warriors of Lincoln County*. Outlaw-Lawmen Research Series 3. Laramie, Wyo.: National Association for Outlaw and Lawman History, 1998.

———. "Amende Honorable—The Life and Death of Billy Wilson." *West Texas Historical Association Year Book* 34 (October 1958): 97–111.

———. "Billy Wilson." In *Trailing Billy the Kid*. Outlaw-Lawmen Research Series 1. Laramie, Wyo.: National Association for Outlaw and Lawman History, 1995.

———. "The Governor Meets the Kid." In *Gunsmoke in Lincoln County*. Outlaw-Lawmen Research Series 2. Laramie, Wyo.: National Association for Outlaw and Lawman History, 1997.

———. *Gunfire in Lincoln County*. In *Gunsmoke in Lincoln County*. Outlaw-Lawmen Research Series 2. Laramie, Wyo.: National Association for Outlaw and Lawman History, 1997.

———. "He Rode with the Kid: The Life of Tom Pickett." English Westerners Tenth Anniversary Publication. London, 1964.

———. "The Men at Fort Stanton." In *Gunsmoke in Lincoln County*. Outlaw-Lawmen Research Series 2. Laramie,

Wyo.: National Association for Outlaw and Lawman History, 1997.

———. "The Murder of Huston I. Chapman." In *Gunsmoke in Lincoln County*. Outlaw-Lawmen Research Series 2. Laramie, Wyo.: National Association for Outlaw and Lawman History, 1997.

———. "They Fought for 'the House.'" In *Warriors of Lincoln County*. Outlaw-Lawmen Research Series 3. Laramie, Wyo.: National Association for Outlaw and Lawman History, 1998.

———. *Violence in Lincoln County*. Carrizozo: Village Press, n.d.

Rayner, Ted. "Ah Nue and The Big City." The Folklore Corner. *Lincoln County News and Carrizozo Outlook*, February 17, 1956.

———. *The Gold Lettered Egg and Other New Mexico Tales*. El Paso: Superior Printing, 1962.

Reily-Branum, Barbara Jeanne, and Roberta Haldane. *Corralled in Old Lincoln County, New Mexico: The Lin Branum Family of Coyote Canyon and the I Bar X*. Alamogordo, N.Mex.: Bennett Printing, 1995.

Rencher, Olive. "The Ups and Downs of Eighty Years." In Eidenbach and Hart, *A Number of Things*.

Reynolds, Bill. *Trouble in New Mexico: The Outlaws, Gunmen, Desperados, Murderers and Lawmen for Fifty Turbulent Years*. Bakersfield, Calif.: privately printed, 1994.

Sherman, James E., and Barbara H. Sherman. *Ghost Towns and Mining Camps of New Mexico*. Norman: University of Oklahoma Press, 1975.

Shinkle, James D. *Reminiscences of Roswell Pioneers*. Roswell: Hall-Poorbaugh Press, 1966.

Simms, J. Denton. *Cowboys, Indians and Pulpits*. Edited by D. Harper Simms. Santa Fe: Parker Books of the West, 1989.

Sinclair, John L. "The Great Bypass." *Westways* 70, no. 9 (September 1998): 18–19.

———. "Little Town of Heart's Desire." *New Mexico Magazine* (December 1940), 18–19, 39.

———. "Mother Julian: She Dished Out Grub and Memories at Her Boarding House." *New Mexico Magazine* (March 1990), 44–49.

Stanley, Francis. *The Chloride, New Mexico Story*. Pep, Tex., 1962.

Stearns, Johnson. *Carrizozo Story*. Carrizozo, N.Mex.: Self-published, 1987.

Stevenson, Robert Louis. *New Poems*. Whitefish, Mont.: Kessinger, 2004.

Taliaferro, James T., comp. *Taliaferro Family History, 1635–1899*. San Diego: Charles Taliaferro, 1955.

Taliaferro, Jones. *Old Hickory Mining Company*. White Oaks: Self-published, n.d. Copy in author's collection.

Tanner, Karen Holliday, and John D. Tanner. *Last of the Old-Time Outlaws: The George West Musgrave Story*. Norman: University of Oklahoma Press, 2002.

———. "The San Antonio–White Oaks Stage robberies of 1896." *Southern New Mexico Historical Review* 8, no. 1 (January 2001): 20–23.

Thorp, N. Howard. *"Along the Rio Grande": Cowboy Jack Thorp's New Mexico*. Edited by Peter White and Mary Ann White. Santa Fe: Ancient City Press, 1988.

Thrapp, Dan L. "Aguayo, Marquis de." In *Encyclopedia of Frontier Biography*. Vol. 1. Glendale, Calif.: Arthur H. Clark, 1988.

———. "Bell, James W., Lawman (c. 1842–Apr. 28, 1881)." In *Encyclopedia of Frontier Biography*. Vol. 1. Glendale, Calif.: Arthur H. Clark, 1988.

Traylor, Herbert Lee, and Louise Coe Runnels. *The Saga of the Sierra Blanca*. Roswell: Old-Time Publications, 1986.

Tucumcari Chamber of Commerce. *Tucumcari: 1910 History and Business Review*. Tucumcari, N.Mex.: Tucumcari Chamber of Commerce, 1910.

———. *Tucumcari: 1912 History and Business Review*. Tucumcari, N.Mex.: Tucumcari Chamber of Commerce, 1910.

Twitchell, Ralph Emerson. *The Leading Facts of New Mexican History*. Cedar Rapids, Iowa: Torch Press, 1912.

Westphall, Victor. *Thomas Benton Catron and His Era*. Tucson: University of Arizona Press, 1973.

Wetzel, Larue Lane. "The House Built for Love." In *The Hoyle House 1893*. White Oaks: Birdsong's Press, n.d.

Whiteman, M. Interview by Georgia B. Redfield. In *American Life Histories: Manuscripts from the Federal Writers' Project, 1936–1940*. Washington, D.C.: Works Progress Administration, November 18, 1938.

"William C. McDonald." In *Illustrated History of New Mexico*. Chicago: Lewis Publishing, 1895.

"William H. Weed." In *Illustrated History of New Mexico*. Chicago: Lewis Publishing, 1895.

Williams, O. W. *Pioneer Surveyor, Frontier Lawyer: The Personal Narrative of O. W. Williams, 1877–1902*. Edited by S. D. Myres. El Paso: Texas Western Press, 1968.

Wilson, John P. *Merchants, Guns, & Money: The Story of Lincoln County and Its Wars*. Santa Fe: Museum of New Mexico Press, 1987.

Wilson, Robert. "The Great Diamond Hoax of 1872." *Smithsonian Magazine* 35, no. 3 (June 2004): 70–79.

Woodard, Bruce A. *Diamonds in the Salt*. Boulder: Pruett Press, 1967.

Wylder, Delbert E. *Emerson Hough*. Edited by David J. Nordloh. Twayne's United States Authors Series 397. Boston: Twayne, 1981.

———. *Emerson Hough*. Southwest Writers Series 19. Austin, Tex.: Steck-Vaughn, 1969.

Zeigler [*sic*], Albert. Interview by Edith Crawford. In *American Life Histories: Manuscripts from the Federal Writers' Project, 1936–1940*. Washington, D.C.: Works Progress Administration, March 1938.

Newspapers and Periodicals

Alamogordo News
Albuquerque Journal
Albuquerque Journal Magazine
Albuquerque Prime Time
Albuquerque Tribune
American Heritage Magazine
Arizona Days and Ways Magazine
Arizonan (Tucson)
Austin Frontier Times
Carrizozo News
Carrizozo Outlook (formerly *White Oaks Outlook* and *Southwestern Outlook*)
Chloride (N.Mex.) Black Range
Chronicles of Oklahoma
Denver Rocky Mountain News
Dodge City (Kans.) Times
Dublin (Tex.) Progress
El Palacio Magazine
El Paso Daily News
El Paso Daily Times
El Paso Herald Post
El Paso Times
Field and Stream
Frontier Times (Austin)
Grant County (N.Mex.) Herald
Iroquois County (Ill.) Times
Larned (Kans.) Eagle-Optic
Larned (Kans.) Press
Las Vegas (N.Mex.) Daily Optic
Las Vegas (N.Mex.) Gazette
Lincoln County Leader
Lincoln County News
Lincoln County News and Carrizozo Outlook
Lincoln County Times
Lincoln Independent
Lincoln Republican
Mesilla (N.Mex.) Southwestern Old Times
Montana, The Magazine of Western History
New Mexico Historical Review
New Mexico Interpreter
New Mexico Magazine
New York Times
Oakland (Calif.) Tribune
Panhandle-Plains Historical Review
Pecos Valley (N.Mex.) Register
Prescott Daily Arizona Miner
Redondo Beach (Calif.) Daily Breeze
Republican Review (Albuquerque)
Richmond (Va.) Whig
Roswell Daily Record
Ruidoso News
Santa Fe New Mexican
Santa Fe Register
Santa Fe Reporter
Silver City (N.Mex.) Enterprise
Smithsonian
Southern New Mexico Historical Review
Southwesterner
Southwestern Outlook (formerly *White Oaks Outlook*, later *Carrizozo Outlook*)
Tombstone (Ariz.) Daily Nugget
True West
Tucson Arizona Citizen
Tularosa Basin News
USA Today
Watseka (Ill.) Republican
White Oaks Eagle
White Oaks Golden Era
White Oaks Outlook (later *Southwestern Outlook*, then *Carrizozo Outlook*)
Willard (N.Mex.) Record

INDEX

Page numbers for photographs and figures are identified by the suffix "p" and "fig.," respectively.

Abraham Lincoln gold claim. *See* Old Abe Mine

African Americans: discrimination and racism, 96, 196–97; lynchings, 219; settlement and land ownership, 33–34. *See also* Jackson, David L. "Happy Jack"

Aguayo, Alejandro ("Alex"), 14, 15, 17p

Aguayo, Amanda, 14, 16p

Aguayo, Araminta, 14

Aguayo, Aristoteles ("Harry"): birth, 14; marriage, 19; recollections of Lincoln County, 17–19; relocation to Douglas, Ariz., 19–20; return to Lincoln County, 20

Aguayo, Edith Sheppard, 19–20

Aguayo, Ester, 14

Aguayo, Frances, 17p

Aguayo, Humphrey, 17p

Aguayo, José María and Francisca Hill: arrival in Tularosa, 13–14; birth and education, 12–13; encounters with Indians, 14–15; later years and deaths, 17; marriage and children, 13–14; relocation to Lincoln, 15–16; relocation to Texas Park, 16

Aguayo, Lucila, 14, 16p, 17p

Aguayo, Manuel María, 12

Aguayo, Sara, 14, 16p, 17p

Aguayo Marquisate: creation of the Mexican dynasty, 11–12; financial decline, 12; Spanish origins of, 11

Ah Nue, John (aka "John Chinaman"): arrival in U.S., 22; arrival in White Oaks, 22–23; death of, 25–26; later life and legends, 24–25; laundry and restaurant, 23–24

Alamogordo, N.Mex., 10, 86, 93, 231, 281, 282

Alaskan Klondike Gold Rush, 295–96

Albuquerque, N.Mex.: Aguayo family relocation, 13, 20; Fairview Park Cemetery, 233; migration to, 298; recollections of, 22; relocations to, 94, 107; Statehood Convention of 1902, 232

Alcock, James A.: arrival in Lincoln County, 27; conflict with sheep raising, 27–28; departure from Lincoln County, 29–30; ownership of Carrizozo Ranch, 27, 29, 180–81, 200; pistol duel with Will Hudgens, 29, 120–21; Salado coalfields, 27

Alcock, Lulu, 30n1

Allen, James M., 107, 230

Alto Light & Power Company, 131, 238–39

American Field (magazine), 101

American Indians: Black Hawk War, 74; filming of *The Covered Wagon*, 105; Navajo Revolt of 1913, 125–26; raids in Mexico, 11–12; raids in N.Mex., 204, 218, 250, 251; raids in Texas, 162–63, 275; relocation, 283; Taos Pueblo, 188p; threat to travelers, 14–15, 98; trading with, 304; white relationships with, 14, 34. *See also* Apache Indians; Comanche Indians; Kiowa Indians

Apache Indians, 14–15, 34, 98, 204, 218, 250–51, 275, 283, 304

Arizona: gold mining, 122; statehood, 232

Armijo, Lola Chávez de, 189–90

Arnold, Philip: diamond mine hoax, 251–53; later years and death, 254; Mexican War service, 249; mining activities, 249–51

"A Romance" (Rencher), 205

Arrastra mining, 4

Askew, Joseph, 118, 122

Atchison, Topeka, and Santa Fe Railroad, 88, 199, 204, 230

Audubon Society, 106n16

Authentic Life of Billy the Kid, The (Fulton), 31

"Baby Doe" Tabor, Silver Queen of Colorado, 183

Baca, B. J., 179

Baca, E. C., 186

Baca, Saturnino ("father of Lincoln County"), 62, 179

Baker, Frank, 61

Ball, Eve, 34, 132

Baptist Church, 31–32, 184, 277, 296

317

Bar W brand, 180fig.
Bar W Ranch: control of Block Ranch, 181–82; McDonald as manager, 180–81; ranch home, 185, 188; Stirling Spencer as manager, 191, 306; Truman Spencer as manager, 190. *See also* Carrizo Ranch (Carrizozo Cattle Ranche Company)
Barber, Eugenia Roberts, 36
Barber, George B., 31–35, 36, 278
Barber, Ralph, 43n40
Barber, Susan McSween: death of, 41–42; death of Alexander McSween, 15–16, 33; development of Three Rivers Ranch, 34; divorce from George Barber, 35, 43n40; early life, 31–32; life and legend in White Oaks, 35–41; marriage to Alexander McSween, 32–33; marriage to George Barber, 33–34
"Bard of Tularosa" (Eugene Manlove Rhodes), 9, 100
Bates, Sebrian, 33–34
Battle of Adobe Walls (1874), 56
Battle of Aibar (Baldillón), 65n1
Baxter, John, 4–5, 283, 288, 290n3
Baxter Gold Mining Company, Patterson vs., 101
Baxter Gulch: Homestake strike, 288, 289; mining activities, 283, 288; Old Abe strike, 107–10; recollections of gold discovery, 117, 246; working placer deposits, 224
Baxter Mountain: early prospecting, 3–4; gold claims, 5, 283; jumping claims, 104; placer diggings, 283
Bell, James W.: arrival in N.Mex., 44–45; as lawman, 44–45, 48n1, 203; capture of Billy the Kid, 45–46; death and burial, 47, 48n29; killed by Billy the Kid, 44, 46–47
Bell, Joseph William, 44
Ben-Hur (Wallace), 9
Berger, William M., 232
Billy the Kid (William H. Bonney): escape from jail, 9, 17–18, 27, 44, 265; fake arrest and pardon, 61, 117, 246; gravesite, 104, 163; killed by Pat Garrett, 104, 116, 258; recollections of encounters with, 22, 31, 39–40, 208, 217, 262, 293p; Regulators gang, 245; shooting of James Bell, 44–47; shootout at Coyote Springs, 98, 119, 247; shootout at Greathouse-Kuch Ranch, 56–59, 119; singing and guitar-playing, 15; at White Oaks, 6, 7
Birth of a Nation (Griffith), 105
Bisbee, Ariz., 153
Black Hawk War of 1833, 74

Black Hills (S.Dak.) gold rush, 88
Blanchard, William, 120, 165–66, 272
Blizzards and snowstorms, 18, 152, 153, 244, 285
Block Ranch, 18, 65p, 125, 164–66, 167n7. *See also* El Capitán Land and Cattle Company (Block Ranch)
Bloom Cattle Company, 18
Bolton family, 33, 288–89
Bonnell, Bertram Jay ("Bert"), 49, 53
Bonnell, Daniel Nelson, 49, 53
Bonnell, Edwin R.: arrival in Kansas, 49; arrival in White Oaks, 50; Civil War service, 49; death of, 52–53; first U.S. tree claim, 49–50; life in White Oaks, 50–52; marriage, 49
Bonnell, Levi Irwin ("Erva"), 49, 51
Bonnell, Viola Albright, 50–53
Bonnell Opera House, 51–52
Bonney, C. D., 19
Bonney, William H. *See* Billy the Kid (William H. Bonney)
Boston Boy Mine, 114p
Brady, William (sheriff), 61
Brands: Bar W brand, 180fig.; Coyote and Red Lake Cattle Company, 120fig.; El Capitán brand, 181fig.; J+H brand, 69; Susan Barber, 34. *See also* Cattle ranching
Branum, Barbara, 38
Branum, Lin, 34
Brent, Jim, 179, 209, 263
Brewer, Dick, 33, 245
Brimberry, J. S., 11
Brimberry, Lorraine Aguayo, 21n3
Brown, John A., 4p, 220
Buffalo/buffalo hunting, 57, 75, 270, 275
Buford, Charles, 265
Bull, Charles (deputy sheriff), 52, 78, 104, 121
Burns, E. W. ("Curly"), 213
Burns, Walter Noble, 31, 40, 41
Burros: acknowledgement of, v; *arrastra* mining, 4; beast of burden, 18, 88–89; transportation, 4p, 202, 224, 237, 285, 286
Bursum, Holm, 72, 186

C. D. Mayer, General Merchandise, 174p, 175p, 176p
Cadenhead, Rufus (Mrs.), 133
Caffrey, William, 246
Calhoun, Jim, 246–47

California: Chinese immigration, 22; emigrant cross-country treks, 83n1, 216, 250; emigrant crossing by Benjamin Gumm, 74–76; Hangtown and gold prospecting, 75–77, 84n2; Jóse Aguayo travels to, 13, 21n3; lure of gold, 98, 208, 241, 249–50, 269–70, 283, 288; mail routes, 216; relocation from Lincoln County to, 40, 64, 114, 127, 141, 142, 143, 160, 175, 185, 206, 225, 228, 239–40, 243–44, 277, 296; travels via Panama, 77–78, 208, 269–70

Campbell, H. C., 230

Campbell, William, 245–46

Canada/Canadian immigration, 6, 62

Canning, John, 161p

Capitán News, 86

Carillo, Urbano and Rosa Montaño, 36

Carlyle, James ("Bermuda Carlisle"): arrival in White Oaks, 56–57; as buffalo hunter, 56; gravesite, 59; shootout at Greathouse-Kuch Ranch, 7, 46, 57–59, 98, 119, 147, 247

Carrizo "flats," 10, 93

Carrizo Mountain: Alto Light & Power Company, 131, 238–39; Cree land boundary, 124; early prospecting, 3–4

Carrizo Ranch (Carrizozo Cattle Ranche Company), 27, 29–30 & n1. *See also* Bar W Ranch

Carrizo Spring, 2, 181, 185

Carrizozo, N.Mex.: arrival of the railroad, 10, 93; churches, 133; Evergreen Cemetery, 20, 64, 175, 247; Exchange Bank, 92; Headlight Saloon, 153; Knights of Pythias, 86; Paden's Drug Store, 208, 211–13; railroad development to, 10

Carrizozo News, 86

Carroll, Henry, 34

Casey, Robert, 62

Catron, Thomas Benton, 27, 29, 231–32

Cattle Queen of New Mexico. *See* Barber, Susan McSween

Cattle ranching: Bar W Ranch, 180–82, 185, 188, 190–91, 306; Block Ranch, 18, 65p, 125, 164–66, 167n7; Bloom Cattle Company, 18; Carrizo Ranch, 27, 29–30 & n1; El Capitán Land and Cattle Company, 9, 38, 147, 181–82; free range grazing, 3, 181; J+H Ranch, 69–72; Lea Cattle Company, 147; mavericking unbranded cattle, 34; "nesters," 165–66; politics, 191n16; railroad shipments, 35; roundups, 165–66; rustlers, 61, 118, 122, 165; sheep-killing cases, 27–29; Three Rivers Land and Cattle Company, 28, 33–37, 38, 43n44; trail drives and winter blizzards, 18; unlawful branding, 164; water rights, 181. *See also* White Oaks, ranches and area buildings

Cedarvale Cemetery, 305p, 306p; Ah Nue, 26; Brent Paden, 213; David and Mary Jackson, 133; dedication as State memorial, 73n11; Edward and May Queen, 160, 243; Edwin Bonnell, 52; Ellyn Queen Whitwell Cady, 244; Felix and Carmalita Tórres, 62; Fred and Maria ("Hattie") Mayer, 169–70; Gallacher family, 72; historical preservation, 167; Jack Winters, 289; James Bell, 47, 48n29; John and Amelia Hewitt, 93, 94, 131; John Slack, 255; Johnnie Weed, 272; Jones Taliaferro, 266; Martha Gumm, 80; Melvin and Brent Paden, 213; Melvin G. and Belle Paden, 213; on National Register of Historic Places, 306; Paul and Ula Mayer, 172; Raymond and Nettie Lemon, 161; Robert and Elizabeth Leslie, 164; Robert Ward Leslie, 167; Sam and Martha Frances Wells, 277; Susan McSween Barber, 131; Urbain Ozanne, 204; Washington Stewart, 257; William and Frances McDonald, 131, 190–91; William Lane, 142; Young children exhumation, 296

Chapman, Huston, 245

Chávez, José Chávez y, 17

Chavéz y Sánchez, Dionicio ("Dio"), 28

Chester, E. S., 220

China/Chinese immigration, 6, 298. *See also* Ah Nue, John (aka "John Chinaman")

Chinese Exclusion Act of 1882, 22

Chisum, John H., 33, 34, 245, 291

Chisum, Pitser M., 34

Chloride, N.Mex., Koch photography business, 134–36

Church, Amelia Bolton, 34

Church of England, 150

Churches, Lincoln County: Baptist Church, 184, 277; Congregational Church, 41, 69, 79, 169, 179, 185, 221–22, 225, 277; Episcopal Church, 102, 191, 259; in Carrizozo, 133; in Chloride, 134; in Tucumcari, 136; Methodist Church, 159, 160, 236, 240; Presbyterian Church, 32, 136; St. Frances de Sales, 142; White Oaks town hall used as, 7

Cinert, Norma Coe, 278n1

Civil rights issues: Chinese citizenship, 22; sexual discrimination, 189–90; voting rights for women, 232. *See also* Racism/racial discrimination

Civil War: Kansas-Missouri border conflict, 144, 145–46; migration to White Oaks following, 7; post-war Reconstruction, 196–97, 291; prison camps, 139–40;

Civil War *(continued)*
   recollections of, 179–80; service by White Oaks residents, 49, 88, 116, 129, 168, 215, 235, 238, 256, 288; State militia, 78
"Clarence Darrow of New Mexico." *See* Prichard, George Worth (Colonel)
Clifford, Frank ("Big Foot Wallace"), 288–89
Cloudcroft, N.Mex., 228
Coahuila, Mex., 11
Coe, Frank: children of, 53; naming of Glencoe, N.Mex., 54; recollections of Susan Barber, 42
Coe, George, 50, 277, 280
Coe, Jap N., 50, 53, 54
Coe, Ross, 53
Coe, Wilbur, 53, 54, 280, 302
Coe, William, 277
Collier, J. B. (judge), 29, 121
Collinson, Frank, 56–57
Colorado: "Baby Doe," 183; Cripple Creek gold rush, 29–30, 225; diamond mine hoax, 252–53; gold mines, 236
Colorado School of Mines, 225
Comanche Indians, 162–63, 275
Comstock Mine, 5, 230, 235
Congregational Church, 41, 69, 79, 169, 179, 185, 221–22, 225, 277
Connor, Jim, 301
Cooper, J. B., 251, 253
Cooper, Tom, 6, 54
Copper mining: Arizona and Mexico, 239; Douglas, Ariz., 259; Ely, Nev., 122; King and Queen Copper Company, 227; Old Hickory Mining Company, 265; Oscura Mountains, 142; stamping mills, 8p
Corbet, Sam, 17
Corn, Betty, 192n44
Corona, Oracio ("Rasho"). *See* Gallacher, Oracio Corona
Covered Wagon, The (Hough), 103, 105
Cox, Fred Nathan and Mary Annie Lee, 151, 156
Coyote and Red Lake Cattle Company, 119–20
Cree family of Scotland (VV Ranch), 53, 94, 124–25
Cullom, Virgil, 118, 122
Curry, George: congressman, 187; county clerk, 24fig., 203, 263; Lincoln County politics, 93, 205; sheriff, 173; territorial governor, 17, 185, 187p, 189; White Oaks businessman, 118, 124

Davis, Alfred T., 136
Davis, Edna Alice Koch, 7p, 134–36, 137n15
Davis, Merle Koch, 7p, 134–35
de Guevara, Anamaría Tórres, 60
de Guevara, Delores Cayales, 60
de Guevara, Felix and Carmelita Tórres, 62
de Guevara, Librada Barela, 61
de Guevara, María Sánches, 60–61
de Guevara, Maximiano ("Max") and Felicita Montoya, 61–62
de Guevara, Plácido, 60–61
de Guevara, Sancho, 65n1
de Guevara, Ygnacio Niño Ladrón, 60
de Guevara family, origins and arrival in White Oaks, 60, 65n1
Dempsey, John (Governor), 126
Desert Land Act of 1877, 34
Diamonds in the Salt (Woodard), 255n2
Díaz, Porfirio, 227
Doe, W. Harvey and "Lizzie" McCourt, 183
Dolan, Caroline Fritz, 288
Dolan, Jimmy, 5, 31, 61, 97, 119, 245–46, 263, 288
Douglas, Ariz., 19, 154, 155, 158–59, 239, 259–60
Douglas, James, 8
Dow, Ina, 127
Dow, Roy, 213, 306
Dunkard Brethren Church, 31–32
Dunn, Elton, 136
Durr, Emil, 35, 36
Dye, B. B., 104, 108p

Eaker, Elmer S. ("Red"), 213
Eastwood, Rich, 65n3
Échevers, Augustin de (1st Marquis of Aguayo), 11
Eddy, Charles B., 10, 92–93, 124
Eddy, John, 92–93, 124
Eddy, N.Mex., 272
Eddy County, 3, 35, 231, 263, 270
El Capitán brand, 181fig.
El Capitán Land and Cattle Company (Block Ranch), 9, 38, 147, 181–82. *See also* Block Ranch
Elkins, Stephen and Hallie, 35
El Paso and Northeastern Railroad, 10, 27, 92–93

El Paso and Southwestern Railroad, 211

England: administration of Samoa, 149–51; cattle ranching, 27, 29, 180–81; presence in Mexico, 12; U.S. immigration, 6, 49, 67

Episcopal Church, 102, 191, 259

Evans, Jesse, 61

Evergreen Cemetery (Carrizozo), 20, 64, 175, 247

Evergreen Cemetery (El Paso), 17, 228

Exchange Bank, 8, 90p, 91–92, 113

F. B. Coe Ranch, 54

Fairview Ranch, 180. *See also* Bar W Ranch

Fall, Albert Bacon, 93, 231–32

"Far from the Maddening Crowds" (Hough), 100

Farr, Jack, 163

Farr Ranch (Felix Guebara Ranch), 62, 163

Ferguson, Wally and Ann, 72

Fergusson, Harvey B., 90, 101, 107, 108p, 186–87, 289

Field and Stream (magazine), 100, 101, 103–104, 106n16

54-40 or Fight (Hough), 103

Fitzpatrick, George and Elizabeth Forsythe, 273n21, 277–78

Fleming, Elvis, 201

Floods/flooding, 131–32

Foley, John and Mary Lane, 142

Forsythe, Lena Weed, 271–72, 273n21

Forsythe, Robert, 273n21, 277–78

Fort Bascom, N.Mex., 18, 217, 229n4

Fort Scott, Kan., 9, 78, 84n4, 168, 178, 202

Fort Stanton, N.Mex., 2fig., 10, 47, 54, 116–17, 147, 198–200, 221, 271

Fort Sumner, N.Mex., 2fig., 18, 46, 48n23, 104, 163, 208, 246, 291

Fountain, Albert Jennings, 233

Fowler, Joel, 59, 118, 246–47

France/French ancestry: Bonnell family, 49; de Guevara family, 65n1; Ozanne family, 196–97; in Mexico, 11; in White Oaks, 6

Freight routes: Central New Mexico map, 2; Santa Fe Trail, 49, 250, 269; shipments to White Oaks, 10. *See also* Stagecoach lines

French, James H., 61–62

Fulton, Maurice G., 31

Gallacher, Edna Eva Gray, 71

Gallacher, Elizabeth Truman, 71

Gallacher, Harry, 67p, 69, 71–72

Gallacher, Ida Collier, 71

Gallacher, Jane Malcolm: arrival in America, 66–67; death of, 72; marriage and children, 67–68; purchase of Sheep Spring Ranch, 69; relocation to Carrizozo, 71

Gallacher, John, 67p, 69, 71

Gallacher, Oracio Corona ("Rasho"), 70, 71–72

Gallacher, William, 67–68, 73n6

Gallacher, William Wilson ("Willy"), 67p, 69, 71, 72, 303

Gallinas Mountains, 265

Galusha, J. R., 126

Garrett, Elizabeth, 189

Garrett, Pat F.: arrival in Lincoln County, 7; as sheriff of Lincoln County, 45, 59, 98; capture of Billy the Kid, 45–46, 48n23, 119; killing of Billy the Kid, 258; life in White Oaks, 50; Lincoln County politics, 263; White Oaks reunion, 104

Gauss, Gottfried Georg, 27, 46

Georgia Placer Mine, 48n12

German ancestry/immigration, 6, 27, 31, 74, 96, 134, 168–69, 236, 245

Germany, administration of Samoa, 149–51

Geronimo (Apache), 204

Gilbert, Bob M., 245

Gildea, Gus, 117, 123n7

Gillett, J. B., 44

Glencoe, N.Mex., 53, 54, 280, 302

Glenn, W. S., 57

Goff, Jacob, 75, 77–78

Gold discovery: Alaska, 279; Baxter Gulch, 246; Black Hills, S.Dak., 88; California, 74–78, 83n1, 249; Carbonateville, N.Mex., 45; Colorado, 29; finding a buried gold brick, 69; Homestake Mine, 5; Jicarilla Mountains, 45, 48n12, 283, 288; Klondike Gold Rush, 295–96; La Paz, Ariz., 250; Nogal Mountains, 45; White Oaks, 4–8, 25, 34, 78, 81, 88–89, 97, 100, 117–18, 123n13, 142, 168, 218, 283. *See also* Silver/silver mining

Gold mining: annual assessment work, 90, 107–108, 127, 206; *arrastra* mining, 4; assaying the "dump," 108; "cleanup," 77, 84n3, 160, 238; cyanide plant/process, 108p, 130, 226; glory hole, 229n21; head frame, 109p; mine block diagram, 220p; mine fires, 8, 110–11,

Gold mining *(continued)*
 221, 224; miners' disease (silicosis), 158, 160, 235, 236, 242, 246; stamping mills, 8p, 108p, 110p, 221; transport of bullion, 170–71, 202; unpatented claims, 107; vertical longitudinal sections, 108p. *See also* Mines/claims

Goodman, Herman, 298–99
Goodman, Ziegler & Company, 299–300
Grand Canyon Park, 106n16
Grant, N.Mex., 250
Grant, Ulysses S., 216
Great Depression, 160, 237, 241–43
Great Diamond Hoax, The (Harpending), 255n2
Great Diamond Hoax of 1872, 250–54, 255n9
Greathouse, Jim ("Whiskey Jim"), 57, 147
Greathouse-Kuch Ranch, 2fig., 8, 46, 50, 56–59, 119
Grey, Zane, 260
Griffith, D. W., 105
Grinnell, George Bird, 106n16
Guebara, Felis (Felix) and Carmelita, 65n3, 163
Guevara, Arturo ("Bud") and Pat, 65n3
Guillermo, Sancho, 65n1
Gumm, Benjamin Franklin: arrival in U.S., 74; arrival in White Oaks, 78–79; en route to California, 74–76; gold rush days in California, 76–78, 83n1; later life and legacy, 79–83
Gumm, Edith Lavenia ("Vena"), 78, 80, 185
Gumm, John, 7p
Gumm, Martha Goff, 78
Gumm, Wallace and Elizabeth Austin, 79, 80, 223p
Gumm House, 37, 39p, 79, 81, 160, 236, 305

Hagerman, Herbert (territorial governor), 186, 232
Haley, J. Evetts, 31
Haley, John Allen ("Judge Haley"): arrival in White Oaks, 85; as schoolteacher, 82; later life and legacy, 87; marriages, 85; political and civic activities, 86–87, 186; purchase of *White Oaks Eagle*, 85–86
Hall, John, 94
Hangings/lynchings, 18, 46, 204, 209, 219, 258. *See also* Murders and killings
Hangtown, Calif., and gold prospecting, 75–77, 84n2
Harpending, Asbury, 250–53, 255n2
Harpending & Roberts Exploration Company, 250–51
Harper, Monroe, 36, 37

Hawkins, William A., 93
Hayes, Bob (alias John West), 18–19, 173, 203
Hayes, Charles, 65n3
Heart's Desire (Hough), 9, 100, 104, 105, 230
Helen Rae Mine, 129, 228
Herman, T. W., 220
Hewitt, Amelia Rawlins, 89, 93
Hewitt, John Y. ("Judge"): arrival in White Oaks, 88–89; Civil War service, 88; later life and legacy, 93–94; life in White Oaks, 8, 24, 25, 50, 89, 90–92; mining interests, 90, 107; political and civic activities, 92; railroad development to White Oaks, 92–93; White Oaks reunion, 104
Hill, James Alexander, and Fernanda Quintero, 14
Hillsboro, N.Mex., 121
Hilton, A. H., 202
Hockradle, Jerry ("Jolly Jerry"): arrival in White Oaks, 96; Dry Gulch mine, 97; Horrell War, 96–97; Lincoln County War, 97; relocation to California, 98; selling water to travelers, 97–98
Homestake Mine: discovery and naming, 5, 123n13; division into North and South, 284, 288. *See also* North Homestake Mine; South Homestake Mine
Homestead Act of 1862, 34
Homesteading: Alfred Ozanne, 201, 206; Frederick Mayer, 168; Gallacher family, 69–70, 72; George Lee, 156; Horrell brothers, 96; Lish Leslie, 165; Robert Leslie, 48n29, 164; Samuel Wells, 276; Susan Barber, 33
Horrell War, 96–97, 98
Hough, Emerson: arrival in White Oaks, 9, 100–110; beginnings of legal career, 101; childhood and education, 100; death of, 105; launch of writing career, 89, 101–103, 230; recollections of White Oaks, 31, 104–105, 247, 306; wildlife preservation and national parks, 103–104, 106n16. *See also* titles of Hough's works
House of Murphy (Murphy-Dolan faction), 27, 31, 33, 57, 180, 245. *See also* Murphy, Lawrence G.
Hoyle, Ida, 111, 221p
Hoyle, Matthew Watson ("Watt"): arrival in White Oaks, 81; death, 114; departure from White Oaks, 82, 113; marriage, 109, 113–14, 115n21; Old Abe assay, 90, 107–108; Old Abe mine fire, 110–11
Hoyle, Sarah Bell Dingey, 113–14
Hoyle House ("Hoyle's Folly"): construction of, 81–82, 111–13; history and legends, 113; ownership today, 114
Hoyt, Henry F. (Dr.), 216–17, 229n4

Hudgens Brothers: Coyote and Red Lake Cattle Company, 119–20; later life and legends, 120–22, 179, 246; operation of "Old Brewery" Saloon, 116–17; operation of Pioneer Saloon, 5–6, 29, 117–18, 246; shootout at Greathouse-Kuch Ranch, 119

Hudgens, John Newton: arrival in Lincoln County, 116; cattle ranch, 119–20; as horse thief, 117; as lawman, 120; marriage, 116

Hudgens, John William ("Jack"), 118, 123n2

Hudgens, Margaret Taylor, 116

Hudgens, Mary Taylor, 116

Hudgens, William Harrison, Jr. ("Will"): arrival in Lincoln County, 116; as deputy sheriff, 45–46, 119, 247; friendship with Billy the Kid, 117; killing of I. T. McCray, 118–19; marriage, 116; pistol duel with James Alcock, 29

Hudspeth, Andrew H. ("Andy"): arrival in Lincoln County, 124; as U.S. Marshal, 125–26; bringing railroad to White Oaks, 93; election to N.Mex. Supreme Court, 126–27; employment on VV Ranch, 124–25; entering into White Oaks politics, 125; later life and legacy, 127; purchase of Hoyle House, 82

Hummer, Susanna Ellen (aka Sue Ellen Homer). *See* Barber, Susan McSween

Hunter, Bertha Mayer, 40, 43n56, 136

Illustrated History of New Mexico (1895), 34

Immigration: Chinese exclusion, 22; lack of restrictions, 104; N.Mex. regulation, 86. *See also* Racism/racial discrimination

Ireland/Irish ancestry, 6, 29, 30n1, 74, 153, 180, 245, 274

Jackson, David L. ("Happy Jack"): Alto Light & Power Company, 131, 238–39; arrival in White Oaks, 129–30; burial of John Ah Nue, 26; caring for Judge Hewitt, 94; marriage and life in White Oaks, 130, 131–32; recollections of, 112, 113, 114, 115n21; waning years and death, 132–33; White Oaks *arrastra*, 4p; White Oaks residence, 111p; Wild Cat Leasing Company, 40, 130–31

Jackson, Mary, 131–32

Jackson, Richard, 219

James, Alfred, 130, 238

James, Jesse, 145

Jap Coe Ranch, 54

Jerome, Ariz., 8p, 122, 123n2

Jews/Jewish ancestry, discrimination, 104. *See also* Whiteman, Marcus and Mary Levy; Ziegler, Albert

J+H Ranch, 69–72, 201

Jicarilla Mountains, 45, 48n12, 97, 218, 283, 288

"John Chinaman." *See* Ah Nue, John (aka "John Chinaman")

Keleher, William, 30n1, 45

Kelley, O. D., 107

Kelley, Troy, 123n2

Kelt, Herman, 115n21

Kelt, John J., Sr., 73n6

King, Clarence, 253, 255n9

King, Israel, 263

King and Queen Copper Company, 227

Kinney, John ("King of the Rustlers"), 118, 122

Kiowa Indians, 275

Kirkpatrick, L. E. (Mrs.), 113, 114, 115n21

Klasner, Lily Casey, 47, 62

Klondike Gold Rush, 295–96

Knights of Pythias, 86–87, 203, 233, 265

Knights of Pythias Cemetery, 272

Koch, Alice, 7p, 135–36

Koch, Edna Alice. *See* Davis, Edna Alice Koch

Koch, Max H.: business in Chloride, 134; later life and legacy, 136; photographic legacy, 65n3; relocation to Tucumcari, 135–36; relocation to White Oaks, 135; White Oaks undertaker, 254

Koch, Merle. *See* Davis, Merle Koch

Krattinger, Mary and Jack, 38–39

Kuch, Fred, 46, 58, 59. *See also* Greathouse-Kuch Ranch

Kudner, Arthur, 125

L. G. Murphy & Co. Mercantile, 15, 31, 33, 97. *See also* Murphy, Lawrence G.

Lacey, Peter Elijah ("Doc") and Fanny LaLone, 63–65

LaLonde, Theophilus. *See* LaLone, Teófilo

LaLone, Estanislada Padilla ("Lada"), 60, 62–63

LaLone, Teófilo, 60, 62–63

La Marr, Belle. *See* Madam Varnish (Belle La Marr)

Lane, Alexander Gallatin (Dr.): as schoolteacher, 82; death of, 142; early life and education, 138; life in White Oaks, 141–42; marriage, 140; post-war difficulties, 140–41; relocation to Lincoln County, 141; service in the Civil War, 138–39

Lane, Allen A., 130–31, 142, 238
Lane, Ida. *See* Sligh, Ida Lane
Lane, John Allen, 138, 142
Lane, Mary Caroline Corbin, 140
Lane, William ("Billy"), 142, 202, 203
Langston (deputy sheriff), 203
Lavash, Donald, 45
Lawrence (Kans.) Massacre of 1863, 146
Lea, Franklin Houston (Judge): arrival in White Oaks, 146–47; family origins, 144–45; life in White Oaks, 147; relocation to Roswell, 147–48; riding with Quantrill's Raiders, 145–46
Lea, Joseph ("Father of Roswell"), 144–48
Lea Cattle Company, 147
Lee, Arizona Tallulah Smith, 158–59
Lee, Edward ("Ned"), 151, 152, 154–55
Lee, George, 151, 152, 156–57
Lee, James, 151, 152, 157–59
Lee, Jeanette ("Nettie"). *See* Lemon, Raymond E. and Nettie Lee
Lee, John (Captain): arrival in U.S., 149; arrival in White Oaks, 153; children of, 151, 154–61; later life and legacy, 154, 239; life at sea, 149–50; Little Casino Saloon, 268n2; marriage and life in Samoan Islands, 150–51; return from Samoa, 151–53
Lee, John (son), 151, 152, 155–56
Lee, Mary Magdalene Purcell, 150–51
Lee, May. *See* Queen, Edward Lawrence ("Ed") and May Lee
Lee, Mel, 158–59
Lee, Oliver, 93, 124
Lee, Robert, 151, 152, 156
Leighner, Laura Stewart, 38, 39, 40p, 259
Lemon, Raymond E. and Nettie Lee: 91p; arrival in U.S. (Lee family), 151–52; latter years in Carrizozo, 243–44; marriage and children, 160–61; recollections of White Oaks, 115n11, 281
Leslie, Bessie, 47
Leslie, Dorothy, 39, 40–41
Leslie, Elizabeth Ward, 163
Leslie, Lish: arrest and imprisonment, 164–65; cattle ranching as a "nester," 165–66; later years and death, 166–67
Leslie, Minnie English, 165
Leslie, Robert (father), 161–62
Leslie, Robert, Sr. (son): arrival in N.Mex., 163; arrival in White Oaks, 48n29; death of, 164; ranching at Texas Park, 163; return to White Oaks, 164; White Oaks recollections, 25–26, 39, 68, 79
Leslie, Robert Ward and Dorothy Pratt, 164, 167, 306
Lesnet, Frank, 263
Lesnet, Irvin, 53
Lewellyn, W. H. H., 198
Light of Western Stars, The (Grey), 260
Lincoln, Abraham, 3, 215, 249
Lincoln County: *1913 Year Book*, 87; history and legacy, 3; last legal hanging, 209; politics, 86, 92, 179, 231, 277, 293, 296; territorial divisions, 3, 263
Lincoln County News, 86
Lincoln County War: Billy the Kid, 117; Five-Day Battle, 31, 33; House of Murphy, 27, 31, 33, 57, 180, 245; impact on population, 33; investigating, 40p; Jerry Hockradle involvement, 97; Jesse Evans gang, 61; murder of Huston Chapman, 245; participants and survivors, 15–16, 42, 53, 57, 60, 62, 262, 283, 291; Susan McSween involvement, 31, 41p
Lincoln National Forest, 121, 124
Littell, Will, 246
Little Casino Saloon, 24–26, 90p, 153, 268n3. *See also* Madam Varnish (Belle La Marr)
Livingston, A. P., 107
Lloyd, Frank, 289
Lone Mountain, 3, 239, 269
Longworth, John, 247
Longworth, Thomas ("Pinto Tom"), 50, 57, 117
Lynch, Evelyn and "Boots," 83n1

Mabry, Thomas, 126
Mack, Fran and Jim, 81
Mackel, Peter, 246
Madam Varnish (Belle La Marr): arrival in White Oaks, 267; Little Casino Saloon, 6, 268; marriage, 267–68; origin of name, 267, 268n1
Madero, Francisco, 225, 227
Mail routes: Central New Mexico map, 2; closure of, 202; delivery to White Oaks, 10, 129, 246–47; government Star Routes, 216–18; robberies, 142, 216
Malcolm, Margaret, 67–68, 69
Malcom, Jane. *See* Gallacher, Jane Malcolm
Man Next Door, The (Hough), 103

Marquisate of Aguayo. *See* Aguayo Marquisate

Mason, Barney, 58

Masons/Masonry: Albert Ziegler, 304; Benjamin Gumm, 78–79, 84n6; E. W. Parker, 222, 228, 254; Franklin Lea, 147; James Bell, 45; John Slack, 249, 255; Jones Taliaferro, 265; Marcus Whiteman, 279; Max and Alice Koch, 136; Roswell, N.Mex. lodge, 147, 254p; William Gallacher, 72; William Gallacher, Sr., 67p. *See also* Order of the Eastern Star

Mayer, Amanda Aguayo Treat, 213

Mayer, Charles and Ina Wauchope, 127, 286

Mayer, Charles D.: arrival in White Oaks, 172–73; as deputy sheriff, 173–74; marriage, 174; relocation to Carrizozo, 174–75

Mayer, Frederick, 168–70

Mayer, Ina Jame Wauchope, 174–75

Mayer, Maria Louise Schultz ("Hattie"), 169

Mayer, Paul: arrival in White Oaks, 98; livery stable and gold deliveries, 36, 62, 170–71; relocation to Carrizozo, 171–72; stage line business, 171, 202; White Oaks reunion, 104

Mayer, Ula Gilmore, 171, 172p

McCourt, Genevieve, 184, 185

McCourt, John and Vena Gumm, 79p, 80, 184, 185

McCourt, Margaret ("Margie"), 184, 185

McCourt, Paul, 183

McCourt, Paul and Louise, 184, 185

McCourt, Thomas B., Jr. ("Tom"), 182–84

McCray, I. T., 118–19

McCutcheon (postmaster), 246–47

McDonald, Frances Tarbell McCourt: arrival in White Oaks, 183; children, 184, 185; later life and legacy, 191; marriage and divorce from McCourt, 182, 184; marriage to McDonald, 184–85; as N.Mex. "First Lady," 188–89

McDonald, William Calhoun (governor): arrival in White Oaks, 9, 178; death and burial, 41, 190; early law practice, 90p, 91, 101, 178–79; entry into politics, 179–80, 185–86; gold mining days, 78; as governor, 86, 125, 186–90; later life and legacy, 190–91; manager of Bar W Ranch, 29, 180–81, 185; manager of Block Ranch, 181–82; marriage, 184–85; recollections of, 210, 223

McSween, Alexander, 16, 32–33

McSween, Susan. *See* Barber, Susan McSween

McSween Mercantile, 15–16

Merrill, Charles Madison: early life and family, 193–94; in Alaska, 193; in N.Mex., 194; return to Texas, 194–95

Methodist Church, 69, 159, 160, 236, 240, 256, 291

Mexican Central Railroad, 227

Mexican Expedition of 1916, 126

Mexican Revolution (1910-1920), 225–27, 259

Mexico: Aguayo latifundios, 11–12; French presence in, 11; independence from Spain, 12

Mills, "Ham" (sheriff), 96–97

Mills, William J. (territorial governor), 186–87, 232, 306

Miners' disease (silicosis), 158, 160, 235, 236, 242, 246

Mines/claims: Abraham Lincoln. *See* Old Abe Mine; Boston Boy, 114p; Bully Bueno Mine (Ariz.), 250; Burro Mines, 250, 251; Captain Kidd, 230; Comstock (N.Mex.), 5, 230; Comstock (Virginia City), 235; Congress, 242; Dark Cloud, 205; Emma Mine (Calif.), 250; Georgia Placer, 48n12; Great Diamond Hoax of 1872, 250–54; Helen Rae, 129, 228; Large Hopes, 118; Little Nell, 118; Matchless Mine (Colo.), 183; Miguel Otero, 78; Miner's Cabin, 286; Mountains of Silver, 250; New River (Calif.), 241–42, 243p; No Man's Friend, 118; Potosí Mines (Nev.), 250; Pyramid Mines, 250–51; Robert E. Lee, 90; Santa Rita, 205; Shattuck Den, 153; Sierra del Oro ("Hill of Gold"), 60; Silver Cliff, 286; Smuggler, 242; White Oaks, 90; Yeager mines (Ariz.), 122. *See also* Gold mining; Homestake Mine; Old Abe Mine; Silver/silver mining

Monjeau, Louis, 121

"Mother Whiteman," 280–81; relocation to Roswell, 281; stores destroyed by fire, 281–82

Mountain Station Ranch, 201

Mules, Mines and Me in Mexico (Parker), 228

Mullin, Robert N., 31

Murders and killings: bushwhacking, 166; Chávez shoots prisoner, 17; disappearance of Albert Fountain, 233; Horrell War, 96; Kansas-Missouri border conflict, 145–46; killing of Robert Hurt, 231; Lincoln County War, 97; mavericking unbranded cattle, 34; murder of Wah Sing, 24; poisoning of Harry Gallacher, 71–72; shooting of Askew and Cullom, 118; shooting of George Fitspatrick, 273n21, 277–78; shooting of George Parker, 18–19, 173, 203; shooting of Huston Chapman, 245–46; shooting of I. T. McCray, 118–19; shooting of Ike Smith, 301; shooting of Irvin Lesnet, 53; shooting of Johnny Mace, 117; shooting of Louis Monjeau, 121. *See also* Hangings/lynchings; Suicides

Murphy, Lawrence G.: Carrizozo Ranch, 27, 120, 180–81, 185; Lincoln County War, 57, 245; mercantile store, 15, 97; "Old Brewery" Saloon, 116. *See also* House of Murphy (Murphy-Dolan faction)

Musgrave, George, 18–19, 173, 203

National Mail Company, 198, 216–18

National Register of Historic Places, Cedarvale Cemetery, 306

Navajo Revolt of 1913, 125–26

Nevada: Comstock Mine, 235–36; copper mining, 122; emigrant trains, 75; filming of *The Covered Wagon*, 105

New, Ann and Michael, 81

New Mexico: copper mining, 8, 19, 122, 142, 227, 239, 259, 265; first indoor bathroom, 113; First Territorial Infantry, 79–80; official state song, 189; Quinquennial Census of 1885, 60, 65n3, 293; right of public domain, 27–28; Santa Fe Ring, 180, 231–32; statehood, 186–87, 232, 238; Statehood Convention, 232; support of women's rights, 189–90, 232

New Mexico Agricultural and Mechanical College (New Mexico State University), 69, 72

New Mexico Cattle Growers' Association, 181

New Mexico Military Institute, 82

New Mexico Normal College, 80

New Mexico Republican Reform League, 231–32

New Mexico Sheriffs' and Police Association, 47

New Mexico State Supreme Court, 126–27

Newton, Howard, 39

1910 History and Business Review of Tucumcari, 136

1913 Year Book of Lincoln County (Haley), 87

Niño Ladrón de Guevara. *See* de Guevara family

Nolan, Frederick, 17, 31

No Scum Allowed Saloon, 305

North Homestake Mine: closure, 110; legends, 113; ownership and gold output, 108p, 289; Wild Cat leasing and operation, 131, 238–39. *See also* Homestake Mine; South Homestake Mine

North of 36 (Hough), 103

"O Fair New Mexico" (state song), 189

O'Falliard, Tom, 46, 245, 247

Oklahoma, 232

Old Abe Eagle, 9p, 85

Old Abe Mine: decline and closure, 10, 82, 113, 130; discovery and development, 5, 8, 107; mine fire, 110–11; miners, 63, 129–30, 186; operations and gold output, 85, 90, 94, 108–11, 170; ownership, 88, 107–108; stamping mill, 110

Old Hickory Mining Company, 265

Olinger, Robert, 18, 27, 45, 46–47

Order of the Eastern Star, 136, 160. *See also* Masons/Masonry

Orozco, Pascual, 227

Ortiz Mountains, 60

Osell, Arvid E. (Dr.), 259

Osell, Harry, 257, 259

Otero, Miguel Antonio (territorial governor), 17, 31–33, 40, 79, 231–33

"Over The Range" (Hough), 101–102

Ozanne, Alfred and Olive Rencher, 196, 201, 205–206, 294

Ozanne, Emile, 196, 204–205

Ozanne, Henry, 196, 204

Ozanne, Paul, 196, 206

Ozanne, Urbain: arrival in U.S., 196; arrival in White Oaks, 197; career in Mississippi, 196–97; children, 204–206; death and burial, 203–204; marriages, 196, 202, 203; White Oaks businesses, 197–203

Ozanne & Son (restaurant and bakery), 197

Ozanne Hotel, 67, 197p, 202, 204

Ozanne Passenger & Stage Line, 198–202, 206

Paden, Belle Williams, 209, 213

Paden, Melvin, Jr., 210p, 213

Paden, Melvin G. (Dr.), 7p; arrival in White Oaks, 208–209; as hunter and game warden, 210–11; later life and legacy, 213; marriage and children, 209–10; officiated at last legal hanging, 209; practice of medicine, 52, 70, 71, 212–13; relocation to Carrizozo, 211–12; White Oaks reunion, 104

Paden, Morgan ("Brent"), 82, 210p, 213

Paden hospital, 212–13

Paden's Drug Store, 208, 211–13

"Pancho" Villa. *See* Villa, Francisco ("Pancho")

Panic of 1873, 88, 215

Panic of 1893, 183, 197, 202

Parker, Bruce Lane, 142

Parker, E. W. (Erastus Wells): childhood and early career, 215; children of, 223–24, 225–28; later years and death, 228; life in White Oaks, 221–22; mail route,

198, 216–18; marriage, 215–16; mining interests, 218; relocation to White Oaks, 98; South Homestake Mine, 219–21, 224–25; sued for Star Route frauds, 217

Parker, Emmeline Brown: later years and death, 228; life in White Oaks, 221–22; marriage and children, 215–16

Parker, Frank Wells, 142, 216, 225

Parker, George, 18–19, 173, 203

Parker, James Henry ("Jamie"), 216, 225

Parker, James W., 217, 219, 223p

Parker, Morris B., 79p, 219p; arrival in White Oaks, 218–19; childhood, 82, 223, 225; gold mining, 108, 129, 224–25, 226–27; growing up in White Oaks, 223–24; later life and legacy, 228; marriage, 185; recollections of White Oaks, 30n18, 98, 113

Parker, O. N., 203

Parker, Olive Genevieve McCourt, 225, 228

Paso por Aquí (Rhodes), 100

Patterson, C. Ewing, 220

Patterson, Henry, 101

Patterson, Will, 284

Patterson vs. Baxter Gold Mining Company, 101

Paynter, David C., 45

Peaker, Oliver, 109p, 278

Peppin, George, 61

Perea, Demetrio, 17

Perrin, James M., 142n1

Pershing, John J. ("Black Jack"), 126

Petty, Joseph Franklin (Mr. and Mrs.), 304

Philippine Insurrection, 79–80

Pickwick Corporation, 54

Pioneer Saloon, 5–6, 29, 117–18, 246

Placerville, Calif. ("Old Dry Diggin's"), 75–77, 84n2

Poe, Sophie, 148n1

Pool, J. Landly, 53

Presbyterian Church, 5, 32, 66, 136, 228

Prichard, George Worth (Colonel): arrival in White Oaks, 230; as lawyer, 230–31; as politician, 231–32; Knights of Pythias, 233; death, 233; life in White Oaks, 104, 203, 210; marriage, 233; Statehood Convention, 232; White Oaks reunion, 104

Prichard, Maude H. Hancock, 233

Pridemore, Tom, 18, 165

Prince, L. Bradford, 231–32

Pyramid Mountains, 250–51

Quantrill's Raiders, 145–46

Queen, Donald, 306

Queen, Donald John ("Don") and Dorothy Bricker, 109p, 236–37, 240–41, 244

Queen, Edward Lawrence ("Ed") and May Lee: 109p, 159p; arrival in U.S. (Lee family), 151–52; arrival in White Oaks, 213; later years and death, 243; living through the Depression, 160, 241–43; marriage, 159–60; ownership of Wild Cat leasing, 130–31, 238–39; relocation to Arizona, Mexico and California, 239–40

Queen, Ellyn Allie. *See* Whitwell, Earl Leonard and Ellyn Queen

Queen, Forrest, 109p

Queen, John and Sophia ("Sophie") Mary Turner, 235–36

Queen, Lawrence Lee and Vera Lynn Tracy, 236–38, 239–40, 244

Racism/racial discrimination: African American, 129; Hispanic-Anglo relations, 96, 98, 298; post-Civil War, 196–97; toward Mexicans, 65n8. *See also* Immigration

Railroads: arrival in N.Mex., 88, 197; arrival in Roswell, 148; construction to White Oaks, 10, 92–93, 230, 282; Salado coalfields, 10, 27, 93; White Oaks land speculation, 35

Ralston, William C., 250–53

Ralston City, N.Mex., 250–51

Range fires, 18

Rasch, Philip J., 56

Reasoner, J. C., 115n21

Redman, James B. (aka Redmond, Readman, W. J. Woodland): aliases, 245; arrival in White Oaks, 117, 246; as deputy sheriff, 247; encounters with Billy the Kid, 46, 246; Greathouse-Kuch shootout, 57; involvement in Lincoln County War, 245–46; later years and death, 247; mining interests, 246

Reeder, O. G., 136

Regulators gang, 61, 245. *See also* Billy the Kid (William H. Bonney)

"Report of San Miguel County, The" (Prichard), 230

Reynolds, Bill, 44, 48n1

Rhodes, Eugene Manlove ("Bard of Tularosa"), 9, 100

Richardson place. *See* Block Ranch

Ridgway, David L. and Ella Young, 296

Robert E. Lee gold claim, 90

Roberts, Albert, 17

Roberts, George D., 250–51, 255n2

"Romance, A" (Rencher), 205
Roosevelt, Franklin D., 237, 242
Roosevelt, Theodore, 80, 231–32
Roswell, N.Mex.: development and settlement, 146–48, 281–82; Masonic Lodge, 254p; New Mexico Military Institute, 82, 210, 265; stage line connections, 201–202
Rothschild, L. G. ("the Baron"), 232
Rudabaugh, Dave, 46, 56, 58, 119
Rudiselle, Lee H., 104, 203, 262
Rugee, John, 35, 36
Ruidoso, N.Mex., 72, 85, 94, 124–25
Ruidoso River, 54, 79, 96
Russell, Baldy, 70

Sacramento Mountains, 10, 35, 119, 153, 194, 271
Safford, N.Mex., 296
Saga of Billy the Kid, The (Burns), 31, 41p
Sagebrusher (Hough), 104
Sager, Frank, 91, 104, 115n10
Salado coalfields, 10, 27, 93
Samoan Islands, 149–51
San Pedro, N.Mex., 300
Sánchez Navarro, Carlos, 12
Sánchez Navarro, José Miguel, 12, 21nn2–3
Santa Fe Ring, 180, 231–32
Saturday Evening Post, 103, 105
Scharf, Laura Sullivan, 213
Schinzing, Augustus ("Gus"), 186, 268n3, 268p
Scott, Hugh (General), 126
Scott, James, 29, 30n1
Searchlight, Nev., 122
Selman, John (aka John Gunter), 116–17, 123n7
Shafer, R. M. ("Bob") and Jane Gallacher, 70, 72
Shafer, R. M., Jr. and Joy, 72
Sharpe, Leonard, 4p
Sharpe, Marlow, 4p
Shattuck Den Mine, 153
Sheep Killing Cases, 27–29
Sheep ranching: Aguayo latifundios, 11; Bar W Ranch, 190; conflicts with cattle, 27–29, 120; Hudgens Brothers, 119; J+H Ranch, 69–70; Lincoln County ranches, 236, 254, 263, 266; politics, 191n16
Sheep Spring Ranch (Patos Mountain), 69–70
Shield, David and Elizabeth, 33

Shield, Edgar P., 39, 41
Sierra del Oro ("Hill of Gold") Mine, 60
Sigafus, B. H., 108p
Sigafus, James A., 226, 238, 289
Silver/silver mining: Chloride, N.Mex., 134; Comstock Mine (Virginia City), 235; Durango, Mexico, 60; Hillsboro, N.Mex., 121; Nogal, N.Mex., 97; Panic of 1893, 183, 197, 202; Pyramid Mountains, 250–51; Sherman Silver Purchase Act, 202
Simms, Carrie, 221p
Simms, J. Denton, 37
Six Years with the Texas Rangers (Gillett), 44
Slack, John Burchem: diamond mine hoax, 251–53; emigration to California gold fields, 249–50; later years and death, 254–55; Mexican War service, 249; mining interests in Arizona and Nevada, 250–51; relocation to White Oaks, 253
Slaves/slavery, 138, 144, 145, 179, 238
Sligh, George, 142–43
Sligh, Ida Lane, 138, 142–43
Sligh, J. E. ("Holy Jim"), 28–29
Smith, Ike, 300–301
Snow, Clara, 113
Sorrells, James, 260n1
South Homestake Mine: block diagram, 220p; building of cyanide plant, 226; development and operation, 110, 218–21, 224–25; miner deaths, 68, 156; operations and gold output, 224, 226; Parker arrival for management of, 98; Parker sale of, 228; stamping mill, 221; Wild Cat leasing and operation, 131, 142–43, 238–39. *See also* Homestake Mine; North Homestake Mine
Southern Pacific Railroad, 259
Southwestern Stage Company, 198
Spanish colonization in America, 11
Spanish-American War, 16, 79–80, 185, 231
Spencer, A. N., 189, 190
Spencer, Frances McDonald, 38, 185, 187, 189
Spencer, Stirling, 191, 306
Spencer, Truman ("Big"), 38–39, 41, 189, 190
Spencer, Truman, Jr., 190
Spring River Ranch, 34
St. Frances de Sales, 142
Stagecoach lines: beginning service to White Oaks, 10, 198; "Buck Board Line," 201; map, 2; Ozanne

Passenger & Stage Line, 198–202, 206; passenger recollections, 278n5; Pickwick Corporation, 54; robberies, 203; White Oaks Passenger Line, 171; winter travel, 153. *See also* Freight routes; Mail routes

Statehood Convention of 1902, 232

Steck, Joe, 57–58, 59, 119

Steins, N.Mex., 227

Stewart, Eugene ("Judge Stewart"), 104, 223p, 256p, 258–60

Stewart, Levin Washington and Louella Vandevort: arrival in White Oaks, 256–57; businesses in Mexico and Arizona, 259; businesses in White Oaks, 257–58; early life and Civil War service, 256; goat herd, 258–59; later years and death, 259–60; marriage, 256

Stewart, Mabel, 257, 259

Stewart, William, 256

Story of the Cowboy, The (Hough), 101, 103

Strausner hack line (stage line), 198

Suicides, 78, 103, 182, 265. *See also* Murders and killings

Sweet, J. A., 118, 246

Sweet, Sallie, 41

Tabor, George D. and Jane Wilcox, 182, 185

Tabor, Horace A. W. and "Baby Doe," 183

Taft, William H., 187

Taliaferro, Eliza ("Lida"), 261, 265

Taliaferro, Jones and Ella Thomson, 7p, 8p; arrival in U.S., 261; arrival in White Oaks, 261–62; businesses and mining interests, 262–63; mercantile store, 263–64, 292p; purchase of *Golden Era*, 262; White Oaks reunion, 104

Taliaferro, Jones, Jr., 8p, 135p

Taliaferro, Richard, 82

Taliaferro, Stanley, 261, 264

Taliaferro Mercantile, 170p, 174, 292p

Terrazas, Luís (Don), 227

Texas: Battle of Adobe Walls, 56; Spanish colonization, 11

Texas County Sheriffs (Tise), 44

Texas Park, N.Mex., 16, 162–66, 272, 275

Texas Rangers, 44–45, 48n1, 203

"The Report of San Miguel County" (Prichard), 230

"The West After Twenty Years" (Hough), 104

Thorp, N. Howard ("Cowboy Jack"), 9, 27, 100

Three Rivers Land and Cattle Company, 28, 33–37, 38, 43n44

Thurber, H. K., 18

Timber Claim Act of 1873, 49–50

Townsend family, 93–94

Tres Ríos Ranch. *See* Three Rivers Land and Cattle Company

Tucumcari, N.Mex., 135–36, 305

Tularosa, N.Mex., 6, 13–14, 33, 54, 93

Tungsten mining, 131, 142, 238, 276, 289

Tunstall, John, 33, 34, 97, 245

Turner, Jane Spencer, 190

Twitchell, Ralph Emerson, 86, 186

Ulrick, George, 21, 91, 101, 104, 120, 178

Union Pacific Railroad, 252, 253

"Unlucky Jims," 56, 58–59

U.S. Census of 1850, 62, 256

U.S. Census of 1860, 60–61, 256

U.S. Census of 1870, 32, 60, 62, 288

U.S. Census of 1880, 6–7, 16, 33, 44, 56, 60, 96, 168, 245, 256, 274, 291

U.S. Census of 1890, 29, 63

U.S. Census of 1900, 22, 30n1, 33, 61, 63, 155, 193, 272

U.S. Census of 1910, 63, 259, 272

U.S. Census of 1920, 272

Valdivielso, José Francisco (3rd Marquis of Aguayo), 11

Valdivielso, José María (4th Marquis of Aguayo), 12

Victory, Mary, 189–90

Villa, Francisco ("Pancho"), 126, 227

Vitro de Vera, José Ramón de Azlor (2nd Marquis of Aguayo), 11

VV Ranch, 53, 94, 124–25

Wah Sing, 22–24

Wallace, Lewis (Governor), 9, 35, 58, 61, 117, 119, 246

Walz, Edward, 29, 30n1

Watseka, Ill., 264

Watson, William, 90, 91, 107–108, 220, 284

Watt, Carol Queen, 306

Watts (Colonel), 230

Wauchope, John and Ione Wilson: arrival in Texas, 284–85; arrival in White Oaks, 285; later years and death, 286; mining interests in South America, 285–86

Weather: blizzards and snowstorms, 18; drought of the 1920s, 182; floods, 131–32; heat/heatstroke, 296

Webb, Charles, 161

Weed, N.Mex., 153, 163, 194

Weed, William H.: arrival in Texas, 270; departure from White Oaks, 272; early trading days, 269–70; marriage, 271–72; operation of the "Big Store," 270–71; recollections of, 225, 295

Welch, Mary Lou Townsend, 93–94

Wells, Erastus, 215, 216, 219

Wells, Samuel and Martha Frances: arrival in Texas and marriage, 274–75; arrival in White Oaks, 276; later years and death, 277; life in White Oaks, 276–77; surviving Indian raids, 275–76

"West After Twenty Years, The" (Hough), 104

Wetzel, Charles, 114

Wetzel, Charles L., 82, 114

Wetzel, Larue Lane ("Rudy"), 37, 82, 114, 126–27, 305

Wheelock, Gilbert and Myra Young, 296

White Caps Gang (Society of Bandits), 17

Whiteman, Marcus and Mary Levy: arrival in N.Mex. and marriage, 279; later years and deaths, 282; Pioneer Store, 279–80; recollections of "Mother Whiteman," 280–81; relocation to Roswell, 281; stores destroyed by fire, 281–82

Whiteman, Mary. *See* "Mother Whiteman"; Whiteman, Marcus and Mary Levy

White Oaks: arrival of the railroad, 10, 35, 92–93; baseball team, 158, 164, 179, 205; census of 1890, 8; "Dave Jackson Day," 133; first election of officers, 246; first electric lights, 131; first hotel, 147; first indoor bathroom, 113; as ghost town, 305–306; locale for *Heart's Desire*, 230; mail delivery, 10, 129, 246–47; map location, 2; Masonic Lodge No. 20, 79, 84n6, 222, 249, 255; origin of town name, 276; politics, 125, 257; reunion of "the boys" in 1904, 104–105. *See also* Cedarvale Cemetery; Churches, Lincoln County

White Oaks, buildings/businesses: Bond and Stewart store, 257p; Bonnell Opera House, 51–52, 79; Brown building, 23, 85, 90, 305; Building and Lumber Company, 82, 142; C. D. Mayer, General Merchandise, 174p; Carizo Hotel, 68–69; Congregational Church, 222p; Exchange Bank, 8, 90p, 91–92, 113; Homestake Saloon, 121; Littell stable, 246; Little Casino Saloon, 6, 24, 25–26, 90p, 153, 267–68, 268n1, 268n3; Mayer General Merchandise store, 170p, 175p, 176p; Mayer livery stable, 36, 62, 98, 170–71; No Scum Allowed Saloon, 305; Ozanne Hotel, 67, 197p, 202, 204; Pioneer Saloon, 5–6, 29, 117–18, 246; Pioneer Store, 279–80; Taliaferro Brothers store, 170p, 264p, 265p, 292p; Taylor blacksmith building, 132p; Teachers' Normal School, 82; town hall, 7, 52; Watson-Lund law office, 305; Weed's "Big Store," 270–71; West and Dedrick stable, 6, 57, 119; White Oaks Schoolhouse, 82–83, 305; Young & Taliaferro Mercantile, 174p, 263, 293–94; Young Hides and Wool, 292p

White Oaks, homes: Ah Nue's log house, 23p; Canning house, 161p, 174; Dave Jackson house, 111p; Gumm House, 79, 81, 236, 305; Hoyle House, 81–82, 111–14, 305; Leslie, Sr. house, 164p; Parker (E. W.) home, 222p; Parker (Morris) home, 226p; Taylor house, 241, 305; Watson House, 107p; Ziegler house, 302p

White Oaks (Parker), 228, 229n1

White Oaks, ranches and area buildings: Jicarilla store, 205p; Lacey dugout home, 64p; Mal Pais stagecoach station, 200p; McDonald ranch house, 185p; Ozanne Mountain Station, 201p

White Oaks Cemetery Association, 47

White Oaks Eagle, 85–86, 90–91

"White Oaks 8," 79–80

White Oaks Golden Era, 9p, 103, 262

White Oaks Historical Association, 167, 305

White Oaks Passenger Line, 171, 202

White Oaks Record of Deaths, 135, 254

White Oaks Schoolhouse, 82–83, 142

White Oaks Schoolhouse Museum, 4p, 83, 167, 305–306

White Oaks Spring, 276

Whitwell, Earl Leonard and Ellyn Queen, 236, 237p, 240, 242p, 244

Wicks, James ("Stumpy"), 101–102

Wild, Azariah F., 45

Wild Cat Leasing Company, 40, 130–31, 238–39

Williams, Ernestine, 201

Williams, Oscar W., 45

Wilson, Billy, 6, 46, 56–58, 119

Wilson, George, 4–5, 156–57, 283–84, 288

Wilson, John A., 220

Wilson, John E., Jr., 286

Wilson, John E. ("Uncle Johnny"), 4, 123n13, 174, 283–86, 288

Wilson, John P., 31

Wilson, Juan Batista (judge), 117

Wilson & the Kid (Lavash), 45
Winters, John V. ("Old Jack," "Old Blue Face"): arrival in N.Mex., 288; death of, 288–89; gold discovery at White Oaks, 4, 123n13, 283–84; gold prospecting, 117–18, 246; Homestake strike, 284–85, 288; sale of North Homestake Mine, 289
Wise, Jacob, 262
Woodland, W. J.. *See* Redman, James B. (aka Redmond, Readman, W. J. Woodland)
World War I, 70–71, 103, 174, 190, 289, 303
World War II, 22, 72, 160, 243

Yellowstone Park, 103–104, 106n16
Young, George Richard ("Dick") and Kate Thomson: arrival in White Oaks, 291–92; Klondike Gold Rush, 295–96; later years and death, 296; loss of children and relocation to Roswell, 295; marriage and children, 292–93; mercantile store, 174p, 263, 293–94; White Oaks reunion of "the boys," 104
Younger Brothers (Cole, Jim, and Bob), 145

Ziegler Brothers: grocery store, 280, 302–303; later life and legacy, 303–304; move to Colorado and Carrizozo, 303; "Quit Business Sale," 304; White Oaks store, 299–301
Ziegler, Albert: 7p; arrival in U.S., 298; arrival in White Oaks, 299; courtship and marriage, 301–302; death of, 304; later years and death, 304; recollections of, 302–303
Ziegler, Carrie Leon, 301–303
Ziegler, Jacob ("Jake"): arrival in White Oaks, 299; later years and death, 303; operation of Manzano store, 298–99
Zimmerman, Minnie Shield, 38p, 39, 42
Zollars, I. W., 91

www.ingramcontent.com/pod-product-compliance
Lightning Source LLC
Chambersburg PA
CBHW081209230426
43666CB00015B/2686